Pro Hadoop

Jason Venner

Apress®

Pro Hadoop

ISBN-13 (pbk): 978-1-4302-1942-2

ISBN-13 (electronic): 978-1-4302-1943-9

Printed and bound in the United States of America 9 8 7 6 5 4 3 2 1

Lead Editor: Matthew Moodie
Technical Reviewer: Steve Cyrus
Editorial Board: Clay Andres, Steve Anglin, Mark Beckner, Ewan Buckingham, Tony Campbell, Gary Cornell, Jonathan Gennick, Michelle Lowman, Matthew Moodie, Duncan Parkes, Jeffrey Pepper, Frank Pohlmann, Douglas Pundick, Ben Renow-Clarke, Dominic Shakeshaft, Matt Wade, Tom Welsh
Project Manager: Richard Dal Porto
Copy Editors: Marilyn Smith, Nancy Sixsmith
Associate Production Director: Kari Brooks-Copony
Production Editor: Laura Cheu
Compositor: Linda Weidemann, Wolf Creek Publishing Services
Proofreader: Linda Seifert
Indexer: Becky Hornyak
Artist: Kinetic Publishing Services
Cover Designer: Kurt Krames
Manufacturing Director: Tom Debolski

Distributed to the book trade worldwide by Springer-Verlag New York, Inc., 233 Spring Street, 6th Floor, New York, NY 10013. Phone 1-800-SPRINGER, fax 201-348-4505, e-mail `orders-ny@springer-sbm.com`, or visit `http://www.springeronline.com`.

For information on translations, please contact Apress directly at 2855 Telegraph Avenue, Suite 600, Berkeley, CA 94705. Phone 510-549-5930, fax 510-549-5939, e-mail `info@apress.com`, or visit `http://www.apress.com`.

Apress and friends of ED books may be purchased in bulk for academic, corporate, or promotional use. eBook versions and licenses are also available for most titles. For more information, reference our Special Bulk Sales–eBook Licensing web page at `http://www.apress.com/info/bulksales`.

The source code for this book is available to readers at `http://www.apress.com`. You may need to answer questions pertaining to this book in order to successfully download the code.

This book is dedicated to Joohn Choe.
He had the idea, walked me through much of the process,
trusted me to write the book, and helped me through the rough spots.

Contents at a Glance

Contents

About the Author

JASON VENNER is a software developer with more than 20 years of experience developing highly scaled, high-performance systems. Earlier, he worked primarily in the financial services industry, building high-performance check-processing systems. His more recent experience has been building the infrastructure to support highly utilized web sites. He has an avid interest in the biological sciences and is an FAA certificated flight instructor.

About the Technical Reviewer

SIA CYRUS's experience in computing spans many decades and areas of software development. During the 1980s, he specialized in database development in Europe. In the 1990s, he moved to the United States, where he focused on client/server applications. Since 2000, he has architected a number of middle-tier business processes. And most recently, he has been specializing in Web 2.0, Ajax, portals, and cloud computing.

Sia is an independent software consultant who is an expert in Java and development of Java enterprise-class applications. He has been responsible for innovative and generic software, holding a U.S. patent in database-driven user interfaces. Sia created a very successful configuration-based framework for the telecommunications industry, which he later converted to the Spring Framework. His passion could be entitled "Enterprise Architecture in Open Source."

When not experimenting with new technologies, Sia enjoys playing ice hockey, especially with his two boys, Jason and Brandon.

Acknowledgments

I would like to thank the people of Attributor.com, as they provided me the opportunity to learn Hadoop. They gracefully let my mistakes pass—and there were some large-scale mistakes—and welcomed my successes.

I would also like to thank Richard M. Stallman, one of the giants who support the world. I remember the days when I couldn't afford to buy a compiler, and had to sneak time on the university computers, when only people who signed horrible NDAs and who worked at large organizations could read the Unix source code. His dedication and yes, fanaticism, has changed our world substantially for the better. Thank you, Richard.

Hadoop rides on the back, sweat, and love of Doug Cutting, and many people of Yahoo! Inc. Thank you Doug and Yahoo! crew. All of the Hadoop users and contributors who help each other on the mailing lists are wonderful people. Thank you.

I would also like to thank the Apress staff members who have applied their expertise to make this book into something readable.

Introduction

This book is a concise guide to getting started with Hadoop and getting the most out of your Hadoop clusters. My early experiences with Hadoop were wonderful and stressful. While Hadoop supplied the tools to scale applications, it lacked documentation on how to use the framework effectively. This book provides that information. It enables you to rapidly and painlessly get up to speed with Hadoop. This is the book I wish was available to me when I started using Hadoop.

Who This Book Is For

This book has three primary audiences: developers who are relatively new to Hadoop or MapReduce and must scale their applications using Hadoop; system administrators who must deploy and manage the Hadoop clusters; and application designers looking for a detailed understanding of what Hadoop will do for them. Hadoop experts will learn some new details and gain insights to add to their expertise.

How This Book Is Structured

This book provides step-by-step instructions and examples that will take you from just beginning to use Hadoop to running complex applications on large clusters of machines. Here's a brief rundown of the book's contents:

Chapter 1, Getting Started with Hadoop Core: This chapter introduces Hadoop Core and MapReduce applications. It then walks you through getting the software, installing it on your computer, and running the basic examples.

Chapter 2, The Basics of a MapReduce Job: This chapter explores what is involved in writing the actual code that performs the map and the reduce portions of a MapReduce job, and how to configure a job to use your map and reduce code.

Chapter 3, The Basics of Multimachine Clusters: This chapter walks you through the basics of creating a multimachine Hadoop cluster. It explains what the servers are, how the servers interact, basic configuration, and how to verify that your cluster is up and running successfully. You'll also find out what to do if a cluster doesn't start.

Chapter 4, HDFS Details for Multimachine Clusters: This chapter covers the details of the Hadoop Distributed File System (HDFS) and provides detailed guidance on the installation, running, troubleshooting, and recovery of your HDFS installations.

Chapter 5, MapReduce Details for Multimachine Clusters: This chapter gives you a detailed understanding of what a MapReduce job is and what the Hadoop Core framework actually does to execute your MapReduce job. You will learn how to set your job classpath and use shared libraries. It also covers the input and output formats used by MapReduce jobs.

Chapter 6, Tuning Your MapReduce Jobs: In this chapter, you will learn what you can tune, how to tell what needs tuning, and how to tune it. With this knowledge, you will be able to achieve optimal performance for your clusters.

Chapter 7, Unit Testing and Debugging: When your job is run across many machines, debugging becomes quite a challenge. Chapter 7 walks you through how to debug your jobs. The examples and unit testing framework provided in this chapter also help you know when your job is working as designed.

Chapter 8, Advanced and Alternate MapReduce Techniques: This chapter demonstrates how to use several advanced features of Hadoop Core: map-side joins, chain mapping, streaming, pipes, and aggregators. You will also learn how to configure your jobs to continue running when some input is bad. Streaming is a particularly powerful tool, as it allows scripts and other external programs to be used to provide the MapReduce functionality.

Chapter 9, Solving Problems with Hadoop: This chapter describes step-by-step development of a nontrivial MapReduce job, including the whys of the design decisions. The sample MapReduce job performs range joins, and uses custom comparator and partitioner classes.

Chapter 10, Projects Based on Hadoop and Future Directions: This chapter provides a summary of several projects that are being built on top of Hadoop Core: distributed column-oriented databases, distributed search, matrix manipulation, and machine learning. There are also references for training and support and future directions for Hadoop Core. Additionally, this chapter provides a short summary of my favorite tools in the examples: a zero-configuration, two-node virtual cluster.

Appendix, The JobConf Object in Detail: The `JobConf` object is the heart of the application developer's interaction with Hadoop. This book's appendix goes through each method in detail.

Prerequisites

For those of you who are new to Hadoop, I strongly urge you to try Cloudera's open source Distribution for Hadoop (`http://www.cloudera.com/hadoop`). It provides the stable base of Hadoop 0.18.3 with bug fixes and some new features back-ported in and added-in hooks to the support scribe log file aggregation service (`http://scribeserver.wiki.sourceforge.net/`). The Cloudera folks have Amazon machine images (AMIs), Debian and RPM installer files, and an online configuration tool to generate configuration files. If you are struggling with Hadoop 0.19 issues, or some of the 0.18.3 issues are biting you, please shift to this distribution. It will reduce your pain.

The following are the stock Hadoop Core distributions at the time of this writing:

- Hadoop 0.18.3 is a good distribution, but has a couple of issues related to file descriptor leakage and reduce task stalls.
- Hadoop 0.19.0 should be avoided, as it has data corruption issues related to the append and sync changes.
- Hadoop 0.19.1 looks to be a reasonably stable release with many useful features.
- Hadoop 0.20.0 has some major API changes and is still unstable.

The examples in this book will work with Hadoop 0.19.0, and 0.19.1, and most of the examples will work with the Cloudera 0.18.3 distribution. Separate Eclipse projects are provided for each of these releases.

Downloading the Code

All of the examples presented in this book can be downloaded from the Apress web site (`http://www.apress.com`). You can access the source code from this book's details page or find the source code at the following URL (search for Hadoop): `http://www.apress.com/book/sourcecode`.

The sample code is designed to be imported into Eclipse as a complete project. There are several versions of the code, each a designated version of Hadoop Core that includes that Hadoop Core version.

The `src` directory has the source code for the examples. The bulk of the examples are in the package `com.apress.hadoopbook.examples`, and subpackages are organized by chapter: `ch2`, `ch5`, `ch7`, and `ch9`, as well as `jobconf` and `advancedtechniques`. The test examples are under `test/src` in the corresponding package directory. The directory `src/config` contains the configuration files that are loaded as Java resources.

Three directories contain JAR or zip files that have specific licenses. The directory `apache_licensed_lib` contains the JARs and source zip files for Apache licensed items. The directory `bsd_license` contains the items that are provided under the BSD license. The directory `other_licenses` contains items that have other licenses. The relevant license files are also in these directories.

A `README.txt` file has more details about the downloadable code.

Contacting the Author

Jason Venner can be contacted via e-mail at `jvenner@prohadoopbook.com`. Also, visit this book's web site at `http://www.prohadoopbook.com`.

CHAPTER 1

Getting Started with Hadoop Core

Applications frequently require more resources than are available on an inexpensive machine. Many organizations find themselves with business processes that no longer fit on a single cost-effective computer. A simple but expensive solution has been to buy specialty machines that have a lot of memory and many CPUs. This solution scales as far as what is supported by the fastest machines available, and usually the only limiting factor is your budget. An alternative solution is to build a high-availability cluster. Such a cluster typically attempts to look like a single machine, and typically requires very specialized installation and administration services. Many high-availability clusters are proprietary and expensive.

A more economical solution for acquiring the necessary computational resources is cloud computing. A common pattern is to have bulk data that needs to be transformed, where the processing of each data item is essentially independent of other data items; that is, using a single-instruction multiple-data (SIMD) algorithm. Hadoop Core provides an open source framework for cloud computing, as well as a distributed file system.

This book is designed to be a practical guide to developing and running software using Hadoop Core, a project hosted by the Apache Software Foundation. This chapter introduces Hadoop Core and details how to get a basic Hadoop Core installation up and running.

Introducing the MapReduce Model

Hadoop supports the MapReduce model, which was introduced by Google as a method of solving a class of petascale problems with large clusters of inexpensive machines. The model is based on two distinct steps for an application:

- *Map*: An initial ingestion and transformation step, in which individual input records can be processed in parallel.
- *Reduce*: An aggregation or summarization step, in which all associated records must be processed together by a single entity.

The core concept of MapReduce in Hadoop is that input may be split into logical chunks, and each chunk may be initially processed independently, by a map task. The results of these individual processing chunks can be physically partitioned into distinct sets, which are then sorted. Each sorted chunk is passed to a reduce task. Figure 1-1 illustrates how the MapReduce model works.

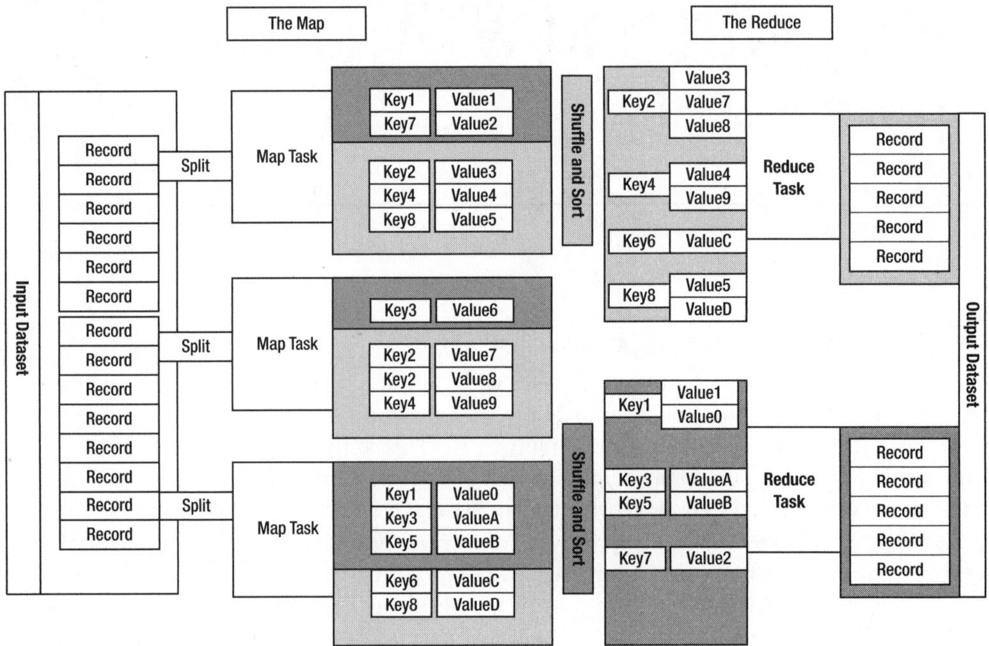

Figure 1-1. *The MapReduce model*

A map task may run on any compute node in the cluster, and multiple map tasks may be running in parallel across the cluster. The map task is responsible for transforming the input records into key/value pairs. The output of all of the maps will be partitioned, and each partition will be sorted. There will be one partition for each reduce task. Each partition's sorted keys and the values associated with the keys are then processed by the reduce task. There may be multiple reduce tasks running in parallel on the cluster.

The application developer needs to provide only four items to the Hadoop framework: the class that will read the input records and transform them into one key/value pair per record, a map method, a reduce method, and a class that will transform the key/value pairs that the reduce method outputs into output records.

My first MapReduce application was a specialized web crawler. This crawler received as input large sets of media URLs that were to have their content fetched and processed. The media items were large, and fetching them had a significant cost in time and resources. The job had several steps:

1. Ingest the URLs and their associated metadata.
2. Normalize the URLs.
3. Eliminate duplicate URLs.

4. Filter the URLs against a set of exclusion and inclusion filters.
5. Filter the URLs against a do not fetch list.
6. Filter the URLs against a recently seen set.
7. Fetch the URLs.
8. Fingerprint the content items.
9. Update the recently seen set.
10. Prepare the work list for the next application.

I had 20 machines to work with on this project. The previous incarnation of the application was very complex and used an open source queuing framework for distribution. It performed very poorly. Hundreds of work hours were invested in writing and tuning the application, and the project was on the brink of failure. Hadoop was suggested by a member of a different team.

After spending a day getting a cluster running on the 20 machines, and running the examples, the team spent a few hours working up a plan for nine map methods and three reduce methods. The goal was to have each map or reduce method take less than 100 lines of code. By the end of the first week, our Hadoop-based application was running substantially faster and more reliably than the prior implementation. Figure 1-2 illustrates its architecture. The fingerprint step used a third-party library that had a habit of crashing and occasionally taking down the entire machine.

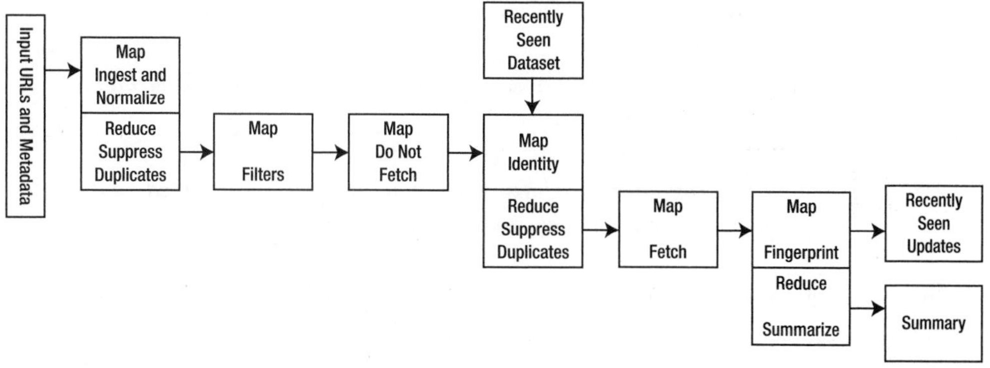

Figure 1-2. *The architecture of my first MapReduce application*

The ease with which Hadoop distributed the application across the cluster, along with the ability to continue to run in the event of individual machine failures, made Hadoop one of my favorite tools.

Both Google and Yahoo handle applications on the petabyte scale with MapReduce clusters. In early 2008, Google announced that it processes 20 petabytes of data a day with MapReduce (see `http://googleblog.blogspot.com/2008/11/sorting-1pb-with-mapreduce.html`).

Introducing Hadoop

Hadoop is the Apache Software Foundation top-level project that holds the various Hadoop subprojects that graduated from the Apache Incubator. The Hadoop project provides and supports the development of open source software that supplies a framework for the development of highly scalable distributed computing applications. The Hadoop framework handles the processing details, leaving developers free to focus on application logic.

Note The Hadoop logo is a stuffed yellow elephant. And Hadoop happened to be the name of a stuffed yellow elephant owned by the child of the principle architect.

The introduction on the Hadoop project web page (`http://hadoop.apache.org/`) states:

The Apache Hadoop project develops open-source software for reliable, scalable, distributed computing, including:

Hadoop Core, our flagship sub-project, provides a distributed filesystem (HDFS) and support for the MapReduce distributed computing metaphor.

HBase builds on Hadoop Core to provide a scalable, distributed database.

Pig is a high-level data-flow language and execution framework for parallel computation. It is built on top of Hadoop Core.

ZooKeeper is a highly available and reliable coordination system. Distributed applications use ZooKeeper to store and mediate updates for critical shared state.

Hive is a data warehouse infrastructure built on Hadoop Core that provides data summarization, adhoc querying and analysis of datasets.

The Hadoop Core project provides the basic services for building a cloud computing environment with commodity hardware, and the APIs for developing software that will run on that cloud. The two fundamental pieces of Hadoop Core are the MapReduce framework, the cloud computing environment, and he Hadoop Distributed File System (HDFS).

Note Within the Hadoop Core framework, MapReduce is often referred to as `mapred`, and HDFS is often referred to as `dfs`.

The Hadoop Core MapReduce framework requires a shared file system. This shared file system does not need to be a system-level file system, as long as there is a distributed file system plug-in available to the framework. While Hadoop Core provides HDFS, HDFS is not required. In Hadoop JIRA (the issue-tracking system), item 4686 is a tracking ticket to separate HDFS into its own Hadoop project. In addition to HDFS, Hadoop Core supports the Cloud-Store (formerly Kosmos) file system (`http://kosmosfs.sourceforge.net/`) and Amazon Simple Storage Service (S3) file system (`http://aws.amazon.com/s3/`). The Hadoop Core framework comes with plug-ins for HDFS, CloudStore, and S3. Users are also free to use any distributed file system that is visible as a system-mounted file system, such as Network File System (NFS), Global File System (GFS), or Lustre.

When HDFS is used as the shared file system, Hadoop is able to take advantage of knowledge about which node hosts a physical copy of input data, and will attempt to schedule the task that is to read that data, to run on that machine. This book mainly focuses on using HDFS as the file system.

Hadoop Core MapReduce

The Hadoop Distributed File System (HDFS)MapReduce environment provides the user with a sophisticated framework to manage the execution of map and reduce tasks across a cluster of machines. The user is required to tell the framework the following:

- The location(s) in the distributed file system of the job input
- The location(s) in the distributed file system for the job output
- The input format
- The output format
- The class containing the map function
- Optionally. the class containing the reduce function
- The JAR file(s) containing the map and reduce functions and any support classes

If a job does not need a reduce function, the user does not need to specify a reducer class, and a reduce phase of the job will not be run. The framework will partition the input, and schedule and execute map tasks across the cluster. If requested, it will sort the results of the map task and execute the reduce task(s) with the map output. The final output will be moved to the output directory, and the job status will be reported to the user.

MapReduce is oriented around key/value pairs. The framework will convert each record of input into a key/value pair, and each pair will be input to the map function once. The map output is a set of key/value pairs—nominally one pair that is the transformed input pair, but it is perfectly acceptable to output multiple pairs. The map output pairs are grouped and sorted by key. The reduce function is called one time for each key, in sort sequence, with the key and the set of values that share that key. The reduce method may output an arbitrary number of key/value pairs, which are written to the output files in the job output directory. If the reduce output keys are unchanged from the reduce input keys, the final output will be sorted.

The framework provides two processes that handle the management of MapReduce jobs:

- TaskTracker manages the execution of individual map and reduce tasks on a compute node in the cluster.
- JobTracker accepts job submissions, provides job monitoring and control, and manages the distribution of tasks to the TaskTracker nodes.

Generally, there is one JobTracker process per cluster and one or more TaskTracker processes per node in the cluster. The JobTracker is a single point of failure, and the JobTracker will work around the failure of individual TaskTracker processes.

■ **Note** One very nice feature of the Hadoop Core MapReduce environment is that you can add TaskTracker nodes to a cluster while a job is running and have the job spread out onto the new nodes.

The Hadoop Distributed File System

HDFS is a file system that is designed for use for MapReduce jobs that read input in large chunks of input, process it, and write potentially large chunks of output. HDFS does not handle random access particularly well. For reliability, file data is simply mirrored to multiple storage nodes. This is referred to as *replication* in the Hadoop community. As long as at least one replica of a data chunk is available, the consumer of that data will not know of storage server failures.

HDFS services are provided by two processes:

- NameNode handles management of the file system metadata, and provides management and control services.
- DataNode provides block storage and retrieval services.

There will be one NameNode process in an HDFS file system, and this is a single point of failure. Hadoop Core provides recovery and automatic backup of the NameNode, but no hot failover services. There will be multiple DataNode processes within the cluster, with typically one DataNode process per storage node in a cluster.

■ **Note** It is common for a node in a cluster to provide both TaskTracker services and DataNode services. It is also common for one node to provide the JobTracker and NameNode services.

Installing Hadoop

As with all software, you need some prerequisite pieces before you can actually use Hadoop. It is possible to run and develop Hadoop applications under Windows, provided that Cygwin is installed. It is strongly suggested that nodes in a production Hadoop cluster run a modern Linux distribution.

■ **Note** To use Hadoop, you'll need a basic working knowledge of Linux and Java. All the examples in this book are set up for the bash shell.

The Prerequisites

The examples in this book were developed with the following:

- Fedora 8
- Sun Java 1.6
- Hadoop 0.19.0 or later

Hadoop versions prior to 0.18.2 make much less use of generics, and the book examples are unlikely to compile with those versions. Java versions prior to 1.6 will not support all of the language features that Hadoop Core requires. In addition, Hadoop Core appears to run most stably with the Sun Java Development Kits (JDKs); there are periodic requests for help from users of other vendors' JDKs. The examples in later chapters of this book are based on Hadoop 0.19.0, which requires JDK 1.6.

Any modern Linux distribution will work. I prefer the Red Hat Package Manager (RPM) tool that is used by Red Hat, Fedora, and CentOS, and the examples reference RPM-based installation procedures.

The wonderful folks of the Fedora project provide *torrents* (downloaded with BitTorrent) for most of the Fedora versions at `http://torrent.fedoraproject.org/`. For those who want to bypass the update process, the people of Fedora Unity provide distributions of Fedora releases that have the updates applied, at `http://spins.fedoraunity.org/spins`. These are referred to as *re-spins*. They do not provide older releases. The re-spins require the use of the custom download tool Jigdo.

For the novice Linux user who just wants to play around a bit, the Live CD and a USB stick for permanent storage can provide a simple and quick way to boot up a test environment. For a more sophisticated user, VMware Linux installation images are readily available at `http://www.vmware.com/appliances/directory/cat/45?sort=changed`.

Hadoop on a Linux System

After Linux is installed, it is necessary to work out where the JDK is installed on the computer so that the `JAVA_HOME` environment variable and the `PATH` environment variable may be set correctly.

The rpm command has options that will tell you which files were in an RPM package: -q to query, -l to list, and -p to specify that the path to package you are querying is the next argument. Look for the string '/bin/javac$', using the egrep program, which searches for simple regular expressions in its input stream:

```
cloud9: ~/Downloads$ rpm -q -l -p ~/Downloads/jdk-6u7-linux-i586.rpm ➥
| egrep '/bin/javac$'
```

```
/usr/java/jdk1.6.0_07/bin/javac
```

■Note The single quotes surrounding the /bin/javac$ are required. If you don't use quotes, or use double quotes, the shell may try to resolve the $ character as a variable.

This assumes a working directory of ~/Downloads when running the JDK installation program, as the installer unpacks the bundled RPM files in the current working directory.

The output indicates that the JDK was installed in /usr/java/jdk1.6.0_07 and the java programs are in the directory /usr/java/jdk1.6.0_07/bin.

Add the following two lines to your .bashrc or .bash_profile:

```
export JAVA_HOME=/usr/java/jdk1.6.0_07
export PATH=${JAVA_HOME}/bin:${PATH}
```

The update_env.sh script, shown in Listing 1-1, will attempt to do this setup for you (this script is provided along with the downloadable code for this book). This script assumes you downloaded the RPM installer for the JDK.

Listing 1-1. *The update_env.sh Script*

```
#! /bin/sh

# This script attempts to work out the installation directory of the jdk,
# given the installer file.
# The script assumes that the installer is an rpm based installer and
# that the name of the downloaded installer ends in
# -rpm-bin
#
# The script first attempts to verify there is one argument and the
# argument is an existing file
# The file may be either the installer binary, the -rpm.bin
# or the actual installation rpm that was unpacked by the installer
#
```

```
# The script will use the rpm command to work out the
# installation package name from the rpm file, and then
# use the rpm command to query the installation database,
# for where the files of the rpm were installed.

# This query of the installation is done rather than
# directly querying the rpm, on the off
# chance that the installation was installed in a different root
# directory than the default.

# Finally, the proper environment set commands are appended
# to the user's .bashrc and .bash_profile file, if they exist, and
# echoed to the standard out so the user may apply them to
# their currently running shell sessions.

# Verify that there was a single command line argument
# which will be referenced as $1
if [ $# != 1 ]; then
    echo "No jdk rpm specified"
    echo "Usage: $0 jdk.rpm" 1>&2
    exit 1
fi

# Verify that the command argument exists in the file system
if [ ! -e $1 ]; then
    echo "the argument specified ($1) for the jdk rpm does not exist" 1>&2
    exit 1
fi

# Does the argument end in '-rpm.bin' which is the suggested install
# file, is the argument the actual .rpm file, or something else
# set the variable RPM to the expected location of the rpm file that
# was extracted from the installer file
if echo $1 | grep -q -e '-rpm.bin'; then
    RPM=`dirname $1`/`basename $1 -rpm.bin`.rpm
elif echo $1 | grep -q -e '.rpm'; then
    RPM=$1
else
    echo -n "$1 does not appear to be the downloaded rpm.bin file or" 1>&2
    echo " the extracted rpm file" 1>&2
    exit 1
fi

# Verify that the rpm file exists and is readable
```

```
if [ ! -r $RPM ]; then
    echo -n "The jdk rpm file (${RPM}) does not appear to exist" 1>&2
    echo -n " have you run "sh ${RPM}" as root?" 1>&2
    exit 1
fi

# Work out the actual installed package name using the rpm command
#. man rpm for details
INSTALLED=`rpm -q --qf %{Name}-%{Version}-%{Release} -p ${RPM}`
if [ $? -ne 0 ]; then
    (echo -n "Unable to extract package name from rpm (${RPM}),"
     Echo " have you installed it yet?") 1>&2
    exit 1
fi

# Where did the rpm install process place the java compiler program 'javac'
JAVAC=`rpm -q -l ${INSTALLED} | egrep '/bin/javac$'`

# If there was no javac found, then issue an error
if [ $? -ne 0 ]; then
    (echo -n "Unable to determine the JAVA_HOME location from $RPM, "
     echo "was the rpm installed? Try rpm -Uvh ${RPM} as root.") 1>&2
    exit 1
fi

# If we found javac, then we can compute the setting for JAVA_HOME
JAVA_HOME=`echo $JAVAC | sed -e 's;/bin/javac;;'`


echo "The setting for the JAVA_HOME environment variable is ${JAVA_HOME}"

echo -n "update the user's .bashrc if they have one with the"
echo " setting for JAVA_HOME and the PATH."
if [ -w ~/.bashrc ]; then
    echo "Updating the ~/.bashrc file with the java environment variables";
    (echo export JAVA_HOME=${JAVA_HOME} ;
        echo export PATH='${JAVA_HOME}'/bin:'${PATH}' ) >> ~/.bashrc
    echo
fi
```

```
echo -n "update the user's .bash_profile if they have one with the"
echo " setting for JAVA_HOME and the PATH."
if [ -w ~/.bash_profile ]; then
    echo "Updating the ~/.bash_profile file with the java environment variables";
    (echo export JAVA_HOME=${JAVA_HOME} ;
        echo export PATH='${JAVA_HOME}'/bin:'${PATH}' ) >> ~/.bash_profile
    echo
fi

echo "paste the following two lines into your running shell sessions"
echo export JAVA_HOME=${JAVA_HOME}
echo export PATH='${JAVA_HOME}'/bin:'${PATH}'
```

Run the script in Listing 1-1 to work out the JDK installation directory and update your environment so that the JDK will be used by the examples:

```
update_env.sh "FULL_PATH_TO_DOWNLOADED_JDK"

./update_env.sh ~/Download/jdk-6u7-linux-i586-rpm.bin
```

```
The setting for the JAVA_HOME environment variable is /usr/java/jdk1.6.0_07

update the user's .bashrc if they have one with the setting ➥
for JAVA_HOME and the PATH.

Updating the ~/.bashrc file with the java environment variables


update the user's .bash_profile if they have one with the setting ➥
for JAVA_HOME and the PATH.

Updating the ~/.bash_profile file with the java environment variables


paste the following two lines into your running shell sessions

export JAVA_HOME=/usr/java/jdk1.6.0_07

export PATH=${JAVA_HOME}/bin:${PATH}
```

Hadoop on a Windows System: How To and Common Problems

To use Hadoop on a Windows system, you will need to install the Sun JDK and the Cygwin environment (available from `http://sources.redhat.com/cygwin`).

Run a Cygwin bash shell by clicking the icon shown in Figure 1-3. You will need to make a symbolic link in the ~ directory between the JDK installation directory and `java`, so that `cd ~/java` will change the working directory to the root of the JDK directory. The appropriate setting for `JAVA_HOME` becomes `export JAVA_HOME=~/java`. This will set up your default process environment to have the `java` programs in your path and let programs, such as Hadoop, know where the Java installation is on your computer.

Figure 1-3. *The Cygwin bash shell icon*

I was unable to make the `bin/hadoop` script work if the path in the `JAVA_HOME` environment variable contained space characters, and the normal installation is in `C:\Program Files\java\jdkRELEASE_VERSION`. When a symbolic link is made and the `JAVA_HOME` set to the symbolic link location, `bin/hadoop` works well enough to use. For my Cygwin installation, I have the following:

```
$ echo $JAVA_HOME
```

```
/home/Jason/jdk1.6.0_12/
```

```
$ ls -l /home/Jason/jdk1.6.0_12
```

```
lrwxrwxrwx 1 Jason None 43 Mar 20 16:32 /home/Jason/jdk1.6.0_12 ➥
 -> /cygdrive/c/Program Files/Java/jdk1.6.0_12/
```

Cygwin maps Windows drive letters to the path `/cygdrive/X`, where `X` is the drive letter, and the Cygwin path element separator character is `/`, compared to Windows use of `\`.

You must keep two views of your files in mind, particularly when running Java programs via the `bin/hadoop` script. The `bin/hadoop` script and all of the Cygwin utilities see a file system that is a subtree of the Windows file system, with the Windows drives mapped to the `/cygdrive` directory. The Windows programs see the traditional `C:\` file system. An example of this is `/tmp`. In a standard Cygwin installation, the `/tmp` directory is also the `C:\cygwin\tmp` directory. Java will parse `/tmp` as `C:\tmp`, a completely different directory. When you receive File Not Found errors from Windows applications launched from Cygwin, the common problem is that the Windows application (Java being a Windows application) is looking in a different directory than you expect.

■ **Note** You will need to customize the Cygwin setup for your system. The exact details change with different Sun JDK releases and with different Windows installations. In particular, the username will probably not be `Jason`, the JDK version may not be `1.6.0_12`, and the Java installation location may not be `C:\Program Files\Java`.

Getting Hadoop Running

After you have your Linux or Cygwin under Windows environment set up, you're ready to download and install Hadoop.

Go to the Hadoop download site at `http://www.apache.org/dyn/closer.cgi/hadoop/core/`. From there, find the tar.gz file of the distribution of your choice, bearing in mind what I said in the introduction, and download that file.

If you are a cautious person, go to the backup site and get the PGP checksum or the MD5 checksum of the download file.

Unpack the `tar` file in the directory where you would like your test installation installed. I typically unpack this in a `src` directory, off my personal home directory:

```
~jason/src.
mkdir ~/src
cd ~/src
tar zxf ~/Downloads/hadoop-0.19.0.tar.gz
```

This will create a directory named `hadoop-0.19.0` in my `~/src` directory.

Add the following two lines to your `.bashrc` or `.bash_profile` file and execute them in your current shell:

```
export HADOOP_HOME=~/src/hadoop-0.19.0
export PATH=${HADOOP_HOME}/bin:${PATH}
```

If you chose a different directory than `~/src`, adjust these `export` statements to reflect your chosen location.

Checking Your Environment

After installing Hadoop, you should check that you have updated your shell environment with the `JAVA_HOME` and `HADOOP_HOME` environment variables correctly; that your `PATH` environment variable has `${JAVA_HOME}/bin` and `${HADOOP_HOME}/bin` to the left of any other Java or Hadoop installations in your path, preferably as the first to elements of your `PATH`; and that your shell's current working directory is `${HADOOP_HOME}`. These settings are required to run the examples in this book.

The shell script `check_basic_env.sh`, shown in Listing 1-2, will verify your runtime environment (this script is provided along with the other downloadable code for this book).

Listing 1-2. *The check_basic_env.sh Script*

```
#! /bin/sh

# This block is trying to do the basics of checking to see if
# the HADOOP_HOME and the JAVA_HOME variables have been set correctly
# and if they are not been set, suggest a setting in line with the earlier examples

# The script actually tests for:
# the presence of the java binary and the hadoop script,
# and verifies that the expected versions are present
# that the version of java and hadoop is as expected (warning if not)
# that the version of java and hadoop referred to by the
# JAVA_HOME and HADOOP_HOME environment variables are default version to run.
#
#
# The 'if [' construct you see is a shortcut for 'if test' ....
# the -z tests for a zero length string
# the -d tests for a directory
# the -x tests for the execute bit
# -eq tests numbers
# = tests strings
# man test will describe all of the options

# The '1>&2' construct directs the standard output of the
# command to the standard error stream.

if [ -z "$HADOOP_HOME" ]; then
    echo "The HADOOP_HOME environment variable is not set" 1>&2
    if [ -d ~/src/hadoop-0.19.0 ]; then
        echo "Try export HADOOP_HOME=~/src/hadoop-0.19.0" 1>&2
    fi
    exit 1;
fi

# This block is trying to do the basics of checking to see if
# the JAVA_HOME variable has been set
# and if it hasn't been set, suggest a setting in line with the earlier examples


if [ -z "$JAVA_HOME" ]; then
    echo "The JAVA HOME environment variable is not set" 1>&2
    if [ -d /usr/java/jdk1.6.0_07 ]; then
        echo "Try export JAVA_HOME=/usr/java/jdk1.6.0_07" 1>&2
    fi
    exit 1
fi
```

```
# We are now going to see if a java program and hadoop programs
# are in the path, and if they are the ones we are expecting.
# The which command returns the full path to the first instance
# of the program in the PATH environment variable
#
JAVA_BIN=`which java`
HADOOP_BIN=`which hadoop`

# Check for the presence of java in the path and suggest an
# appropriate path setting if java is not found
if [ -z "${JAVA_BIN}" ]; then
    echo "The java binary was not found using your PATH settings" 1>&2
    if [ -x ${JAVA_HOME}/bin/java ]; then
        echo 'Try export PATH=${JAVA_HOME}/bin' 1>&2
    fi
    exit 1
fi

# Check for the presence of hadoop in the path and suggest an
# appropriate path setting if java is not found
if [ -z "${HADOOP_BIN}" ]; then
    echo "The hadoop binary was not found using your PATH settings" 1>&2
    if [ -x ${HADOOP_HOME}/bin/hadoop ]; then
        echo 'Try export PATH=${HADOOP_HOME}/bin:${PATH}' 1>&2
    fi
    exit 1
fi


# Double check that the version of java installed in ${JAVA_HOME}
# is the one stated in the examples.
# If you have installed a different version your results may vary.
#
if ! ${JAVA_HOME}/bin/java -version 2>&1 | grep -q 1.6.0_07; then
    (echo -n "Your JAVA_HOME version of java is not the"
     echo -n " 1.6.0_07 version, your results may vary from"
     echo " the book examples.") 1>&2
fi

# Double check that the java in the PATH is the expected version.
if ! java -version 2>&1 | grep -q 1.6.0_07; then
    (echo -n "Your default java version is not the 1.6.0_07 "
     echo -n "version, your results may vary from the book"
     echo " examples.") 1>&2
fi
```

```
# Try to get the location of the hadoop core jar file
# This is used to verify the version of hadoop installed
HADOOP_JAR=`ls -1 ${HADOOP_HOME}/hadoop-0.19.0-core.jar`
HADOOP_ALT_JAR=`ls -1 ${HADOOP_HOME}/hadoop-*-core.jar`
# If a hadoop jar was not found, either the installation
# was incorrect or a different version installed
if [ -z "${HADOOP_JAR}" -a -z "${HADOOP_ALT_JAR}" ]; then
    (echo -n "Your HADOOP_HOME does not provide a hadoop"
     echo -n " core jar. Your installation probably needs"
     echo -n " to be redone or the HADOOP_HOME environment"
     echo variable needs to be correctly set.") 1>&2
    exit 1
fi

if [ -z "${HADOOP_JAR}" -a ! -z "${HADOOP_ALT_JAR}" ]; then
    (echo -n "Your hadoop version appears to be different"
     echo -n " than the 0.19.0 version, your results may vary"
     echo " from the book examples.") 1>&2
fi

if [ `pwd` != ${HADOOP_HOME} ]; then
    (echo -n 'Please change your working directory to"
     echo -n " ${HADOOP_HOME}. cd ${HADOOP_HOME} <Enter>") 1>&2
    exit 1
fi

echo "You are good to go"
echo -n "your JAVA_HOME is set to ${JAVA_HOME} which "
echo "appears to exist and be the right version for the examples."
echo -n "your HADOOP_HOME is set to ${HADOOP_HOME} which "
echo "appears to exist and be the right version for the examples."
echo "your java program is the one in ${JAVA_HOME}"
echo "your hadoop program is the one in ${HADOOP_HOME}"
echo -n "The shell current working directory is ${HADOOP_HOME} "
echo "as the examples require."


if [ "${JAVA_BIN}" = "${JAVA_HOME}/bin/java" ]; then
    echo "Your PATH appears to have the JAVA_HOME java program as the default java."
else
    echo -n "Your PATH does not appear to provide the JAVA_HOME"
    echo " java program as the default java."
fi
```

```
if [ "${HADOOP_BIN}" = "${HADOOP_HOME}/bin/hadoop" ]; then
    echo -n "Your PATH appears to have the HADOOP_HOME"
    echo " hadoop program as the default hadoop."
else
    echo -n "Your PATH does not appear to provide the the HADOOP_HOME "
    echo "hadoop program as the default hadoop program."
fi


exit 0
```

Run the script as follows:

```
[scyrus@localhost ~]$ ./check_basic_env.sh
```

```
Please change your working directory to ${HADOOP_HOME}. cd ➥
${HADOOP_HOME} <Enter>
```

```
[scyrus@localhost ~]$ cd $HADOOP_HOME
[scyrus@localhost hadoop-0.19.0]$

[scyrus@localhost hadoop-0.19.0]$ ~/check_basic_env.sh
```

```
You are good to go
your JAVA_HOME is set to /usr/java/jdk1.6.0_07 which appears to exist
and be the right version for the examples.
your HADOOP_HOME is set to /home/scyrus/src/hadoop-0.19.0 which appears
to exist and be the right version for the examples.
your java program is the one in /usr/java/jdk1.6.0_07
your hadoop program is the one in /home/scyrus/src/hadoop-0.19.0
The shell current working directory is /home/scyrus/src/hadoop-0.19.0 as
the examples require.
Your PATH appears to have the JAVA_HOME java program as the default
java.
Your PATH appears to have the HADOOP_HOME hadoop program as the default
hadoop.
```

Running Hadoop Examples and Tests

The Hadoop installation provides JAR files with sample programs and tests that you can run. Before you run these, you should have verified that your installation is complete and that your runtime environment is set up correctly. As discussed in the previous section, the check_basic_env.sh script will help verify your installation and suggest corrections if any are required.

Hadoop Examples

The hadoop-0.19.0-examples.jar file includes ready-to-run sample programs. Included in the JAR file are the programs listed in Table 2-1.

Table 2-1. *Examples in hadoop-0.19.0-examples.jar*

Program	Description
aggregatewordcount	An aggregate-based MapReduce program that counts the words in the input files
aggregatewordhist	An aggregate-based MapReduce program that computes the histogram of the words in the input files
grep	A MapReduce program that counts the matches of a regular expression in the input
join	A job that performs a join over sorted, equally partitioned datasets
multifilewc	A job that counts words from several files
pentomino	A MapReduce tile-laying program to find solutions to pentomino problems
pi	A MapReduce program that estimates pi using the Monte Carlo method
randomtextwriter	A MapReduce program that writes 10GB of random textual data per node
randomwriter	A MapReduce program that writes 10GB of random data per node
sleep	A job that sleeps at each map and reduce task
sort	A MapReduce program that sorts the data written by the random writer
sudoku	A sudoku solver
wordcount	A MapReduce program that counts the words in the input files

To demonstrate using the Hadoop examples, let's walk through running the pi program.

Running the Pi Estimator

The pi example estimates pi using the Monte Carlo method. The web site http://www.chem.unl.edu/zeng/joy/mclab/mcintro.html provides a good discussion of this technique. The number of samples is the number of points randomly set in the square. The larger this value, the more accurate the calculation of pi. For the sake of simplicity, we are going to make a very poor estimate of pi by using very few operations.

The pi program takes two integer arguments: the number of maps and the number of samples per map. The total number of samples used in the calculation is the number of maps times the number of samples per map.

The map task generates a random point in a 1 × 1 area. For each sample where $X^2+Y^2 <=1$, the point is inside; otherwise, the point is outside. The map outputs a key of 1 or 0 and a value of 1 for a point that is inside or outside the circle, diameter 1. The reduce task sums the number of inside points and the number of outside points. The ratio between this is, in the limit, pi.

In this example, to help the job run quicker and with less output, you will choose 2 maps, with 10 samples each, for a total of 20 samples.

To run the example, change the working directory of your shell to HADOOP_HOME (via cd ${HADOOP_HOME}) and enter the following:

```
jason@cloud9:~/src/hadoop-0.19.0$ hadoop jar hadoop-0.19.0-examples.jar pi 2 10
```

The bin/hadoop jar command submits jobs to the cluster. The command-line arguments are processed in three steps, with each step consuming some of the command-line arguments. We'll see this in detail in Chapter 5, but for now the hadoop-0.19.0-examples.jar file contains the main class for the application. The next three arguments are passed to this class.

Examining the Output: Input Splits, Shuffles, Spills, and Sorts

Your output will look something like that shown in Listing 2-3.

Listing 2-3. *Output from the Sample Pi Program*

```
Number of Maps = 2 Samples per Map = 10
Wrote input for Map #0
Wrote input for Map #1
Starting Job
jvm.JvmMetrics: Initializing JVM Metrics with processName=JobTracker, sessionId=
mapred.FileInputFormat: Total input paths to process : 2
mapred.FileInputFormat: Total input paths to process : 2
mapred.JobClient: Running job: job_local_0001
mapred.FileInputFormat: Total input paths to process : 2
mapred.FileInputFormat: Total input paths to process : 2
mapred.MapTask: numReduceTasks: 1
mapred.MapTask: io.sort.mb = 100
mapred.MapTask: data buffer = 79691776/99614720
mapred.MapTask: record buffer = 262144/327680
mapred.JobClient:  map 0% reduce 0%
mapred.MapTask: Starting flush of map output
mapred.MapTask: bufstart = 0; bufend = 32; bufvoid = 99614720
mapred.MapTask: kvstart = 0; kvend = 2; length = 327680
mapred.LocalJobRunner: Generated 1 samples
mapred.MapTask: Index: (0, 38, 38)
mapred.MapTask: Finished spill 0
mapred.LocalJobRunner: Generated 1 samples.
mapred.TaskRunner: Task 'attempt_local_0001_m_000000_0' done.
mapred.TaskRunner: Saved output of task 'attempt_local_0001_m_000000_0' ➥
to file:/home/jason/src/hadoop-0.19.0/test-mini-mr/outmapred.
MapTask: numReduceTasks: 1
mapred.MapTask: io.sort.mb = 100
mapred.JobClient:  map 0% reduce 0%
mapred.LocalJobRunner: Generated 1 samples
mapred.MapTask: data buffer = 79691776/99614720
```

```
mapred.MapTask: record buffer = 262144/327680
mapred.MapTask: Starting flush of map output
mapred.MapTask: bufstart = 0; bufend = 32; bufvoid = 99614720
mapred.MapTask: kvstart = 0; kvend = 2; length = 327680
mapred.JobClient:  map 100% reduce 0%
mapred.MapTask: Index: (0, 38, 38)
mapred.MapTask: Finished spill 0
mapred.LocalJobRunner: Generated 1 samples.
mapred.TaskRunner: Task 'attempt_local_0001_m_000001_0' done.
mapred.TaskRunner: Saved output of task 'attempt_local_0001_m_000001_0' ➥
to file:/home/jason/src/hadoop-0.19.0/test-mini-mr/out
mapred.ReduceTask: Initiating final on-disk merge with 2 files
mapred.Merger: Merging 2 sorted segments
mapred.Merger: Down to the last merge-pass, with 2 segments left of ➥
total size: 76 bytes
mapred.LocalJobRunner: reduce > reduce
mapred.TaskRunner: Task 'attempt_local_0001_r_000000_0' done.
mapred.TaskRunner: Saved output of task 'attempt_local_0001_r_000000_0' ➥
to file:/home/jason/src/hadoop-0.19.0/test-mini-mr/out
mapred.JobClient: Job complete: job_local_0001
mapred.JobClient: Counters: 11
mapred.JobClient:   File Systems
mapred.JobClient:     Local bytes read=314895
mapred.JobClient:     Local bytes written=359635
mapred.JobClient:   Map-Reduce Framework
mapred.JobClient:     Reduce input groups=2
mapred.JobClient:     Combine output records=0
mapred.JobClient:     Map input records=2
mapred.JobClient:     Reduce output records=0
mapred.JobClient:     Map output bytes=64
mapred.JobClient:     Map input bytes=48
mapred.JobClient:     Combine input records=0
mapred.JobClient:     Map output records=4
mapred.JobClient:     Reduce input records=4
Job Finished in 2.322 seconds
Estimated value of PI is 3.8
```

Note The Hadoop projects use the Apache Foundation's log4j package for logging. By default, all output by the framework will have a leading date stamp, a log level, and the name of the class that emitted the message. In addition, the default is only to emit log messages of level `INFO` or higher. For brevity, I've removed the data stamp and log level from the output reproduced in this book.

Of particular interest here is that the last line of output states something of the form "Estimated value of PI is...". In that case, you know that your local installation of Hadoop is ready for you to play with.

Now we will go through the output in Listing 2-3 chunk by chunk, so that you have an understanding of what is going on and can recognize when something is wrong.

The first section is output by the pi estimator as it is setting up the job. Here, you requested 2 maps and 10 samples:

```
Number of Maps = 2 Samples per Map = 10
Wrote input for Map #0
Wrote input for Map #1
```

The framework has taken over at this point and sets up input splits (each fragment of input is called an *input split*) for the map tasks.

The following line provides the job ID, which you could use to refer to this job with the job control tools:

```
Running job: job_local_0001
```

The following lines let you know that there are two input files and two input splits:

```
jvm.JvmMetrics: Initializing JVM Metrics with processName=JobTracker, sessionId=
mapred.FileInputFormat: Total input paths to process : 2
mapred.FileInputFormat: Total input paths to process : 2
mapred.JobClient: Running job: job_local_0001
mapred.FileInputFormat: Total input paths to process : 2
mapred.FileInputFormat: Total input paths to process : 2
```

The map output key/value pairs are partitioned, and then the partitions are sorted, which is referred to as the *shuffle*. The file created for each sorted partition is called a *spill*. There will be one spill file for each configured reduce task. For each reduce task, the framework will pull its spill from the output of each map task, and merge-sort these spills. This step is referred to as the *sort*.

In Listing 2-3, the next block provides detailed information on the map task and shuffle process that was run. The framework is expecting to produce output for one reduce task (`numReduceTasks: 1`), which receives all of the map task output records. Also, it expects that the map outputs have been partitioned and sorted and stored in the file system (`Finished spill 0`). If there were multiple reduce tasks specified, you would see a `Finished spill` *N* for each reduce task. The rest of the lines primarily have to do with output buffering and may be ignored.

Next, you see the following:

```
mapred.MapTask: numReduceTasks: 1
...
mapred.MapTask: Finished spill 0
mapred.LocalJobRunner: Generated 1 samples.

mapred.TaskRunner: Task 'attempt_local_0001_m_000000_0' ➥
done.mapred.TaskRunner: Saved output of task ➥
'attempt_local_0001_m_000000_0' ➥
to file:/home/jason/src/hadoop-0.19.0/test-mini-mr/out
```

`Generated 1 samples` is the output of the ending status of the map job. The Hadoop framework is telling you that the first map task is done via `Task 'attempt_local_0001_m_000000_0' done`, and that the output was saved to the default file system at `file:/home/jason/src/hadoop-0.19.0/test-mini-mr/out`.

The following block handles the sort:

```
mapred.ReduceTask: Initiating final on-disk merge with 2 files
mapred.Merger: Merging 2 sorted segments
mapred.Merger: Down to the last merge-pass, with 2 segments left of ➥
total size: 76 bytes
```

Listing 2-3 has exactly two map tasks, per your command-line instructions to the task, and one reduce task, per the job design.. With a single reduce task, each map task's output is placed into a single partition and sorted. This results in two files, or *spills*, as input to the framework sort phase. Each reduce task in a job will have its output go to the output directory and be named `part-0`*N*, where *N* is the ordinal number starting from zero of the reduce task. The numeric portion of the name is traditionally five digits, with leading zeros as needed.

The next block describes the single reduce task that will be run:

```
mapred.LocalJobRunner: reduce > reduce
mapred.TaskRunner: Task 'attempt_local_0001_r_000000_0' done.
mapred.TaskRunner: Saved output of task 'attempt_local_0001_r_000000_0' to ➥
file:/home/jason/src/hadoop-0.19.0/test-mini-mr/out
```

The output of this reduce task is written to `attempt_local_0001_r_000000_0`, and then will be renamed to `part-00000` in the job output directory.

The next block of output provides summary information about the completed job:

```
mapred.JobClient: Job complete: job_local_0001
mapred.JobClient: Counters: 11
mapred.JobClient:   File Systems
mapred.JobClient:     Local bytes read=314895
mapred.JobClient:     Local bytes written=359635
mapred.JobClient:   Map-Reduce Framework
mapred.JobClient:     Reduce input groups=2
mapred.JobClient:     Combine output records=0
mapred.JobClient:     Map input records=2
mapred.JobClient:     Reduce output records=0
mapred.JobClient:     Map output bytes=64
mapred.JobClient:     Map input bytes=48
mapred.JobClient:     Combine input records=0
mapred.JobClient:     Map output records=4
mapred.JobClient:     Reduce input records=4
```

The final two lines are printed by the `PiEstimator` code, not the framework.

```
Job Finished in 2.322 seconds
Estimated value of PI is 3.8
```

Hadoop Tests

Hadoop provides a JAR that contains tests `hadoop-0.19.0-test.jar`, which are primarily for testing the distributed file system or MapReduce jobs on top of the distributed file system. Table 2-2 lists the tests provided

Table 2-2. *Tests in hadoop-0.19.0-test.jar*

Test	Description
`DFSCIOTest`	Distributed I/O benchmark of `libhdfs`, a shared library that provides HDFS file services for C/C++ applications
`DistributedFSCheck`	Distributed checkup of the file system consistency
`TestDFSIO`	Distributed I/O benchmark
`clustertestdfs`	A pseudo distributed test for the distributed file system
`dfsthroughput`	Measures HDFS throughput
`filebench`	Benchmark `SequenceFileInputFormat` and `SequenceFileOutputFormat`, with `BLOCK` compression, `RECORD` compression, and no compression; and `TextInputFormat` and `TextOutputFormat`, compressed and uncompressed
`loadgen`	Generic MapReduce load generator
`mapredtest`	A MapReduce test check
`mrbench`	A MapReduce benchmark that can create many small jobs
`nnbench`	A benchmark that stresses the NameNode
`testarrayfile`	A test for flat files of binary key/value pairs
`testbigmapoutput`	A MapReduce program that works on a very big nonsplittable file and does an identity MapReduce
`testfilesystem`	A test for file system read/write
`testipc`	A test for Hadoop Core interprocess communications
`testmapredsort`	A MapReduce program that validates the MapReduce framework's sort
`testrpc`	A test for remote procedure calls
`testsequencefile`	A test for flat files of binary key/value pairs
`testsequencefileinputformat`	A test for sequence file input format
`testsetfile`	A test for flat files of binary key/value pairs
`testtextinputformat`	A test for text input format
`threadedmapbench`	A MapReduce benchmark that compares the performance of maps with multiple spills over maps with one spill

Troubleshooting

The issues that can cause problems in running the examples in this book will most likely be due to environment differences. You may also experience problems if you have space shortages on your computer.

The following environment variables are important:

JAVA_HOME: This is the root of the Java installations. All of the examples assume that the JAVA_HOME environment variable contains the root of the Sun JDK 1.6_07 installation, which is expected to be installed into the directory /usr/java/jdk1.6.0_07. So, this environment variable is set as follows: export JAVA_HOME=/usr/java/jdk1.6.0_07.

HADOOP_HOME: This is the root of the Hadoop installations. You should have unpacked the hadoop-0.19.0.tar.gz downloaded file with a parent directory of ~/src, such that the Hadoop program is available as ~/src/hadoop-0.19.0/bin/hadoop. The HADOOP_HOME environment variable is expected to be set to the root of the Hadoop installation, which in the examples is ~/src/hadoop-0.19.0. This environment variable is set as follows: export HADOOP_HOME=~/src/hadoop-0.19.0.

PATH: The user's path is expected to have ${JAVA_HOME}/bin and ${HADOOP_HOME}/bin as the first two elements. This environment variable is set as follows: export PATH=${JAVA_HOME}/bin:${HADOOP_HOME}/bin:${PATH}.

For Windows users, C:\cygwin\bin;C:\cygwin\usr\bin must be added to the system environment Path variable, or the Hadoop Core servers will not start. You can set this system variable through the System Control Panel. In the System Properties dialog box, click the Advanced tab, and then click the Environment Variables button. In the System Variables section of the Environment Variables dialog box, select Path, click Edit, and add the following string:

```
C:\cygwin\bin;C:\cygwin\usr\bin
```

The semicolon (;) is the separator character.

While not critical, the current working directory for the shell session used for running the examples is expected to be ${HADOOP_HOME}.

If you see the message java.lang.OutOfMemoryError: Java heap space in your output, your computer either has insufficient free RAM or the Java heap is set too small. The PiEstimator example with 2 maps and 100 samples will run with a Java Virtual Machine (JVM) maximum heap size of 128MB (-Xmx128m). To force this, you may execute the following:

```
HADOOP_OPTS="-Xmx128m" hadoop  jar hadoop-0.19.0-examples.jar pi 2 100
```

Summary

Hadoop Core provides a robust framework for distributing tasks across large numbers of general-purpose computers. Application developers just need to write the map and reduce methods for their data, and use one of the existing input and output formats. The framework provides a rich set of input and output handlers, and you can create custom handlers, if necessary.

Getting over the installation hurdle can be difficult, but it is getting simpler as more people and organizations understand the issues and refine the processes and procedures. Cloudera (`http://www.cloudera.com`) now provides a self-installing Hadoop distribution in RPM format.

New features and functionality are being tried. Read the information on the `http://hadoop.apache.org/core` web site, join the mailing lists referenced there (to join the Core user mailing list, send an e-mail to `core-user-subscribe@hadoop.apache.org`), and have fun writing your applications.

The chapters to come will guide you through the trouble spots as you develop your own applications with Hadoop.

CHAPTER 2

The Basics of a MapReduce Job

This chapter walks you through what is involved in a MapReduce job. You will be able to write and run simple stand-alone MapReduce programs by the end of the chapter.

The examples in this chapter assume the setup as described in Chapter 1. They should be explicitly run in a special local mode configuration for executing on a single machine, with no requirements for a running the Hadoop Core framework. This single machine (local) configuration is also ideal for debugging and for unit tests. The code for the examples is available from this book's details page at the Apress web site (`http://www.apress.com`). The downloadable code also includes a JAR file you can use to run the examples.

Let's start by examining the parts that make up a MapReduce job.

The Parts of a Hadoop MapReduce Job

The user configures and submits a MapReduce job (or just *job* for short) to the framework, which will decompose the job into a set of map tasks, shuffles, a sort, and a set of reduce tasks. The framework will then manage the distribution and execution of the tasks, collect the output, and report the status to the user.

The job consists of the parts shown in Figure 2-1 and listed in Table 2-1.

Table 2-1. *Parts of a MapReduce Job*

Part	Handled By
Configuration of the job	User
Input splitting and distribution	Hadoop framework
Start of the individual map tasks with their input split	Hadoop framework
Map function, called once for each input key/value pair	User
Shuffle, which partitions and sorts the per-map output	Hadoop framework
Sort, which merge sorts the shuffle output for each partition of all map outputs	Hadoop framework
Start of the individual reduce tasks, with their input partition	Hadoop framework
Reduce function, which is called once for each unique input key, with all of the input values that share that key	User
Collection of the output and storage in the configured job output directory, in *N* parts, where *N* is the number of reduce tasks	Hadoop framework

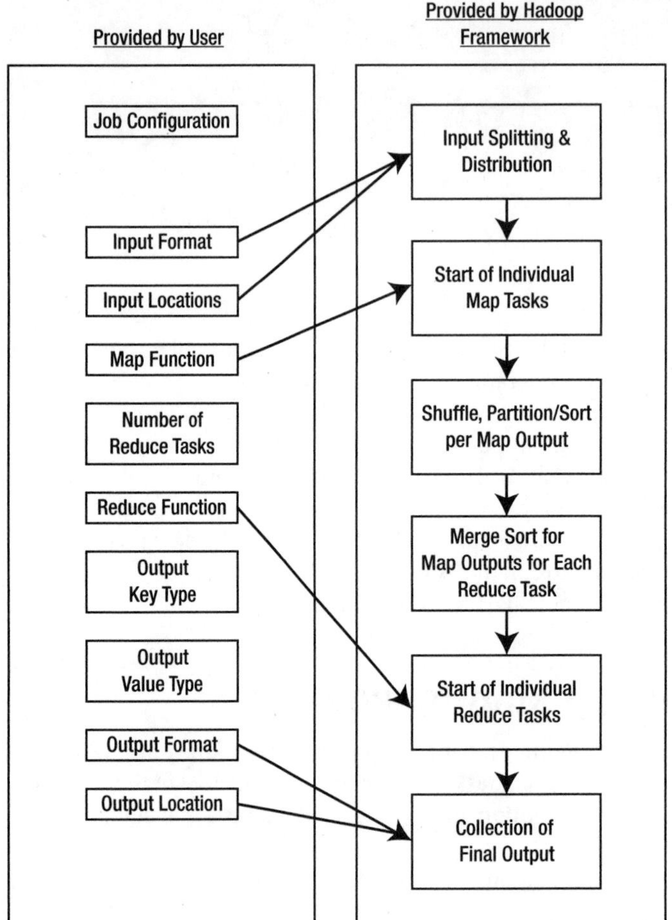

Figure 2-1. *Parts of a MapReduce job*

The user is responsible for handling the job setup, specifying the input location(s), specifying the input, and ensuring the input is in the expected format and location. The framework is responsible for distributing the job among the TaskTracker nodes of the cluster; running the map, shuffle, sort, and reduce phases; placing the output in the output directory; and informing the user of the job-completion status.

All the examples in this chapter are based on the file `MapReduceIntro.java`, shown in Listing 2-1. The job created by the code in `MapReduceIntro.java` will read all of its textual input line by line, and sort the lines based on that portion of the line before the first tab character. If there are no tab characters in the line, the sort will be based on the entire line. The `MapReduceIntro.java` file is structured to provide a simple example of configuring and running a MapReduce job.

Listing 2-1. *MapReduceIntro.java*

```
package com.apress.hadoopbook.examples.ch2;

import java.io.IOException;

import org.apache.hadoop.io.Text;
import org.apache.hadoop.mapred.FileInputFormat;
import org.apache.hadoop.mapred.FileOutputFormat;
import org.apache.hadoop.mapred.JobClient;
import org.apache.hadoop.mapred.JobConf;
import org.apache.hadoop.mapred.KeyValueTextInputFormat;
import org.apache.hadoop.mapred.RunningJob;
import org.apache.hadoop.mapred.lib.IdentityMapper;
import org.apache.hadoop.mapred.lib.IdentityReducer;
import org.apache.log4j.Logger;

/** A very simple MapReduce example that reads textual input where
 * each record is a single line, and sorts all of the input lines into
 * a single output file.
 *
 * The records are parsed into Key and Value using the first TAB
 * character as a separator. If there is no TAB character the entire
 * line is the Key. *
 *
 * @author Jason Venner
 *
 */
public class MapReduceIntro {
    protected static Logger logger = Logger.getLogger(MapReduceIntro.class);

    /**
     * Configure and run the MapReduceIntro job.
     *
     * @param args
     *            Not used.
     */
    public static void main(final String[] args) {
        try {

/** Construct the job conf object that will be used to submit this job
 * to the Hadoop framework. ensure that the jar or directory that
 * contains MapReduceIntroConfig.class is made available to all of the
 * Tasktracker nodes that will run maps or reduces for this job.
 */
            final JobConf conf = new JobConf(MapReduceIntro.class);
```

```
/**
 * Take care of some housekeeping to ensure that this simple example
 * job will run
 */
MapReduceIntroConfig.
    exampleHouseKeeping(conf,
                        MapReduceIntroConfig.getInputDirectory(),
                        MapReduceIntroConfig.getOutputDirectory());
/**
 * This section is the actual job configuration portion /**
 * Configure the inputDirectory and the type of input. In this case
 * we are stating that the input is text, and each record is a
 * single line, and the first TAB is the separator between the key
 * and the value of the record.
 */
conf.setInputFormat(KeyValueTextInputFormat.class);
FileInputFormat.setInputPaths(conf,
                        MapReduceIntroConfig.getInputDirectory());

/** Inform the framework that the mapper class will be the
 * {@link IdentityMapper}.  This class simply passes the
 * input Key Value pairs directly to its output, which in
 * our case will be the shuffle.
 */
conf.setMapperClass(IdentityMapper.class);

/** Configure the output of the job to go to the output
 * directory.  Inform the framework that the Output Key
 * and Value classes will be {@link Text} and the output
 * file format will {@link TextOutputFormat}. The
 * TextOutput format class joins produces a record of
 * output for each Key,Value pair, with the following
 * format.  Formatter.format( "%s\t%s%n", key.toString(),
 * value.toString() );.
 *
 * In addition indicate to the framework that there will be
 * 1 reduce. This results in all input keys being placed
 * into the same, single, partition, and the final output
 * being a single sorted file.
 */
FileOutputFormat.setOutputPath(conf,
                    MapReduceIntroConfig.getOutputDirectory());
conf.setOutputKeyClass(Text.class);
conf.setOutputValueClass(Text.class);
conf.setNumReduceTasks(1);
```

```
/** Inform the framework that the reducer class will be the {@link
  * IdentityReducer}.  This class simply writes an output record key,
  * value record for each value in the key, valueset it receives as
  * input.  The value ordering is arbitrary.
  */
            conf.setReducerClass(IdentityReducer.class);

            logger .info("Launching the job.");
/** Send the job configuration to the framework and request that the
 * job be run.
 */
            final RunningJob job = JobClient.runJob(conf);
            logger.info("The job has completed.");

            if (!job.isSuccessful()) {
                logger.error("The job failed.");
                System.exit(1);
            }
            logger.info("The job completed successfully.");
            System.exit(0);
        } catch (final IOException e) {
            logger.error("The job has failed due to an IO Exception", e);
            e.printStackTrace();
        }

    }
}
```

Input Splitting

For the framework to be able to distribute pieces of the job to multiple machines, it needs to fragment the input into individual pieces, which can in turn be provided as input to the individual distributed tasks. Each fragment of input is called an *input split*. The default rules for how input splits are constructed from the actual input files are a combination of configuration parameters and the capabilities of the class that actually reads the input records. These parameters are covered in Chapter 6.

An input split will normally be a contiguous group of records from a single input file, and in this case, there will be at least *N* input splits, where *N* is the number of input files. If the number of requested map tasks is larger than this number, or the individual files are larger than the suggested fragment size, there may be multiple input splits constructed of each input file. The user has considerable control over the number of input splits. The number and size of the input splits strongly influence overall job performance.

A Simple Map Function: IdentityMapper

The Hadoop framework provides a very simple map function, called `IdentityMapper`. It is used in jobs that only need to reduce the input, and not transform the raw input. We are going to examine the code of the IdentityMapper class, shown in Listing 2-2, in this section. If you have downloaded a Hadoop Core installation and followed the instructions in Chapter 1, this code is also available in the directory where you installed it, ${HADOOP_HOME}/src/mapred/org/apache/hadoop/mapred/lib/IdentityMapper.java.

Listing 2-2. *IdentityMapper.java*

```
/**
 * Licensed to the Apache Software Foundation (ASF) under one
 * or more contributor license agreements.  See the NOTICE file
 * distributed with this work for additional information
 * regarding copyright ownership.  The ASF licenses this file
 * to you under the Apache License, Version 2.0 (the
 * "License"); you may not use this file except in compliance
 * with the License.  You may obtain a copy of the License at
 *
 *     http://www.apache.org/licenses/LICENSE-2.0
 *
 * Unless required by applicable law or agreed to in writing, software
 * distributed under the License is distributed on an "AS IS" BASIS,
 * WITHOUT WARRANTIES OR CONDITIONS OF ANY KIND, either express or implied.
 * See the License for the specific language governing permissions and
 * limitations under the License.
 */

package org.apache.hadoop.mapred.lib;

import java.io.IOException;

import org.apache.hadoop.mapred.Mapper;
import org.apache.hadoop.mapred.OutputCollector;
import org.apache.hadoop.mapred.Reporter;
import org.apache.hadoop.mapred.MapReduceBase;

/** Implements the identity function, mapping inputs directly to outputs. */
public class IdentityMapper<K, V>
    extends MapReduceBase implements Mapper<K, V, K, V> {

  /** The identify function.  Input key/value pair is written directly to
   * output.*/
  public void map(K key, V val,
                  OutputCollector<K, V> output, Reporter reporter)
    throws IOException {
    output.collect(key, val);
  }
}
```

The magic piece of code is the line `output.collect(key, val)`, which passes a key/value pair back to the framework for further processing.

All map functions must implement the `Mapper` interface, which guarantees that the map function will always be called with a key. The key is an instance of a `WritableComparable` object, a value that is an instance of a `Writable` object, an output object, and a reporter. For now, just remember that the reporter is useful. Reporters are discussed in more detail in the "Creating a Custom Mapper and Reducer" section later in this chapter.

■ **Note** The code for the `Mapper.java` and `Reducer.java` interfaces is available from this book's details page at the Apress web site (`http://www.apress.com`), along with the rest of the downloadable code for this book.

The framework will make one call to your map function for each record in your input. There will be multiple instances of your map function running, potentially in multiple Java Virtual Machines (JVMs), and potentially on multiple machines. The framework coordinates all of this for you.

COMMON MAPPERS

One common mapper drops the values and passes only the keys forward:

```
public void map(K key,
                V val,
                OutputCollector<K, V> output,
                Reporter reporter)
    throws IOException {

    output.collect(key, null); /** Note, no value, just a null */

  }
```

Another common mapper converts the key to lowercase:

```
/** put the keys in lower case. */
public void map(Text key,
                V val,
                OutputCollector<Text, V> output,
                Reporter reporter)
    throws IOException {

     Text lowerCaseKey = new Text( key.toString().toLowerCase());
     output.collect(lowerCaseKey, value);

  }
```

A Simple Reduce Function: IdentityReducer

The Hadoop framework calls the reduce function one time for each unique key. The framework provides the key and the set of values that share that key.

The framework-supplied class `IdentityReducer` is a simple example that produces one output record for every value. Listing 2-3 shows this class.

Listing 2-3. *IdentityReducer.java*

```
/**
 * Licensed to the Apache Software Foundation (ASF) under one
 * or more contributor license agreements.  See the NOTICE file
 * distributed with this work for additional information
 * regarding copyright ownership.  The ASF licenses this file
 * to you under the Apache License, Version 2.0 (the
 * "License"); you may not use this file except in compliance
 * with the License.  You may obtain a copy of the License at
 *
 *     http://www.apache.org/licenses/LICENSE-2.0
 *
 * Unless required by applicable law or agreed to in writing, software
 * distributed under the License is distributed on an "AS IS" BASIS,
 * WITHOUT WARRANTIES OR CONDITIONS OF ANY KIND, either express or implied.
 * See the License for the specific language governing permissions and
 * limitations under the License.
 */

package org.apache.hadoop.mapred.lib;

import java.io.IOException;

import java.util.Iterator;

import org.apache.hadoop.mapred.Reducer;
import org.apache.hadoop.mapred.OutputCollector;
import org.apache.hadoop.mapred.Reporter;
import org.apache.hadoop.mapred.MapReduceBase;

/** Performs no reduction, writing all input values directly to the output. */
public class IdentityReducer<K, V>
    extends MapReduceBase implements Reducer<K, V, K, V> {
```

```
/** Writes all keys and values directly to output. */
public void reduce(K key, Iterator<V> values,
                   OutputCollector<K, V> output, Reporter reporter)
  throws IOException {
  while (values.hasNext()) {
    output.collect(key, values.next());
  }
}
```

If you require the output of your job to be sorted, the reducer function must pass the key objects to the `output.collect()` method unchanged. The reduce phase is, however, free to output any number of records, including zero records, with the same key and different values. This particular constraint is also why the map tasks may be multithreaded, while the reduce tasks are explicitly only single-threaded.

COMMON REDUCERS

A common reducer drops the values and passes only the keys forward:

```
public void map(K key,
                  V val,
                  OutputCollector<K, V> output,
                  Reporter reporter)
    throws IOException {

    output.collect(key, null);

  }
```

Another common reducer provides count information for each key:

```
  protected Text count = new Text();
  /** Writes all keys and values directly to output. */
  public void reduce(K key, Iterator<V> values,
                     OutputCollector<K, V> output, Reporter reporter)
    throws IOException {
    int i = 0;
    while (values.hasNext()) {
        i++
    }
    count.set( "" + i );
    output.collect(key, count);
  }
```

Configuring a Job

All Hadoop jobs have a driver program that configures the actual MapReduce job and submits it to the Hadoop framework. This configuration is handled through the `JobConf` object. The sample class `MapReduceIntro` provides a walk-through for using the `JobConf` object to configure and submit a job to the Hadoop framework for execution. The code relies on a class called `MapReduceIntroConfig`, shown in Listing 2-4, which ensures that the input and output directories are set up and ready.

Listing 2-4. *MapReduceIntroConfig.java*

```
package com.apress.hadoopbook.examples.ch2;

import java.io.IOException;
import java.util.Formatter;
import java.util.Random;

import org.apache.hadoop.fs.FSDataOutputStream;
import org.apache.hadoop.fs.FileStatus;
import org.apache.hadoop.fs.FileSystem;
import org.apache.hadoop.fs.Path;
import org.apache.hadoop.mapred.JobConf;
import org.apache.log4j.Logger;

/** A simple class to handle the housekeeping for the MapReduceIntro
 * example job.
 *
 *
 * <p>
 * This job explicitly configures the job to run, locally and without a
 * distributed file system, as a stand alone application.
 * </p>
 * <p>
 * The input is read from the directory /tmp/MapReduceIntroInput and
 * the output is written to the directory
 * /tmp/MapReduceIntroOutput. If the directory
 * /tmp/MapReduceIntroInput is missing or empty, it is created and
 * some input data files generated. If the directory
 * /tmp/MapReduceIntroOutput is present, it is removed.
 * </p>
 *
 * @author Jason Venner
 */
```

```
public class MapReduceIntroConfig {
    /**
     * Log4j is the recommended way to provide textual information to the user
     * about the job.
     */
    protected static Logger logger =
        Logger.getLogger(MapReduceIntroConfig.class);

    /** Some simple defaults for the job input and job output. */
    /**
     * This is the directory that the framework will look for input files in.
     * The search is recursive if the entry is a directory.
     */
    protected static Path inputDirectory =
        new Path("file:///tmp/MapReduceIntroInput");
    /**
     * This is the directory that the job output will be written to. It must not
     * exist at Job Submission time.
     */
    protected static Path outputDirectory =
        new Path("file:///tmp/MapReduceIntroOutput");

    /**
     * Ensure that there is some input in the <code>inputDirectory</code>,
     * the <code>outputDirectory</code> does not exist and that this job will
     * be run as a local stand alone application.
     *
     * @param conf
     *            The {@link JobConf} object that is required for doing file
     *            system access.
     * @param inputDirectory
     *            The directory the input will reside in.
     * @param outputDirectory
     *            The directory that the output will reside in
     * @throws IOException
     */
    protected static void exampleHouseKeeping(final JobConf conf,
            final Path inputDirectory, final Path outputDirectory)
            throws IOException {
        /**
         * Ensure that this job will be run stand alone rather than relying on
         * the services of an external JobTracker.
         */
        conf.set("mapred.job.tracker", "local");
```

```
        /** Ensure that no global file system is required to run this job. */
        conf.set("fs.default.name", "file:///");
        /**
         * Reduce the in ram sort space, so that the user does not need to
         * increase the jvm memory size. This sets the sort space to 1 Mbyte,
         * which is very small for a real job.
         */
        conf.setInt("io.sort.mb", 1);
        /**
         * Generate some sample input if the <code>inputDirectory</code> is
         * empty or absent.
         */
        generateSampleInputIf(conf, inputDirectory);

        /**
         * Remove the file system item at <code>outputDirectory</code> if it
         * exists.
         */
        if (!removeIf(conf, outputDirectory)) {
            logger.error("Unable to remove " + outputDirectory + "job aborted");
            System.exit(1);
        }
    }

    /**
     * Generate <code>fileCount</code> files in the directory
     * <code>inputDirectory</code>, where the individual lines of the file
     * are a random integer TAB file name.
     *
     * The file names will be file-N where N is between 0 and
     * <code>fileCount</code> - 1. There will be between 1 and
     * <code>maxLines</code> + 1 lines in each file.
     *
     * @param fs
     *            The file system that <code>inputDirectory</code> exists in.
     * @param inputDirectory
     *            The directory to create the files in. This directory must
     *            already exist.
     * @param fileCount
     *            The number of files to create.
     * @param maxLines
     *            The maximum number of lines to write to the file.
     * @throws IOException
     */
```

```
protected static void generateRandomFiles(final FileSystem fs,
        final Path inputDirectory, final int fileCount, final int maxLines)
        throws IOException {

    final Random random = new Random();
    logger .info( "Generating 3 input files of random data," +
                 "each record is a random number TAB the input file name");

    for (int file = 0; file < fileCount; file++) {

        final Path outputFile = new Path(inputDirectory, "file-" + file);
        final String qualifiedOutputFile = outputFile.makeQualified(fs)
                .toUri().toASCIIString();
        FSDataOutputStream out = null;
        try {
            /**
             * This is the standard way to create a file using the Hadoop
             * Framework. An error will be thrown if the file already
             * exists.
             */
            out = fs.create(outputFile);

            final Formatter fmt = new Formatter(out);
            final int lineCount = (int) (Math.abs(random.nextFloat())
                    * maxLines + 1);
            for (int line = 0; line < lineCount; line++) {
                fmt.format("%d\t%s%n", Math.abs(random.nextInt()),
                        qualifiedOutputFile);
            }
            fmt.flush();
        } finally {
            /**
             * It is very important to ensure that file descriptors are
             * closed. The distributed file system code can run out of file
             * descriptors and the errors generated in that case are
             * misleading.
             */
            out.close();
        }
    }
}
```

```
/**
 * This method will generate some sample input, if the
 * <code>inputDirectory</code> is missing or empty.
 *
 * This method also demonstrates some of the basic APIs for interacting
 * with file systems and files. Note: the code has no particular knowledge
 * of the type of file system.
 *
 * @param conf
 *            The Job Configuration object, used for acquiring the
 *            {@link FileSystem} objects.
 * @param inputDirectory
 *            The directory to ensure has sample files.
 * @throws IOException
 */
protected static void generateSampleInputIf(final JobConf conf,
        final Path inputDirectory) throws IOException {

    boolean inputDirectoryExists;
    final FileSystem fs = inputDirectory.getFileSystem(conf);

    if ((inputDirectoryExists = fs.exists(inputDirectory))
            && !isEmptyDirectory(fs, inputDirectory)) {
        if (logger.isDebugEnabled()) {
            logger
                    .debug("The inputDirectory "
                            + inputDirectory
                            + " exists and is either a"
                            + " file or a non empty directory");
        }
        return;
    }

    /**
     * We should only get here if <code>inputDirectory</code> does not
     * exist, or is an empty directory.
     */
    if (!inputDirectoryExists) {
        if (!fs.mkdirs(inputDirectory)) {
            logger.error("Unable to make the inputDirectory "
                    + inputDirectory.makeQualified(fs) + " aborting job");
            System.exit(1);
        }
    }
```

```
        final int fileCount = 3;
        final int maxLines = 100;
        generateRandomFiles(fs, inputDirectory, fileCount, maxLines);
    }

    /**
     * bean access getter to the {@link #inputDirectory} field.
     *
     * @return the value of inputDirectory.
     */
    public static Path getInputDirectory() {
        return inputDirectory;
    }

    /**
     * bean access getter to the {@link outputDirectory} field.
     *
     * @return the value of outputDirectory.
     */
    public static Path getOutputDirectory() {
        return outputDirectory;
    }

    /**
     * Determine if a directory has any non zero files in it or its descendant
     * directories.
     *
     * @param fs
     *            The {@link FileSystem} object to use for access.
     * @param inputDirectory
     *            The root of the directory tree to search
     * @return true if the directory is missing or does not contain at least one
     *         non empty file.
     * @throws IOException
     */
    private static boolean isEmptyDirectory(final FileSystem fs,
            final Path inputDirectory) throws IOException {

        /**
         * This is the standard way to read a directory's contents. This can be
         * quite expensive for a large directory.
         */
        final FileStatus[] statai = fs.listStatus(inputDirectory);
```

```
        /**
         * This method returns null under some circumstances, in particular if
         * the directory does not exist.
         */
        if ((statai == null) || (statai.length == 0)) {
            if (logger.isDebugEnabled()) {
                logger.debug(inputDirectory.makeQualified(fs).toUri()
                        + " is empty or missing");
            }
            return true;
        }
        if (logger.isDebugEnabled()) {
            logger.debug(inputDirectory.makeQualified(fs).toUri()
                    + " is not empty");
        }
        /** Try to find a file in the top level that is not empty. */
        for (final FileStatus status : statai) {
            if (!status.isDir() && (status.getLen() != 0)) {
                if (logger.isDebugEnabled()) {
                    logger.debug("A non empty file "
                            + status.getPath().makeQualified(fs).toUri()
                            + " was found");
                    return false;
                }
            }
        }
        /** Recurse if there are sub directories,
         * looking for a non empty file.
         */
        for (final FileStatus status : statai) {
            if (status.isDir() && isEmptyDirectory(fs, status.getPath())) {
                continue;
            }
            /**
             * If status is a directory it must not be empty or the previous
             * test block would have triggered.
             */
            if (status.isDir()) {
                return false;
            }
        }
        /**
         * Only get here if no non empty files were found in the entire subtree
         * of <code>inputPath</code>.
         */
        return true;
    }
```

```
/**
 * Ensure that the <code>outputDirectory</code> does not exist.
 *
 * <p>
 * The framework requires that the output directory not be present at job
 * submission time.
 * </p>
 * <p>
 * This method also demonstrates how to remove a directory using the
 * {@link FileSystem} API.
 * </p>
 *
 * @param conf
 *            The configuration object. This is needed to know what file
 *            systems and file system plugins are being used.
 * @param outputDirectory
 *            The directory that must be removed if present.
 * @return true if the the <code>outputPath</code> is now missing, or
 *         false if the <code>outputPath</code> is present and was unable
 *         to be removed.
 * @throws IOException
 *             If there is an error loading or configuring the FileSystem
 *             plugin, or other IO error when attempting to access or remove
 *             the <code>outputDirectory</code>.
 */
protected static boolean removeIf(final JobConf conf,
        final Path outputDirectory) throws IOException {

    /** This is standard way to acquire a FileSystem object. */
    final FileSystem fs = outputDirectory.getFileSystem(conf);

    /**
     * If the <code>outputDirectory</code> does not exist this method is
     * done.
     */
    if (!fs.exists(outputDirectory)) {
        if (logger.isDebugEnabled()) {
            logger .debug("The output directory does not exist,"
                          + " no removal needed.");
        }
        return true;
    }
    /**
     * The getFileStatus command will throw an IOException if the path does
     * not exist.
     */
```

```
        final FileStatus status = fs.getFileStatus(outputDirectory);
        logger.info("The job output directory "
                + outputDirectory.makeQualified(fs) + " exists"
                + (status.isDir() ? " and is not a directory" : "")
                + " and will be removed");

        /**
         * Attempt to delete the file or directory. delete recursively just in
         * case <code>outputDirectory</code> is a directory with
         * sub-directories.
         */
        if (!fs.delete(outputDirectory, true)) {
            logger.error("Unable to delete the configured output directory "
                    + outputDirectory);
            return false;
        }

        /** The outputDirectory did exist, but has now been removed. */
        return true;

    }

    /**
     * bean access setter to the {@link inputDirectory} field.
     *
     * @param inputDirectory
     *            The value to set inputDirectory to.
     */
    public static void setInputDirectory(final Path inputDirectory) {
        MapReduceIntroConfig.inputDirectory = inputDirectory;
    }

    /**
     * bean access setter for the {@link outpuDirectory field.
     *
     * @param outputDirectory
     *            The value to set outputDirectory to.
     */
    public static void setOutputDirectory(final Path outputDirectory) {
        MapReduceIntroConfig.outputDirectory = outputDirectory;
    }

}
```

First, you must create a `JobConf` object. It is good practice to pass in a class that is contained in the JAR file that has your map and reduce functions. This ensures that the framework will make the JAR available to the map and reduce tasks run for your job.

```
JobConf conf = new JobConf(MapReduceIntro.class);
```

Now that you have a `JobConfig` object, `conf`, you need to set the required parameters for the job. These include the input and output directory locations, the format of the input and output, and the mapper and reducer classes.

All jobs will have a map phase, and the map phase is responsible for handling the job input. The configuration of the map phase requires you to specify the input locations and the class that will produce the key/value pairs from the input, the mapper class, and potentially, the suggested number of map tasks, map output types, and per-map task threading, as listed in Table 2-2.

Table 2-2. *Map Phase Configuration*

Element	Required?	Default
Input path(s)	Yes	
Class to read and convert the input path elements to key/value pairs	Yes	
Map output key class	No	Job output key class
Map output value class	No	Job output value class
Class supplying the map function	Yes	
Suggested minimum number of map tasks	No	Cluster default
Number of threads to run in each map task	No	1

Most Hadoop Core jobs have their input as some set of files, and these files are either a textual key/value pair per line or a Hadoop-specific binary file format that provides serialized key/value pairs. The class that handles the key/value text input is `KeyValueTextInputFormat`. The class that handles the Hadoop-specific binary file is `SequenceFileInputFormat`.

Specifying Input Formats

The Hadoop framework provides a large variety of input formats. The major distinctions are between textual input formats and binary input formats. The following are the available formats:

- `KeyValueTextInputFormat`: Key/value pairs, one per line.
- `TextInputFormant`: The key is the line number, and the value is the line.
- `NLineInputFormat`: Similar to `KeyValueTextInputFormat`, but the splits are based on *N* lines of input rather than *Y* bytes of input.
- `MultiFileInputFormat`: An abstract class that lets the user implement an input format that aggregates multiple files into one split.
- `SequenceFIleInputFormat`: The input file is a Hadoop sequence file, containing serialized key/value pairs.

`KeyValueTextInputFormat` and `SequenceFileInputFormat` are the most commonly used input formats. The examples in this chapter use `KeyValueTextInputFormat`, as the input files are human-readable.

The following block of code informs the framework of the type and location of the job input:

```
/**
 * This section is the actual job configuration portion /**
 * Configure the inputDirectory and the type of input. In this case
 * we are stating that the input is text, and each record is a
 * single line, and the first TAB is the separator between the key
 * and the value of the record.
 */
conf.setInputFormat(KeyValueTextInputFormat.class);
FileInputFormat.setInputPaths(conf,
                              MapReduceIntroConfig.getInputDirectory());
```

The line `conf.setInputFormat(KeyValueTextInputFormat.class)` informs the framework that all of the files used for input will be textual key/value pairs, one per line.

THE KEYVALUETEXTINPUTFORMAT CLASS

The `KeyValueTextInputFormat` format reads a text file and splits it into records, one record per line. The records are further divided into key/value pairs by splitting the line at the first tab character. If there is no tab character in the line, the entire line is the key, and the value object will contain a zero-length string. There is no way to distinguish an input line that contains a single tab as the last character and the same line without a trailing tab character.

Suppose that an input file has the following three lines, where TAB is replaced by an US-ASCII horizontal tab character (0x09):

```
key1TABvalue1
key2
key3TABvalue3TABvalue4
```

Your mapper would be called with the following key/value pairs:

- `key1`, `value1`
- `key2`
- `key3`, `value3TABvalue4`

The actual order in which the keys are passed to your map function is indeterminate. In a real-world example, the actual machine that ran the map that got a given key would be indeterminate. It is very likely, however, that sets of contiguous records in the input will be processed by the same map task, as each task is given one input split from which to work.

The input bytes are considered to be in the UTF-8 character set. As of Hadoop 0.18.2, there is no configurable way to change the character set interpretation of the input files handled by the `KeyValueTextInputFormat` class.

Now that the framework knows where to look for the input files and the class to use to generate key/value pairs from the input files, you need to inform the framework which map function to use.

```
/** Inform the framework that the mapper class will be the {@link
 * IdentityMapper}. This class simply passes the input key-value
 * pairs directly to its output, which in our case will be the
 * shuffle.
 */
conf.setMapperClass(IdentityMapper.class);
```

■ **Note** The simple example in this chapter does not use the optional configuration parameters. If the map function needs to output a different key or value class than the job output, those classes may be set here. In addition, Hadoop supports threading for map functions. This is ideal if the map function is not able to fully utilize the resources allocated for the map task. A simple case of where this might be beneficial is a map task that performs DNS lookups on the IP addresses in a server log.

Setting the Output Parameters

The framework requires that the output parameters be configured, even if the job will not produce any output. The framework will collect the output from the specified tasks (either the output of the map tasks for a MapReduce job that did not include reduce tasks or the output of the job's reduce tasks) and place them into the configured output directory. To avoid issues with file name collisions when placing the task output into the output directory, the framework requires that the output directory not exist when you start the job.

In our simple example, the `MapReduceIntroConfig` class handles ensuring that the output directory does not exist and provides the path to the output directory. The output parameters are actually a little more comprehensive than just the setting of the output path. The code will also set the output format and the output key and value classes.

The `Text` class is the functional equivalent of a `String`. It implements the `WritableComparable` interface, which is necessary for keys, and the `Writable` interface (which is actually a subset of `WritableComparable`), which is necessary for values. Unlike `String`, `Text` is mutable, and the `Text` class has some explicit methods for UTF-8 byte handling.

The key feature of a `Writable` is that the framework knows how to serialize and deserialize a `Writable` object. The `WritableComparable` adds the `compareTo` interface so the framework knows how to sort the `WritableComparable` objects. The interface references for `Writable Comparable` and `Writable` are shown in Listings 2-5 and 2-6.

The following code block provides an example of the minimum required configuration for the output of a MapReduce job:

```
/** Configure the output of the job to go to the output directory.
 * Inform the framework that the Output Key and Value classes will be
 * {@link Text} and the output file format will {@link
 * TextOutputFormat}. The TextOutput format class produces a record of
 * output for each Key,Value pair, with the following format.
 * Formatter.format( "%s\t%s%n", key.toString(), value.toString() );.
 *
 * In addition indicate to the framework that there will be
 * 1 reduce. This results in all input keys being placed
 * into the same, single, partition, and the final output
 * being a single sorted file.
 */
FileOutputFormat.setOutputPath(conf,
                               MapReduceIntroConfig.getOutputDirectory());
conf.setOutputKeyClass(Text.class);
conf.setOutputValueClass(Text.class);
```

The `FileOutputFormat.setOutputPath(conf, MapReduceIntroConfig.getOutputDirectory())` setting is familiar from the input example discussed earlier in the chapter. The `conf.setOutputKeyClass(Text.class)` and `conf.setOutputValueClass(Text.class)` settings are new. These settings inform the framework of the types of the key/value pairs to expect for the reduce phase. By default, these classes will also be used to set the values the framework will expect from the map output. Unsurprisingly, the method to set the output key class for the map output is `conf.setMapOutputKeyClass(Class<? extends WritableComparable>)`. To set the output value class, the method is `conf.setMapOutputValueClass(Class<? extends Writable>)`.

Listing 2-5. *WritableComparable.java*

```
/**
 * Licensed to the Apache Software Foundation (ASF) under one
 * or more contributor license agreements.  See the NOTICE file
 * distributed with this work for additional information
 * regarding copyright ownership.  The ASF licenses this file
 * to you under the Apache License, Version 2.0 (the
 * "License"); you may not use this file except in compliance
 * with the License.  You may obtain a copy of the License at
 *
 *     http://www.apache.org/licenses/LICENSE-2.0
 *
 * Unless required by applicable law or agreed to in writing, software
 * distributed under the License is distributed on an "AS IS" BASIS,
 * WITHOUT WARRANTIES OR CONDITIONS OF ANY KIND, either express or implied.
 * See the License for the specific language governing permissions and
 * limitations under the License.
 */
```

```
package org.apache.hadoop.io;

/**
 * A {@link Writable} which is also {@link Comparable}.
 *
 * <p><code>WritableComparable</code>s can be compared to each other, typically
 * via <code>Comparator</code>s. Any type which is to be used as a
 * <code>key</code> in the Hadoop Map-Reduce framework should implement this
 * interface.</p>
 *
 * <p>Example:</p>
 * <p><blockquote><pre>
 *     public class MyWritableComparable implements WritableComparable {
 *       // Some data
 *       private int counter;
 *       private long timestamp;
 *
 *       public void write(DataOutput out) throws IOException {
 *         out.writeInt(counter);
 *         out.writeLong(timestamp);
 *       }
 *
 *       public void readFields(DataInput in) throws IOException {
 *         counter = in.readInt();
 *         timestamp = in.readLong();
 *       }
 *
 *       public int compareTo(MyWritableComparable w) {
 *         int thisValue = this.value;
 *         int thatValue = ((IntWritable)o).value;
 *         return (thisValue &lt; thatValue ? -1 : (thisValue==thatValue ? 0 : 1));
 *       }
 *     }
 * </pre></blockquote></p>
 */
public interface WritableComparable<T> extends Writable, Comparable<T> {
}
```

Listing 2-6. *Writable.java*

```
/**
 * Licensed to the Apache Software Foundation (ASF) under one
 * or more contributor license agreements.  See the NOTICE file
 * distributed with this work for additional information
 * regarding copyright ownership.  The ASF licenses this file
```

```
 * to you under the Apache License, Version 2.0 (the
 * "License"); you may not use this file except in compliance
 * with the License.  You may obtain a copy of the License at
 *
 *     http://www.apache.org/licenses/LICENSE-2.0
 *
 * Unless required by applicable law or agreed to in writing, software
 * distributed under the License is distributed on an "AS IS" BASIS,
 * WITHOUT WARRANTIES OR CONDITIONS OF ANY KIND, either express or implied.
 * See the License for the specific language governing permissions and
 * limitations under the License.
 */

package org.apache.hadoop.io;

import java.io.DataOutput;
import java.io.DataInput;
import java.io.IOException;

/**
 * A serializable object which implements a simple, efficient, serialization
 * protocol, based on {@link DataInput} and {@link DataOutput}.
 *
 * <p>Any <code>key</code> or <code>value</code> type in the Hadoop Map-Reduce
 * framework implements this interface.</p>
 *
 * <p>Implementations typically implement a static <code>read(DataInput)</code>
 * method which constructs a new instance, calls {@link #readFields(DataInput)}
 * and returns the instance.</p>
 *
 * <p>Example:</p>
 * <p><blockquote><pre>
 *     public class MyWritable implements Writable {
 *       // Some data
 *       private int counter;
 *       private long timestamp;
 *
 *       public void write(DataOutput out) throws IOException {
 *         out.writeInt(counter);
 *         out.writeLong(timestamp);
 *       }
 *
 *       public void readFields(DataInput in) throws IOException {
 *         counter = in.readInt();
 *         timestamp = in.readLong();
 *       }
 *
```

```
 *       public static MyWritable read(DataInput in) throws IOException {
 *         MyWritable w = new MyWritable();
 *         w.readFields(in);
 *         return w;
 *       }
 *     }
 * </pre></blockquote></p>
 */
public interface Writable {
  /**
   * Serialize the fields of this object to <code>out</code>.
   *
   * @param out <code>DataOuput</code> to serialize this object into.
   * @throws IOException
   */
  void write(DataOutput out) throws IOException;

  /**
   * Deserialize the fields of this object from <code>in</code>.
   *
   * <p>For efficiency, implementations should attempt to re-use storage in the
   * existing object where possible.</p>
   *
   * @param in <code>DataInput</code> to deseriablize this object from.
   * @throws IOException
   */
  void readFields(DataInput in) throws IOException;
}
```

Configuring the Reduce Phase

To configure the reduce phase, the user must supply the framework with five pieces of information:

- The number of reduce tasks; if zero, no reduce phase is run
- The class supplying the reduce method
- The input key and value types for the reduce task; by default, the same as the reduce output
- The output key and value types for the reduce task
- The output file type for the reduce task output

The input and output key and value types, as well as the output file type, are the same as those covered in the previous “Setting the Output Parameters” section. Here, we will look at setting the number of reduce tasks and the reducer class.

The configured number of reduce tasks determines the number of output files for a job that will run the reduce phase. Tuning this value will have a significant impact on the overall performance of your job. The time spent sorting the keys for each output file is a function of the number of keys. In addition, the number of reduce tasks determines the maximum number of reduce tasks that can be run in parallel.

The framework generally has a default number of reduce tasks configured. This value is set by the `mapred.reduce.tasks` parameter, which defaults to 1. This will result in a single output file containing all of the output keys, in sorted order. There will be one reduce task, run on a single machine that processes every key.

The number of reduce tasks is commonly set in the configuration phase of a job.

```
conf.setNumReduceTasks(1);
```

In general, unless there is a significant need for a single output file, the number of reduce tasks is set to roughly the number of simultaneous execution slots in the cluster. In Chapter 9, the class `DataJoinReduceOutput` is provided as a sample for efficiently merging multiple reduce task outputs into a single sorted file.

CLUSTER EXECUTION SLOTS

A typical cluster is composed of *M* TaskTracker machines, with *C* CPUs, each of which supports *T* threads. This would result in *M* * *C* * *T* execution slots in the cluster. In my environment, the machines typically have eight CPUs that support one thread per CPU, and a small cluster might have ten TaskTracker machines. This gives us 10 * 8 * 1 = 80 execution slots in the cluster.

If your tasks tend not to be CPU-bound, you may adjust the number of execution slots configured to optimize the CPU utilization on your TaskTracker machines.

The configuration parameter `mapred.tasktracker.map.tasks.maximum` controls the maximum number of map tasks that will be run simultaneously on a TaskTracker node.

The configuration parameter `mapred.tasktracker.reduce.tasks.maximum` controls the maximum number of reduce tasks that will be run simultaneously on a TaskTracker node.

This requires tuning on a per-job basis and is a weakness in Hadoop at present, as the maximum values are not per-job configurable and instead require a cluster restart.

The reducer class needs to be set only if the number of reduce tasks is not zero. It is very common to not need a reducer, since frequently you do not require sorted output or value grouping by key. The actual setting of the reducer class is straightforward:

```
/** Inform the framework that the reducer class will be the
 * {@link IdentityReducer}. This class simply writes an output record
 * key/value record for each value in the key/value set it receives as
 * input. The value ordering is arbitrary.
 */
conf.setReducerClass(IdentityReducer.class);
```

A COMMON EXCEPTION

The framework relies on the output parameters being set correctly. One of the more common errors is to have each reduce task fail with an exception of the form:

```
java.io.IOException: Type mismatch in key from map: expected
org.apache.hadoop.io.LongWritable, recieved org.apache.hadoop.io.Text
```

This error indicates that output key class has been defaulted by the framework, or was set incorrectly during the job configuration.

To correct this, use the following:

```
conf.setOutputKeyClass( Text.class )
```

Or if your map output is not the same as your job output, use this form:

```
conf.setMapOutputKeyClass( Text.class )
```

This error may occur for the value class as well:

```
java.io.IOException: Type mismatch in value from map: expected
org.apache.hadoop.io.LongWritable, recieved org.apache.hadoop.io.Text
```

The corresponding `setOutputValueClass()` or `setMapOutputValue()` class methods are needed to correct this.

Running a Job

The ultimate aim of all your MapReduce job configuration is to actually run that job. The `MapReduceIntro.java` example (Listing 2-1) demonstrates a common and simple way to run a job:

```
logger .info("Launching the job.");
/** Send the job configuration to the framework
 * and request that the job be run.
*/
final RunningJob job = JobClient.runJob(conf);
logger.info("The job has completed.");
```

The method `runJob()` submits the configuration information to the framework and waits for the framework to finish running the job. The response is provided in the `job` object.

The `RunningJob` class provides a number of methods for examining the response. Perhaps the most useful is `job.isSuccessful()`.

Run `MapReduceIntro.java` as follows (using the `CH2.jar` file provided with this book's downloadable code):

```
hadoop jar DOWNLOAD_PATH/ch2.jar ➥
com.apress.hadoopbook.examples.ch2.MapReduceIntro
```

The response should be as follows:

```
ch2.MapReduceIntroConfig: Generating 3 input files of random data, each record
is a random number TAB the input file name
ch2.MapReduceIntro: Launching the job.
jvm.JvmMetrics: Initializing JVM Metrics with processName=JobTracker, sessionId=
mapred.JobClient: Use GenericOptionsParser for parsing the arguments.
Applications should implement Tool for the same.
mapred.FileInputFormat: Total input paths to process : 3
mapred.FileInputFormat: Total input paths to process : 3
mapred.FileInputFormat: Total input paths to process : 3
mapred.FileInputFormat: Total input paths to process : 3
mapred.JobClient: Running job: job_local_0001
mapred.MapTask: numReduceTasks: 1
mapred.MapTask: io.sort.mb = 1
mapred.MapTask: data buffer = 796928/996160
mapred.MapTask: record buffer = 2620/3276
mapred.MapTask: Starting flush of map output
mapred.MapTask: bufstart = 0; bufend = 664; bufvoid = 996160
mapred.MapTask: kvstart = 0; kvend = 14; length = 3276
mapred.MapTask: Index: (0, 694, 694)
mapred.MapTask: Finished spill 0
mapred.LocalJobRunner: file:/tmp/MapReduceIntroInput/file-2:0+664
mapred.TaskRunner: Task 'attempt_local_0001_m_000000_0' done.
mapred.TaskRunner: Saved output of task 'attempt_local_0001_m_000000_0' to
file:/tmp/MapReduceIntroOutput
mapred.MapTask: numReduceTasks: 1
mapred.MapTask: io.sort.mb = 1
mapred.MapTask: data buffer = 796928/996160
mapred.MapTask: record buffer = 2620/3276
mapred.MapTask: Starting flush of map output
mapred.MapTask: bufstart = 0; bufend = 3418; bufvoid = 996160
mapred.MapTask: kvstart = 0; kvend = 72; length = 3276
mapred.MapTask: Index: (0, 3564, 3564)
mapred.MapTask: Finished spill 0
mapred.LocalJobRunner: file:/tmp/MapReduceIntroInput/file-1:0+3418
mapred.TaskRunner: Task 'attempt_local_0001_m_000001_0' done.
mapred.TaskRunner: Saved output of task 'attempt_local_0001_m_000001_0' to
file:/tmp/MapReduceIntroOutput
mapred.MapTask: numReduceTasks: 1
mapred.MapTask: io.sort.mb = 1
mapred.MapTask: data buffer = 796928/996160
mapred.MapTask: record buffer = 2620/3276
```

```
mapred.MapTask: Starting flush of map output
mapred.MapTask: bufstart = 0; bufend = 3986; bufvoid = 996160
mapred.MapTask: kvstart = 0; kvend = 84; length = 3276
mapred.MapTask: Index: (0, 4156, 4156)
mapred.MapTask: Finished spill 0
mapred.LocalJobRunner: file:/tmp/MapReduceIntroInput/file-0:0+3986
mapred.TaskRunner: Task 'attempt_local_0001_m_000002_0' done.
mapred.TaskRunner: Saved output of task 'attempt_local_0001_m_000002_0' to
file:/tmp/MapReduceIntroOutput
mapred.ReduceTask: Initiating final on-disk merge with 3 files
mapred.Merger: Merging 3 sorted segments
mapred.Merger: Down to the last merge-pass, with 3 segments left of total size:
8414 bytes
mapred.LocalJobRunner: reduce > reduce
mapred.TaskRunner: Task 'attempt_local_0001_r_000000_0' done.
mapred.TaskRunner: Saved output of task 'attempt_local_0001_r_000000_0' to
file:/tmp/MapReduceIntroOutput
mapred.JobClient: Job complete: job_local_0001
mapred.JobClient: Counters: 11
mapred.JobClient:   File Systems
mapred.JobClient:     Local bytes read=230060
mapred.JobClient:     Local bytes written=319797
mapred.JobClient:   Map-Reduce Framework
mapred.JobClient:     Reduce input groups=170
mapred.JobClient:     Combine output records=0
mapred.JobClient:     Map input records=170
mapred.JobClient:     Reduce output records=170
mapred.JobClient:     Map output bytes=8068
mapred.JobClient:     Map input bytes=8068
mapred.JobClient:     Combine input records=0
mapred.JobClient:     Map output records=170
mapred.JobClient:     Reduce input records=170
ch2.MapReduceIntro: The job has completed.
ch2.MapReduceIntro: The job completed successfully.
```

Congratulations, you have run a MapReduce job.

The single output file of the reduce task in the file `/tmp/MapReduceIntroOutput/part-00000` will have a series of lines of the form *`Number`* `TAB file:/tmp/MapReduceIntroInput/file-`*`N`*. The first thing you will notice is that the numbers don't seem to be in order. The code that generates the input produces a random number for the key of each line, but the example tells the framework that the keys are `Text`. Therefore, the numbers have been sorted as text rather than as numbers.

Creating a Custom Mapper and Reducer

As you've seen, your first Hadoop job, in MapReduceIntro, produced sorted output, but the sorting was not suitable, as it sorted lexically rather than numerically, and the keys for the job were numbers. Now, let's work out what is required to sort numerically, using a custom mapper. Then we'll look at a custom reducer that outputs the values in a format that is easy to parse.

Setting Up a Custom Mapper

Sorting numerically doesn't sound difficult. Let's try making the output key class a LongWritable, another class supplied by the framework:

```
conf.setOutputKeyClass(LongWritable.class);
```

instead of:

```
conf.setOutputKeyClass(Text.class);
```

The class with this change is available as MapReduceIntroLongWritable.java. Run this class via this command:

```
hadoop jar DOWNLOAD_PATH/ch2.jar ➥
com.apress.hadoopbook.examples.ch2.MapReduceIntroLongWritable
```

You will see the following in the output:

```
mapred.LocalJobRunner: job_local_0001
java.io.IOException: Type mismatch in key from map: expected
org.apache.hadoop.io.LongWritable, recieved org.apache.hadoop.io.Text
    at org.apache.hadoop.mapred.MapTask$MapOutputBuffer.collect(MapTask.java:415)
    at org.apache.hadoop.mapred.lib.IdentityMapper.map(IdentityMapper.java:37)
    at org.apache.hadoop.mapred.MapRunner.run(MapRunner.java:47)
    at org.apache.hadoop.mapred.MapTask.run(MapTask.java:227)
    at org.apache.hadoop.mapred.LocalJobRunner$Job.run(LocalJobRunner.java:157)
ch2.MapReduceIntroLongWritable: The job has failed due to an IO Exception
```

As you can see, just changing the output key class was insufficient. If you are going to change the output key class to a LongWritable, you also need to modify the map function so that it outputs LongWritable keys.

For the job to actually produce output that is sorted numerically, you must change the job configuration and provide a custom mapper class. This is done by two calls on the JobConf object:

- conf.setOutputKeyClass(LongWritable.class): Informs the framework of the key class for map and reduce output.
- conf.setMapperClass(TransformKeysToLongMapper.class): Informs the framework of the custom class that provides the map method that takes as input Text keys and outputs LongWritable keys.

A demonstration class MapReduceIntroLongWritableCorrect.java provides the configuration for this. This class is identical to MapReduceIntro, except for these two replacement method calls.

Note The job configuration could also provide a custom sort option. One way to do this is to provide a custom class that implements WritableComparable and use that as the key class. Another way is to specify a CustomComparator in the job configuration via the setOutputKeyComparatorClass() method on the JobConf object. An example of implementing a custom comparator is provided in Chapter 9.

You also need to provide a mapper class that performs the transformation. The sample mapper class TransformKeysToLongMapper.java does this. The TransformKeysToLongMapper.java class file has a number of changes from the IdentityMapper class (shown earlier in Listing 2-2).

First, the class declaration is no longer generic; the types have been made concrete:

```
/** Transform the input Text, Text key value
 * pairs into LongWritable, Text key/value pairs.
*/
public class TransformKeysToLongMapperMapper
  extends MapReduceBase implements Mapper<Text, Text, LongWritable, Text>
```

Notice that the code actually provides the types for the key/value pairs for input and for output. The original IdentityMapper class was completely generic. In addition, the identity mapper's declaration was implements Mapper<K, V, K, **V**>. In TransformKeysToLongMapperMapper, the declaration is implements Mapper<Text, Text, **LongWritable**, Text>.

The map() method of TransformKeysToLongMapper is substantially different from the IdentityMapper and introduces the use of the reporter object.

The Reporter Object

The map and reduce methods both take four parameters: the key, the value, the output collector, and the reporter. The reporter object provides a mechanism for informing the framework of the current status of your job.

The reporter object provides three methods:

- incrCounter(): Provides counters that are aggregated and reported at the end of the job.
- setStatus(): Provides a status line for this map or reduce task.
- getInputSplit(): Returns information about the input source for this task. If the input is simple files, this can provide useful information for log messages.

Each call on the reporter object or the output collector provides a heartbeat to the framework, informing it that the task is not deadlocked or otherwise unresponsive. If your map or reduce method takes substantial time, the method must make periodic calls on the reporter object methods, to inform the framework that it is still working. The framework will kill tasks that have not reported in 600 seconds by default.

Listing 2-6 shows the body of the `TransformKeysToLongMapper` mapper that uses the `reporter` object.

Listing 2-6. *The Reporter Object in TransformKeysToLongMapper.java*

```
/** Map input to the output, transforming the input {@link Text}
 * keys into {@link LongWritable} keys.
 * The values are passed through unchanged.
 *
 * Report on the status of the job.
 * @param key The input key, supplied by the framework, a {@link Text} value.
 * @param value The input value, supplied by the framework, a {@link Text} value.
 * @param output The {@link OutputCollector} that takes
 * {@link LongWritable}, {@link Text} pairs.
 * @param reporter The object that provides a way
 * to report status back to the framework.
 * @exception IOException if there is any error.
 */
public void map(Text key, Text value,
                OutputCollector<LongWritable, Text> output, Reporter reporter)
  throws IOException {

    try {
        try {
            reporter.incrCounter( "Input", "total records", 1 );
            LongWritable newKey =
                  new LongWritable( Long.parseLong( key.toString() ) );
            reporter.incrCounter( "Input", "parsed records", 1 );
            output.collect(newKey, value);
        } catch( NumberFormatException e ) {
            /** This is a somewhat expected case and we handle it specially. */
            logger.warn( "Unable to parse key as a long for key,"
                              +" value " + key + " " + value, e );
            reporter.incrCounter( "Input", "number format", 1 );
            return;
        }
    } catch( Throwable e ) {
        /** It is very important to report back if there were
          * exceptions in the mapper.
         * In particular it is very handy to report the number of exceptions.
         * If this is done, the driver can make better assumptions
         * on the success or failure of the job.
         */
```

```
            logger.error( "Unexpected exception in mapper for key,"
                + " value " + key + ", " + value, e );
            reporter.incrCounter( "Input", "Exception", 1 );
            reporter.incrCounter( "Exceptions", e.getClass().getName(), 1 );
            if (e instanceof IOException) {
                throw (IOException) e;
            }
            if (e instanceof RuntimeException) {
                throw (RuntimeException) e;
            }
            throw new IOException( "Unknown Exception", e );
        }
    }
```

This block of code introduces a new object, reporter, and some best practice patterns. The key piece of this is the transformation of the Text key to a LongWritable key.

```
LongWritable newKey = new LongWritable(Long.parseLong(key.toString()));
output.collect(newKey, value);
```

The code in Listing 2-6 is sufficient to perform the transformation, and also includes some additional code for tracking and reporting.

CODE EFFICIENCY

The pattern of creating a new key object in the mapper for the transformation object is not the most efficient pattern. Most key classes provide a `set()` method, which sets the current value of the key. The `output.collect()` method uses the current value of the key, and once the `collect()` method is complete, the key object or the value object is free to be reused.

If the job is configured to multithread the map method, via `conf.setMapRunner(MultithreadedMapRunner.class)`, the map method will be called by multiple threads. Extreme care must be taken in using the mapper class member variables. A `ThreadLocal LongWritable` object could be used to ensure thread safety. To simplify the example, a new `LongWritable` is constructed. In the reduce method; there are no threading issues.

Object churn is a significant performance issue in a map method, and to a lesser extent, in the reduce method. Object reuse can provide a significant performance gain.

The Counters and Exceptions

This example includes two `try`/`catch` blocks and several calls to the `reporter.incrCounter()` method. It is a good practice to wrap your map and reduce methods in a `try` block that catches `Throwables` and reports on the catches.

The JobTracker, the Hadoop Core server process that manages job execution on the cluster, accumulates the counter values and provides a final count in the job output, as well as making the instantaneous count available in the JobTracker web interface (available on http://jobtracker_host:50030/ by default). This interface will be discussed in more detail in Chapter 6, which covers the setup of a multimachine cluster.

You can now run the job:

```
hadoop jar ch2.jar ➥
com.apress.hadoopbook.examples.ch2.MapReduceIntroLongWritableCorrect
```

The output that reflects the counters is as follows:

```
mapred.JobClient: Job complete: job_local_0001
mapred.JobClient: Counters: 13
mapred.JobClient:   File Systems
mapred.JobClient:     Local bytes read=78562
mapred.JobClient:     Local bytes written=157868
mapred.JobClient:   Input
mapred.JobClient:     total records=126
mapred.JobClient:     parsed records=126
mapred.JobClient:   Map-Reduce Framework
mapred.JobClient:     Reduce input groups=126
mapred.JobClient:     Combine output records=0
mapred.JobClient:     Map input records=126
mapred.JobClient:     Reduce output records=126
mapred.JobClient:     Map output bytes=5670
mapred.JobClient:     Map input bytes=5992
mapred.JobClient:     Combine input records=0
mapred.JobClient:     Map output records=126
mapred.JobClient:     Reduce input records=126
```

The first catch block handles exceptions related to reporter.incrCounter("Input", "number format", 1);, which may be thrown during the key transformation:

```
        } catch( NumberFormatException e ) {
            /** This is a somewhat expected case and we handle it specially. */
            reporter.incrCounter( "Input", "number format", 1 );
            return;
        }
```

You expect that some of the keys may not convert correctly into Long values, so you capture the exception. The reporter.incrCounter() call tells the framework to increment a counter in the Input group, of the name number format, by 1. If the counter does not already exist, it will be created.

In the sample input, there are no records that will cause a number format exception. The only counters that are accumulated are Input.total records and Input.parsed records. These two counters will show up in the job output as part of the Input group:

```
mapred.JobClient:   Input
mapred.JobClient:     total records=126
mapred.JobClient:     parsed records=126
```

If one or more keys caused an exception during the conversion to Long, the output might look more like this:

```
mapred.JobClient:   Input
mapred.JobClient:     total records=126
mapred.JobClient:     parsed records=125
mapred.JobClient:    number format=1
```

Note The sum of the parsed records and the number formats should equal the total records. The counters are also available via the RunningJob object, allowing for a more comprehensive check of the success status. The totals for your job will vary from this example.

After the Job Finishes

Once the job finishes, the framework will provide you with a filled-out RunningJob object. This object has information about the framework's opinion on the success status of your job via the conf.isSuccessful() method. The framework will report that the job was unsuccessful if it was unable to complete any single map task or if the job was killed.

This generally doesn't provide enough information to make a determination on the actual success. It may be that there was an exception in the map or method for every key or for most keys. If the map or reduce function provides job counters for these cases, your job driver will be able to make a better determination regarding the actual success or failure of your job.

In the sample mapper, several counters were collected under different circumstances:

- reporter.incrCounter(TransformKeysToLongMapper.INPUT, TransformKeysToLongMapper.TOTAL_RECORDS, 1): Reports the total number of input records seen.
- reporter.incrCounter(TransformKeysToLongMapper.INPUT, TransformKeysToLongMapper.PARSED_RECORDS, 1): Reports the total number of records successfully parsed.
- reporter.incrCounter(TransformKeysToLongMapper.INPUT, TransformKeysToLongMapper.NUMBER_FORMAT, 1): Reports the total number of records where the key could not be parsed.

- `reporter.incrCounter( TransformKeysToLongMapper.INPUT, TransformKeysToLongMapper.EXCEPTION, 1 )`: Reports the number of records that generated an exception when being processed.
- `reporter.incrCounter( TransformKeysToLongMapper.EXCEPTIONS, e.getClass().getName(), 1 )`: Reports the counts of exceptions by type.

Examining the Counters

Once the framework fills in the `RunningJob` object and returns control back to the job driver, the driver is able to examine the values of the various counters, as well as the framework's success or failure status.

Making the counter values available is a multistep process.

```
/** Get the job counters. {@see RunningJob.getCounters()}. */
Counters jobCounters = job.getCounters();

/** Look up the "Input" Group of counters. */
Counters.Group inputGroup = jobCounters.getGroup( TransformKeysToLongMapper.INPUT );

/** The map task potentially outputs 4 counters in the input group.
  * Get each of them.
  */
long total = inputGroup.getCounter( TransformKeysToLongMapper.TOTAL_RECORDS );
long parsed = inputGroup.getCounter( TransformKeysToLongMapper.PARSED_RECORDS );
long format = inputGroup.getCounter( TransformKeysToLongMapper.NUMBER_FORMAT );
long exceptions = inputGroup.getCounter( TransformKeysToLongMapper.EXCEPTION );
```

Now that the job driver has the counters issued by the map method, a much more accurate determination of success can be made.

■ Caution An accurate determination of success is critical. In one of my production clusters, a TaskTracker node was incorrectly configured. The result of this misconfiguration was that none of the computationally intense work could be run in the map task, and the map method would return immediately with an exception. As far as the framework was concerned, this machine was super fast, and it scheduled almost all of the map tasks on this machine. The job was successful as far as the framework was concerned, but totally unsuccessful per the business rules. At that point. the pattern of checking the exception count was not part of the standard practice, and the failure was uncovered only when the consumer of the results noticed there were no valid results. Save yourself much embarrassment—collect information about the successes and failures in the mapper and reducer objects and check those results in your job driver.

Was This Job Really Successful?

The check for success primarily involves ensuring that the number of records output is roughly the same as the number of records input. Hadoop jobs are generally dealing with bulk real-world data, which is never 100% clean, so a small error rate is generally acceptable.

```
if (format != 0) {
    logger.warn( "There were " + format + " keys that were not "
            + "transformable to long values");
}

/** Check to see if we had any unexpected exceptions.
  * This usually indicates some significant problem,
  * either with the machine running the task that had
  * the exception, or the map or reduce function code.
  * Log an error for each type of exception with the count.
  */
if (exceptions > 0 ) {
    Counters.Group exceptionGroup = jobCounters.getGroup(
                    TransformKeysToLongMapper.EXCEPTIONS );
    for (Counters.Counter counter : exceptionGroup) {
        logger.error( "There were " + counter.getCounter()
                + " exceptions of type " + counter.getDisplayName() );
    }
}

if (total == parsed) {
    logger.info("The job completed successfully.");
    System.exit(0);
}

// We had some failures in handling the input records.
// Did enough records process for this to be a successful job?
// is 90% good enough?
if (total * .9 <= parsed) {
    logger.warn( "The job completed with some errors, "
                + (total - parsed) + " out of " + total );
    System.exit( 0 );
}

logger.error( "The job did not complete successfully,"
    +" too many errors processing the input, only "
    + parsed + " of " + total + "records completed" );
System.exit( 1 );
```

In this particular case, you would expect a small number of `NumberFormatExceptions` but no other exceptions. If the total number of input records is roughly the number of parsed input records, and you have no unexpected exceptions, this job is a success.

Creating a Custom Reducer

The reduce method is called once for each key, and passes the key and an iterator to all of the map output values that share that key. The reduce task is an ideal place for summarizing data and for doing basic duplicate suppression.

Note For managing duplicate suppression against a prior seen set, it is usually best to keep the prior seen set in either HBase (the Hadoop database) or in a sorted format, such as a Hadoop map file. If this is not done, then the dataset of *seen* records and the dataset of *input* records must be merged and sorted, which can take considerable time if either dataset is large. In the HBase case, if the input data is already sorted, the duplicate status of an input record can be rapidly determined. With a simple sorted seen set, map-side joins may be performed. HBase is discussed in Chapter 10, and map-side joins are covered in Chapters 8 and 9.

For the sample custom reducer, let's merge the values into a comma-separated values (CSV) form, so you have one output line per key, with all of the values in a simple-to-parse format.

After your work with the custom mapper in the preceding sections, creating a custom reducer will seem familiar. This version is in `MapReduceIntroLongWritableReduce.java`, which is based on `MapReduceIntroLongWritableCorrect.java`. First, the framework needs to be informed of the reducer class. The key piece is, as usual, to inform the framework of the reducer class, so add the following single line:

```
/** Inform the framework that the reducer class will be the
  * {@link MergeValuesToCSV}.
 * This class simply writes an output record key,
 * value record for each value in the key, valueset it receives as
 * input.
 * The value ordering is arbitrary.
 */
conf.setReducerClass(MergeValuesToCSV.class);
```

There have been no changes to the output classes, so no other changes are required to `MapReduceIntroLongWritableCorrect.java`.

The class to actually perform the work is `MergeValuesToCSVReducer.java`. As with the mapper example, `TransformKeysToLongMapper`, you start with your class declaration, which has partially specified the generic types:

```
public class MergeValuesToCSVReducer<K, V>
    extends MapReduceBase implements Reducer<K, V, K, Text> {
```

The reduce method doesn't need to know the incoming value class; it requires only the `toString()` method to work. The reduce method does need to construct a new output value, and for simplicity's sake, given this transformation, the output value is declared to be `Text`.

The actual method declaration also has the same type specification:

```
/** Merge the values for each key into a CSV text string.
 *
 * @param key The key object for this group.
 * @param values Iterator to the set of values that share the <code>key</code>.
 * @param output The {@see OutputCollector} to pass the transformed output to.
 * @param reporter The reporter object to update counters and set task status.
 * @exception IOException if there is an error.
 */
```

```
public void reduce(K key, Iterator<V> values,
       OutputCollector<K, Text> output, Reporter reporter)
    throws IOException {
```

The framework will throw an error if the job is expecting a different output value type than Text. As with the mapper example, you have a method body that employs the reporter.incrCounter() method to make detailed information available to the job and via the web interface. As a performance optimization, to reduce object churn, two class fields are declared. These variables are used in the reduce() method:

```
/** Used to construct the merged value.
  * The {@link Text.set() Text.set} method is used
  * to prevent object churn.
  */
protected Text mergedValue = new Text();
/** Working storage for constructing the resulting string. */
protected StringBuilder buffer = new StringBuilder();
```

The buffer object is used to build the CSV-style line for the output, and mergedValue is the actual object that is sent to the output on each reduce() call. It is safe to declare these as class fields, rather than as local variables, because the individual reduce tasks are run only as single threads by the framework.

Note There may be multiple reduce tasks running simultaneously, but each task is running in a separate JVM, and the JVMs are potentially running on separate physical machines.

The reduce() method is called with the key and an iterator to the values that share that key. Recall that, ideally, a reduce task will make no changes to the key, and will use that key as the key argument to the output.collect() method calls in the reduce() method. The design goal for this reduce() method is to output only a single row for every key, with a comma-separated list of the values that shared that key. The core of the reduce() method has a bit of boilerplate for the object churn optimizations to reset the StringBuilder object, and a loop to process each of the values for this key:

```
buffer.setLength(0);
for (;values.hasNext(); valueCount++) {
    reporter.incrCounter( OUTPUT, MergeValuesToCSVReducer.TOTAL_VALUES, 1 );
    String value = values.next().toString();
    if (value.contains("\"")) { // Perform Excel style quoting
        value.replaceAll( "\"", "\\\"" );
    }
    buffer.append( '"' );
    buffer.append( value );
    buffer.append( "\"," );
}
buffer.setLength( buffer.length() - 1 );
```

It is rare that a reduce() method doesn't have a loop that iterates over the values. It is good form to report on the number of values input. In this example, reporter.incrCounter (OUTPUT, MergeValuesToCSVReducer.TOTAL_VALUES, 1) handles the reporting.

This reducer relies on the toString() method of the value object, which seems reasonable for a textual output job, as the framework would also be using the toString() method to produce the output. The rest of the preceding code block simply builds a comma-separated list of values, with Excel-style CSV quoting.

The actual output block must build a new value for the output. In this case, a class field mergedValue will be used. In a larger job, there may be a billion keys passed through the reduce() method, and by using the class field, the amount of object churn is greatly reduced. In this example, there are also counters for the output records:

```
mergedValue.set(buffer.toString());
reporter.incrCounter( OUTPUT, TOTAL_OUTPUT_RECORDS, 1 );
output.collect( key, mergedValue );
```

The value is set on the mergedValue object, using the mergedValue.set(buffer.toString()) statement, and the value is output using the output.collect(key, mergedValue) line. This example uses Text as the output value class; it is acceptable to use any Writable as the output value class. If the output format is a SequenceFile, there is no need for a functional toString() method on your object.

■ **Note** The framework serializes the key and value into the output stream during the collect() method, leaving the user free to change the objects values when the method returns.

Why Do the Mapper and Reducer Extend MapReduceBase?

The custom mapper class TransformKeysToLongMapper and reducer class MergeValuesToCSVReducer both extend the class org.apache.hadoop.mapred.MapReduceBase. This class provides basic implementations of two additional methods that are required of a mapper or a reducer by the framework. The framework calls the configure() method upon initializing a task, and it calls the close() method when the task has finished processing its input split:

```
/** Default implementation that does nothing. */
public void close() throws IOException {
}

/** Default implementation that does nothing. */
public void configure(JobConf job) {
}
```

The configure Method

The configure() method is the only way to get access to the JobConf object for your task. This method is where any per-task configuration and setup is done. If your application relies on the Spring Framework for setup, the application context would be established here and the relevant beans found.

It is very common for the developer to have a JobConf member variable, which would be initialized in this method with the passed-in JobConf object. (I prefer to issue a logging record with detailed information about the input split.) The configure() method is also the ideal place to open additional files that need to be read or written to during the map() or reduce() method.

The close Method

The close() method is called by the framework when all of the input-split entries have been processed by the applicable map() or reduce() method. It is very important to close any supplemental files here to ensure that they are properly flushed to the file system. Particularly for HDFS, if the file is not closed, data in the last block may be lost.

The following example also makes a reporter call in the close() method:

```
/** Keep track of the maximum number of keys a value had.
 * Report it in the counters so that per task counters can be examined as needed
 * and set the task status to include this maximum count.
 */
@Override
public void close() throws IOException {
    super.close();
    if (reporter!=null) {
        reporter.incrCounter( OUTPUT, MAX_VALUES, maxValueCount );
        reporter.setStatus( "Job Complete, maxixmum ValueCount was "
                  + maxValueCount );
    }
}
```

The reporter field was made a class instance field, via protected Reporter reporter, and set in the reduce() method via this.reporter = reporter. In the reduce() method, the count of values is kept in valueCount, and if it's larger than the instance member field, maxValueCount, maxValueCount is set to it. This enables you to output the maximum number of values that shared a specific key.

In this case, the overall summary value is not particularly useful, as that value is the sum of all of the maximum values, but the per-task value is interesting and available via the web interface. A more useful solution would be to maintain an additional output file and output the key/value counts into that file.

When you select a completed or running task through the web interface (which is on port 50030 on the machine running the JobTracker, by default), you are presented with the counter summary for the job and links to detailed information about the map and reduce tasks. Each map and reduce task will have a link to the counters.

Using a Custom Partitioner

By default, the framework partitions your output based on the hash value of the key, using the HashPartitioner class. There are times when you need your output data partitioned differently. The standard example is a single output file where multiple output files would usually result, which is handled by setting the number of reduce tasks to 1, via `conf.setNumReduces(1)`, or unsorted/unreduced output, which is handled via `conf.setNumReduces(0)`. If you need different partitioning, you have the option of setting a partitioner.

This chapter's example has `Long` keys. Some simple partitioner concepts could be to sort into odd/even or, if the minimum and maximum key values are known, to sort into key range-based buckets. It is also possible to partition by the value.

HOW PARTITIONING IS DONE

When the framework is performing the shuffle, each key output by the mapper is examined, and the following operation is performed:

```
int partition = partitioner.getPartition(key, value, partitions);
```

The value `partitions` is the number of reduce tasks to perform. The key, if actually output by the reducer, will end up in the output file part `partition`, with an appropriate number of leading zeros so that the file names are all the same length.

The critical issues are that the number of partitions is fixed at job start time and the partition is determined in the `output.collect()` method of the map task. The only information the partitioner has is the key, the value, the number of partitions, and whatever data was made available to it when it was instantiated.

The partitioner interface is very simple, as shown in Listing 2-7.

Listing 2-7. *The Partitioner Interface*

```
/**
 * Partitions the key space.
 *
 * <p><code>Partitioner</code> controls the partitioning of the keys of the
 * intermediate map-outputs. The key (or a subset of the key) is used to derive
 * the partition, typically by a hash function. The total number of partitions
 * is the same as the number of reduce tasks for the job. Hence this controls
 * which of the <code>m</code> reduce tasks the intermediate key (and hence the
 * record) is sent for reduction.</p>
 *
 * @see Reducer
 */
public interface Partitioner<K2, V2> extends JobConfigurable {

  /**
   * Get the paritition number for a given key (hence record) given the total
   * number of partitions i.e. number of reduce-tasks for the job.
   *
```

```
   * <p>Typically a hash function on a all or a subset of the key.</p>
   *
   * @param key the key to be paritioned.
   * @param value the entry value.
   * @param numPartitions the total number of partitions.
   * @return the partition number for the <code>key</code>.
   */
  int getPartition(K2 key, V2 value, int numPartitions);
}
```

The `JobConfigurable` interface provides an additional `configure()` method, as the `MapReduceBase` class does.

Summary

This chapter explained what is involved in executing a MapReduce job. You now have a basic understanding of the `JobConf` object and how to use it to inform the framework of the requirements for your jobs.

You've seen how to write mapper and reducer classes, and how the `reporter` object is one of your best friends, because of the wonderful information it can provide about what is happening during the execution of your jobs. Output partitions finally make sense, and you have a sense of when and why you configure your job to reduce, and how many reducers you will use.

As a brilliant Hadoop expert, you are totally prepared to inform people of why the files they open in mapper or reducer classes are empty or short, because you know you need to close files before the framework will flush the last file system block size worth of data to disk.

In the next chapter, you'll learn how to set up of a multimachine cluster.

CHAPTER 3

The Basics of Multimachine Clusters

This chapter explains how to set up a multimachine cluster. You'll learn about the makeup of a cluster, the tools for managing clusters, and how to configure a cluster. Here, we'll walk-through a simple cluster configuration, using the minimum HDFS setup necessary to bring up the cluster. Chapter 4 will go into the details for a high-usage HDFS.

The Makeup of a Cluster

A typical Hadoop Core cluster is made up of machines running a set of cooperating server processes. The machines in the cluster are not required to be homogeneous, and commonly they are not. The cluster machines may even have different CPU architectures and operating systems. But if the machines have similar processing power, memory, and disk bandwidth, cluster administration is a lot easier, because in that case, only one set of configuration files and runtime environments needs to be maintained and distributed.

Figure 3-1 illustrates a typical Hadoop cluster. A cluster will have one JobTracker server, one NameNode server, and one secondary NameNode server, and DataNodes and Task-Trackers. The JobTracker coordinates the activities of the TaskTrackers, and the NameNode manages the DataNodes.

In the context of Hadoop, a node/machine running the TaskTracker or DataNode server is considered a *slave* node. It is common to have nodes that run both the TaskTracker and DataNode servers. The Hadoop server processes on the slave nodes are controlled by their respective masters, the JobTracker and NameNode servers.

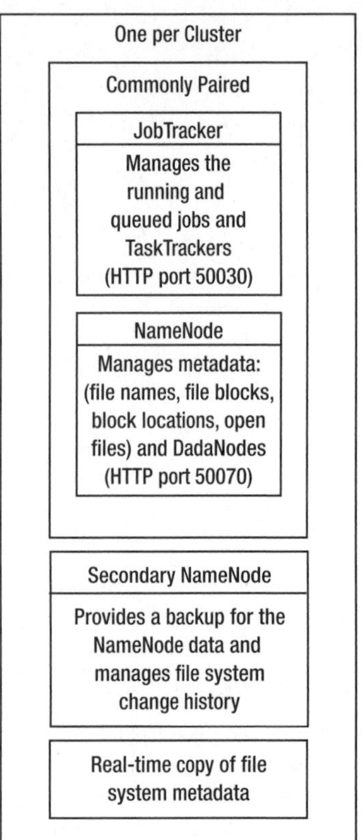

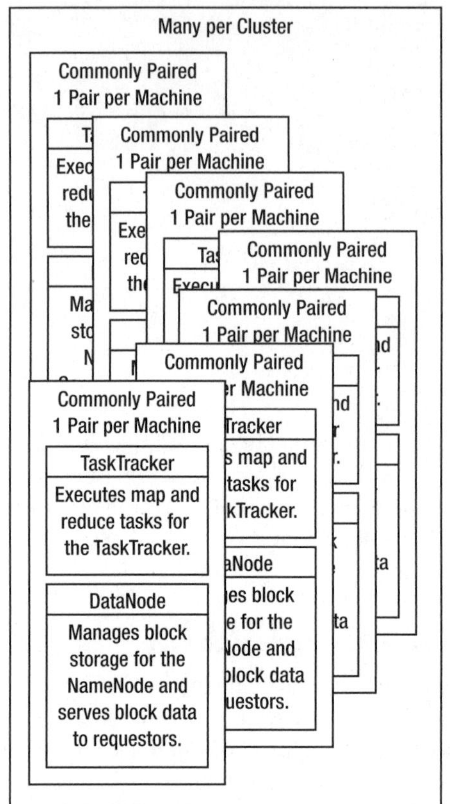

Figure 3-1. *A typical Hadoop cluster*

Let's look at each of the server processes run by the machines in a cluster:

JobTracker: The JobTracker provides command and control for job management. It supplies the primary user interface to a MapReduce cluster. It also handles the distribution and management of tasks. There is one instance of this server running on a cluster. The machine running the JobTracker server is the MapReduce master.

TaskTracker: The TaskTracker provides execution services for the submitted jobs. Each TaskTracker manages the execution of tasks on an individual compute node in the MapReduce cluster. The JobTracker manages all of the TaskTracker processes. There is one instance of this server per compute node.

■ **Note** If your MapReduce jobs utilize external packages or services, it is very important that these external packages and services are identically configured across all of your TaskTracker machines. It is not uncommon for external JAR files to be required for the successful running of a task. If these JAR files differ in version or are absent, unexpected and difficult-to-diagnose errors occur.

NameNode: The NameNode provides metadata storage for the shared file system. The NameNode supplies the primary user interface to the HDFS. It also manages all of the metadata for the HDFS. There is one instance of this server running on a cluster. The metadata includes such critical information as the file directory structure and which DataNodes have copies of the data blocks that contain each file's data. The machine running the NameNode server process is the HDFS master.

Secondary NameNode: The secondary NameNode provides both file system metadata backup and metadata compaction. It supplies near real-time backup of the metadata for the NameNode. There is at least one instance of this server running on a cluster, ideally on a separate physical machine than the one running the NameNode. The secondary NameNode also merges the metadata change history, the edit log, into the NameNode's file system image.

Real-time backup of the NameNode data: Many installations configure the NameNode to store the file system metadata to multiple locations, where at least one of these locations resides on a separate physical machine. Other installations use a tool such as DRBD (`http://www.drbd.org/`) to replicate the host file system in near real time to a separate physical machine.

DataNode: The DataNode provides data storage services for the shared file system. Each DataNode supplies block storage services for the HDFS. The NameNode coordinates the storage and retrieval of the individual data blocks managed by a DataNode. There is one instance of this server process per HDFS storage node.

Balancer: During normal usage, the disk utilization on the DataNode machines may become uneven. This is particularly common if some DataNodes have less disk space available for use by HDFS. The Balancer moves data blocks between DataNodes to even out the per-DataNode available disk space. The Balancer will also rebalance the cluster as new DataNodes are added to an existing cluster. The Balancer is not a started automatically. It must be run by the user via the command `bin/hadoop balancer [-threshold <threshold>]`. The optional argument is the maximum amount of variance in disk space utilization between DataNodes for the cluster to be considered balanced. The default is 10%. As of Hadoop 0.19.0, this is not a configuration parameter.

These server processes are typically started once per cluster instance. DataNodes and TaskTrackers may be dynamically added and removed from a running cluster, as described in the next section.

All of these servers are implemented in Java and require at least Java version 1.6.

Cluster Administration Tools

The Hadoop Core installation provides a number of scripts in the `bin` subdirectory of the installation that are used to stop and start the entire cluster or various pieces of the cluster. There are also administrative scripts for the Hadoop Core servers. Table 3-1 lists the available scripts for administering clusters.

The administrator has the option of starting or stopping the full set of Hadoop Core servers with the `start-all.sh` and `stop-all.sh` scripts. These scripts start all of the server processes on the cluster machines. The NameNode and JobTracker will be started or stopped on the machine on which the script is run, and DataNodes and TaskTracker nodes will be started on the configured slave machines. Any requested secondary NameNodes will also be started on configured machines.

Table 3-1. *Cluster Administration Scripts*

Script	Description
`start-all.sh` and `stop-all.sh`	Start and stop the full set of Hadoop Core servers in the cluster.
`start-mapred.sh` and `stop-mapred.sh`	Start and stop just the MapReduce servers. These scripts start or stop only the JobTracker and TaskTracker nodes. The JobTracker is expected to run on the machine on which these scripts are executed.
`start-dfs.sh` and `stop-dfs.sh`	Start and stop the HDFS servers, in the same way as the `start-mapred.sh` and `stop-mapred` scripts manage the MapReduce servers.
`start-balancer.sh` and `stop-balancer.sh`	Start and stop the Balancer. The Balancer is expected to run on the machine on which these scripts are executed.
`hadoop-daemon.sh`	Starts or stops a single instance of a server on the current machine. The preceding start and stop scripts actually use the `hadoop-daemon.sh` script to start or stop the servers.
`hadoop-daemons.sh`	Starts or stops a set of servers on the relevant set of machines. This script is used by the start and stop scripts to start the DataNodes, TaskTrackers, and secondary NameNodes. This script will use the `hadoop-daemon.sh` script to start the servers on each specific machine in the set of machines on which it operates.
`hadoop-config.sh`	Used by the other scripts to load the Hadoop configuration.
`slaves.sh`	Runs its arguments as a command on each of the hosts listed in the `conf/slaves` file, collecting the output and presenting the output back the user, prefixed with the name of the host on which each output line originated.
`hadoop`	Provides command-level access to the services provided by the HDFS and MapReduce servers.
`rcc`	Provides services for creating RPC interfaces. This feature of Hadoop Core is expected to be discontinued. As of Hadoop 0.17, the Serialization service is the preferred method for handling external data structures.

Cluster Configuration

The Hadoop Core servers load their configuration from files in the `conf` directory of your Hadoop Core installation. As a general rule, identical copies of the configuration files are maintained in the `conf` directory of every machine in the cluster. The current default is to have one set of configuration files that provides the configuration for all of the server processes. It is also common to have the NameNode and the JobTracker servers on the same node, especially in smaller installations.

■Note Many difficult-to-diagnose problems occur when the configuration files or the supporting runtime environments differ between TaskTracker nodes.

Hadoop Configuration Files

The configuration files fall into the following groups:

Hadoop Core configuration: The Hadoop Core is configured by two XML files: `hadoop-default.xml` and `hadoop-site.xml`. `hadoop-default.xml` provides reasonable defaults and comes with the Hadoop distribution. The configuration provided by this default file is suitable for a single machine instance and is the configuration used to run the examples in Chapters 1 and 2. `hadoop-site.xml` is where cluster-specific information is specified by the cluster administrator. In this chapter, we will walk through constructing a `hadoop-site.xml` file for a small cluster.

Slaves and masters: Two files are used by the startup and shutdown commands discussed in the previous section to start and stop the DataNode, TaskTracker, and secondary NameNode servers. The `slaves` file contains a list of hosts, one per line, that are to host DataNode and TaskTracker servers. The `masters` file contains a list of hosts, one per line, that are to host secondary NameNode servers. If the `start-all.sh` script is used, a DataNode and TaskTracker will be started on each host in the `slaves` file, and a secondary NameNode will be started on each host in the `masters` file. `start-mapred.sh` starts only the TaskTracker servers. If the `start-dfs.sh` script is used, DataNodes will be started on the hosts listed in `slaves`, and secondary NameNodes will be started on the hosts listed in `masters`.

Per-process runtime environment: The file `hadoop-env.sh` is responsible for tailoring the per-process environment. In particular, it includes the `JAVA_HOME` environment variable, which provides the JVM installation location. This file also offers a way to provide custom parameters for each of the servers. `hadoop-env.sh` is sourced by all of the Hadoop Core scripts provided in the `conf` directory of the installation.

Reporting: Hadoop Core may be configured to report detailed information about the activities on the cluster. Hadoop Core may report to a file, via Ganglia (`http://ganglia.info/`), which provides a framework for displaying graphical reports summarizing the activities of large clusters of machines. The file `hadoop-metrics.properties` controls the reporting. The default is to not report.

Hadoop Core Server Configuration

The hadoop-default.xml file defines more than 150 parameters, divided into six groups:

- Global properties
- Logging properties
- I/O properties
- File system properties
- MapReduce properties
- IPC properties

■ Note Some of the parameters listed in hadoop-default.xml may be modified on a per-job basis by setting alternate values using the JobConf.set* methods. The administrator may specify that a parameter is final by adding <final>true</final> to the parameter's declaration in the hadoop-site.xml file. In general, the modification of the server configuration parameters by a job have no effect on the servers.

It is customary to consider the hadoop-default.xml file to be read-only, and to make changes only to the hadoop-site.xml file. The framework will load configuration files in order, with the values defined in later files superseding those earlier definitions. The loading order is hadoop-default.xml, hadoop-site.xml, and then any user specified resources.

■ Note Values that have a ${text} are replaced with the system property value or a previously defined value. The search order is system properties, then previously defined values.

Three critical parameters must be configured for any Hadoop cluster: hadoop.tmp.dir, fs.default.name, and mapred.job.tracker. Several other parameters are important to tune but not critical: mapred.tasktracker.map.tasks.maximum, mapred.tasktracker.reduce.tasks.maximum, mapred.child.java.opts, and webinterface.private.actions. The default values for these parameters are suitable for single-machine, single-CPU temporary use only. The following sections discuss the minimum set for cluster configuration. The parameters required for large-scale HDFS installations are covered in Chapters 4 and 5.

■ Note The hadoop-default.xml file includes some documentation on the parameters that control a cluster and a job, although some configuration parameters are not documented here.

Per-Machine Data

The `hadoop.tmp.dir` parameter is critical to configure. If it is not configured, data loss will occur. This parameter is poorly named. It informs the framework of the directory to use for all Hadoop Core server data storage, as follows:

- The NameNode will store file system metadata in a subdirectory.
- The DataNodes will store per-file blocks in a subdirectory.
- The TaskTrackers will store intermediate output in a subdirectory.
- The JobTracker will store per-job data in a subdirectory.
- The secondary NameNode will store the backup metadata in a subdirectory.

In the default configuration, the parameters listed in Table 3-2 use the value of `${hadoop.tmp.dir}` as the leading component of their paths. A high-performance cluster will have values tailored to minimize I/O contention on individual devices. To maximize performance, the I/O for these various functions needs to be distributed over multiple devices.

Table 3-2. *Parameters That Use the hadoop.tmp.dir Value*

Parameter	Description
`fs.checkpoint.dir`	Determines where on the local file system the secondary NameNode will store name data.
`dfs.name.dir`	Determines where on the local file system the NameNode metadata is stored. This may be a comma- or space-separated list of directories. All the provided directories are used for redundant storage. This is of critical importance and should be stored on a low-latency device with redundancy.
`dfs.client.buffer.dir`	Determines where on the local file system data to be written to HDFS is accumulated prior to transmission to the DataNodes. This directory will experience bulk I/O that has a short life.
`dfs.data.dir`	Determines where on the local file system a DataNode stores blocks. This may be a comma- or space-separated list of directories. The data will be distributed among the directories. By default, HDFS replicates data storage blocks to multiple DataNodes. This directory will experience bulk I/O transactions.
`mapred.local.dir`	The local directory where TaskTracker stores intermediate output. This may be a comma-separated list of directories, preferably on different devices. I/O will be spread among the directories for increased performance. This directory will also experience bulk I/O that has a short life.
`mapred.system.dir`	The shared directory where the JobTracker stores control files. This value must be unique per JobTracker if multiple MapReduce clusters share a single HDFS.
`mapred.temp.dir`	A shared directory for temporary files. The default value of this setting is `${hadoop.tmp.dir}/mapred/temp}`.

The cluster administrator must pick a location or locations for these directories that provide the required I/O performance and the required reliability. For example, consider the `dfs.data.dir` parameter.

The HDFS-level redundant block storage reduces the requirements for highly reliable block storage for individual DataNodes. The directory specified by the `dfs.data.dir` parameter will experience bulk I/O transactions and should be optimized for maximum speed. There will be a large number of files and directories created, each file being an HDFS data block.

The default value for `dfs.data.dir` is `${hadoop.tmp.dir}/dfs/data`. If you don't change this default value for `${hadoop.tmp.dir}`, the HDFS data will be stored in `/tmp` and deleted by the system `/tmp` cleaning service. This causes interesting chaos for users when their HDFS file's data blocks start vanishing about a week after the file was created.

The framework will attempt to create the `hadoop.tmp.dir` directory and all of the subdirectories if they do not exist. The user that the relevant server processes are running as must have the required permissions to be able to create these directories and to add and remove files from them.

A perhaps ideal configuration would be for the NameNode to have a RAID 10 array for the `dfs.name.dir`, and for DataNodes and TaskTrackers to have RAID 0 arrays for `dfs.data.dir`, `mapred.local.dir`, and `dfs.client.buffer.dir`. This configuration assumes that HDFS is doing redundant block storage at the HDFS level.

Default Shared File System URI and NameNode Location for HDFS

The `fs.default.name` parameter is critical to configure. If it is not configured, there is no shared file system. The URI specified here informs the Hadoop Core framework of the default file system. The default value is `file:///`, which instructs the framework to use the local file system.

An example of an HDFS URI is `hdfs://NamenodeHost[:8020]/`. The file system protocol is `hdfs`, the host to contact for services is `NamenodeHost`, and the port to connect to is `8020`, which is the default port for HDFS. If the default 8020 port is used, the URI may be simplified as `hdfs://NamenodeHost/`. This value may be altered by individual jobs. You can choose an arbitrary port for the `hdfs` NameNode.

JobTracker Host and Port

The `mapred.job.tracker` parameter is critical to configure. If it is not configured, only a single machine will be used for task execution The URI specified in this parameter informs the Hadoop Core framework of the JobTracker's location. The default value is `local`, which indicates that no JobTracker server is to be run, and all tasks will be run from a single JVM.

The appropriate value for a cluster is `JobtrackerHost:8021`. The `JobtrackerHost` is the host on which the JobTracker server process will be run. This value may be altered by individual jobs.

Maximum Concurrent Map Tasks per TaskTracker

The `mapred.tasktracker.map.tasks.maximum` parameter sets the maximum number of map tasks that may be run by a TaskTracker server process on a host at one time. The default value is 2. This value is read only when the TaskTracker is started. Changes made by a job will not be honored or persist.

This parameter should be tuned to ensure that the CPU resources of the TaskTracker nodes are fully utilized. If the machine hosts only the TaskTracker, it is common to set this value to the effective number of CPUs on the node. This may result in a large memory footprint, as each of the JVMs executing tasks will have a full memory allocation.

Many administrators set this value to 1 and require that individual jobs specify that the class `MultiThreadedMapRunner` is to be used via the `JobConf.setMapRunner(MultiThreadedMapRunner.class)` method, and specify the number of threads to use per map task. The default number of threads as of Hadoop 0.18.2 is 10, and this value may be altered by setting the number of threads via the following:

```
JobConf.set("mapred.map.multithreadedrunner.threads", threadCount);
```

This latter choice is preferred, as the number of threads may be set on a per-job basis, allowing the job to customize its CPU consumption.

The following sample snippet demonstrates a common pattern for per-job management of map task parallelism. The choice of 100 was made for demonstration purposes and is not suitable for a CPU-intensive map task.

```
if (conf.getInt("mapred.tasktracker.map.tasks.maximum", 2)==1) {
    conf.setMapRunnerClass(MultithreadedMapRunner.class);
    conf.setInt("mapred.map.multithreadedrunner.threads", 100);
}
```

Maximum Concurrent Reduce Tasks per TaskTracker

The `mapred.tasktracker.reduce.tasks.maximum` parameter sets the maximum number of reduce tasks that may be run by an individual TaskTracker server at one time. Unlike in a map task, the output key ordering is critical for a reduce tasks, which precludes running multi-threaded reduce tasks. This value also determines the number of parts in which your job output is placed. The default value, 2, is specified in the `conf/hadoop-default.xml` file.

Reduce tasks tend to be I/O bound, and it is not uncommon to have the per-machine maximum reduce task value set to 1 or 2. This value is utilized when the cluster is started. Changes made by a job will not be honored or persist.

JVM Options for the Task Virtual Machines

The `mapred.child.java.opts` parameter is commonly used to set a default maximum heap size for tasks. The default value is `-Xmx200m`. Most installation administrators immediately change this value to `-Xmx500m`. A significant and unexpected influence on this is the heap requirements (`io.sort.mb`), which by default will cause 100MB of space to be used for sorting.

During the run phase of a job, there may be up to `mapred.tasktracker.map.tasks.maximum` map tasks and `mapred.tasktracker.reduce.tasks.maximum` reduce tasks running simultaneously on each TaskTracker node, as well the TaskTracker JVM. The node must have sufficient virtual memory to meet the memory requirements of all of the JVMs. JVMs have non-heap memory requirements; for simplicity, 20MB is assumed.

A cluster that sets the map task maximum to 1, the reduce task maximum to 8, and the JVM heap size to 500MB would need a minimum of (1 + 8 + 1) * (500+20) = 10 * 520 = 5200MB, or 5GB, of virtual memory available for the JVMs on each TaskTracker host. This 5GB value does not include memory for other processes or servers that may be running on the node.

The `mapred.child.java.opts` parameter is configurable by the job.

Enable Job Control Options on the Web Interfaces

Both the JobTracker and the NameNode provide a web interface for monitoring and control. By default, the JobTracker provides web service on `http://JobtrackerHost:50030` and the NameNode provides web service on `http://NamenodeHost:50070`. If the `webinterface.private.actions` parameter is set to `true`, the JobTracker web interface will add Kill This Job and Change Job Priority options to the per-job detail page. The default location of these additional options is the bottom-left corner of the page (so you usually need to scroll down the page to see them).

A Sample Cluster Configuration

In this section, we will walk through a simple configuration of a six-node Hadoop cluster. The cluster will be composed of six machines: `master01`, `slave01`, `slave02`, `slave03`, `slave04`, and `slave05`. The JobTracker and NameNode will reside on the machine `master01`, and a secondary NameNode will be placed on `slave01`. The DataNodes and TaskTrackers will be colocated on the same machines, and the nodes will be named `slave01` through `slave05`. Figure 3-2 shows this setup.

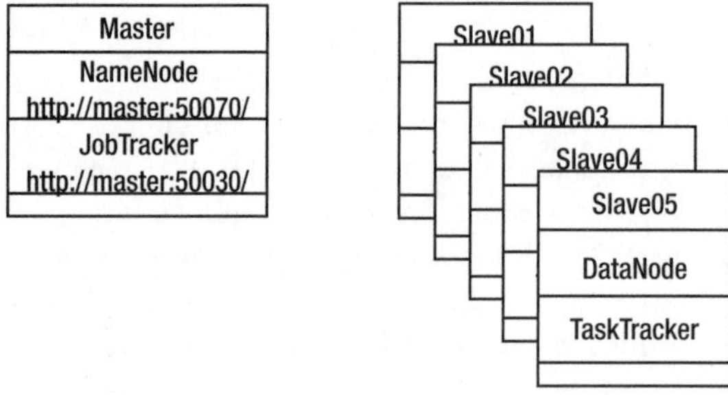

Figure 3-2. *A simple six-node cluster*

The standard machine configuration is usually an eight-CPU machine with 8GB of RAM, and a hardware RAID controller presenting a single partition to the operating system. This configuration is the favorite of IT departments. The single partition presentation is not ideal for Hadoop Core, as there is no opportunity to segregate the I/O for MapReduce and for HDFS.

Configuration Requirements

This configuration will require the customization of several files in the `conf` directory and compliance with some simple network requirements. The user that will own the Hadoop processes must also be determined.

Network Requirements

Hadoop Core uses Secure Shell (SSH) to launch the server processes on the slave nodes. Hadoop Core requires that passwordless SSH work between the master machines and all of the slave and secondary machines.

For example, for OpenSSH, you can generate an unencrypted key for the user that will own the Hadoop Core server processes. The user will need to have a directory, `~/.ssh`, that only the user has access permissions for, on all machines in the cluster. The following command generates a `dsa` key with an empty password in the file `~/.ssh/id_dsa`:

```
ssh-keygen -t dsa -P '' -f ~/.ssh/id_dsa
```

This command will generate two files in the `~/.ssh` directory: `id_dsa` and `id_dsa.pub`. The quotes in the command are a pair of single quote characters, side by side.

For each machine in the cluster, append the contents of the `~/.ssh/id_dsa.pub` file to the `~/.ssh/authorized_keys` file. If required, create the `~/.ssh` directory and the `~/.ssh/authorized_keys` file.

Execute the command `chmod og-rwx ~/.ssh` on each machine in the cluster.

You should now be able to run `bin/slaves.sh uptime` and receive the output of uptime from each of the machines listed in the `conf/slaves` file, as follows:

```
bin/slaves.sh uptime | sort
```

```
slave01:  21:18:41 up 47 days, 22:22,  9 users,  load average: 1.22, 1.23, 1.26
. . .
slave06:  21:18:22 up 20 days, 11:59,  3 users,  load average: 0.02, 0.09, 0.13
```

These slave servers will need to contact their specific master (either the NameNode or the JobTracker), and this will require that several ports in the low 50000 range be unblocked and available. Table 3-3 lists the default ports that must be unfiltered and available.

Table 3-3. *Default Ports Used by Hadoop Core*

Port	Setting	Description
50030	`mapred.job.tracker.http.address`	JobTracker administrative web GUI
50070	`dfs.http.address`	NameNode administrative web GUI
50010	`dfs.datanode.address`	DataNode control port (each DataNode listens on this port and registers it with the NameNode on startup)
50020	`dfs.datanode.ipc.address`	DataNode IPC port, used for block transfer
50060	`mapred.task.tracker.http.address`	Per TaskTracker web interface
50075	`dfs.datanode.http.address`	Per DataNode web interface
50090	`dfs.secondary.http.address`	Per secondary NameNode web interface
50470	`dfs.https.address`	NameNode web GUI via HTTPS
50475	`dfs.datanode.https.address`	Per DataNode web GUI via HTTPS

■ **Note** Hadoop Core uses a number of TCP ports in the low 50000 range. If other applications, such as Squid, are also using ports in this range, difficult-to-diagnose problems may occur.

Advanced Networking: Support for Multihomed Machines

In installations with more complex network topologies, it can become important to control which network interface is used by Hadoop for interprocess communications (IPC). Some installations have multiple network interfaces per machine. Hadoop provides two parameters to control this:

- `dfs.datanode.dns.interface`: If set, this parameter is the name of the network interface to be used for HDFS transactions to the DataNode. The IP address of this interface will be advertised by the DataNode as its contact address.
- `dfs.datanode.dns.nameserver`: If set, this parameter is the hostname or IP address of a machine to use to perform a reverse host lookup on the IP address associated with the specified network interface.

An example of a more complex network setup is for the machines in the Amazon cloud. Each Amazon cloud machine has two network interfaces: one for Internet access and one for intracloud traffic. It is helpful to set `dfs.datanode.dns.interface` to the name of the intracloud network interface.

Machine Configuration Requirements

In the simplest case, all of the machines in the cluster will be identically configured. They will have the same number of CPUs, the same amount of physical RAM, and the same disk capacity and configuration, with the same file system mount points. The same version of the JVM will be installed in the same location, and the Hadoop installation directory will be the same on all of the machines.

Hadoop Core maintains process ID files in the directory `/tmp`, by default. This directory must exist and be writable by the user that will own the Hadoop server processes. This directory is configurable by editing the `conf/hadoop-env.sh` file and uncommenting and optionally altering the setting for the environment variable `HADOOP_PID_DIR`.

Caution Remember that the parent of the directory to be used for `hadoop.tmp.dir` must exist and be writable by the Hadoop server user, or the `hadoop.tmp.dir` directory must exist and be writable by the Hadoop server user.

Configuration Files for the Sample Cluster

The examples provided in this section were run using the VMware images provided by Cloudera as part of its boot camp (`http://www.cloudera.com/hadoop-training-basic`). The boot camp VMware image 0.2 is used.

The hadoop-site.xml File

The `hadoop-site.xml` file contains the XML-based configuration data for our sample cluster. The source code for this file is provided with the rest of this book's downloadable code.

The partition /hadoop is used as the value for hadoop.tmp.dir, and is assumed to be on the hardware RAID partition.

```
<?xml version="1.0" encoding="UTF-8"?>
<?xml-stylesheet type="text/xsl" href="configuration.xsl"?>

<configuration>

    <property>
        <name>hadoop.tmp.dir</name>
        <value>/hadoop</value>
        <description>A base for other temporary directories. Set to a
            directory off of the user's home directory for the simple test.
        </description>
    </property>
```

The NameNode is configured by setting fs.default.name to hdfs://master01. The port of 8020 is implicit in the protocol declaration.

```
    <property>
        <name>fs.default.name</name>
        <value>hdfs://master01</value>
        <description>The name of the default file system. A URI whose
            scheme and authority determine the FileSystem implementation. The
            uri's scheme determines the config property (fs.SCHEME.impl) naming
            the FileSystem implementation class. The uri's authority is used to
            determine the host, port, etc. for a filesystem. Pass in the hostname
            via the -Dhadoop.namenode=NAMENODE_HOST java option.
         </description>
    </property>
```

The JobTracker is configured by setting mapred.job.tracker to master01:8011. The value for mapred.job.tracker is not interpreted as a URI, but rather as a host:port pair. The port must be specified explicitly.

```
    <property>
        <name>mapred.job.tracker</name>
        <value>master01:8012</value>
        <description>The host and port that the MapReduce job tracker runs
            at. If "local", then jobs are run in-process as a single map
            and reduce task.
            Pass in the jobtracker hostname via the
            -Dhadoop.jobtracker=JOBTRACKER_HOST java option.
        </description>
    </property>
```

For setting the maximum number of map and reduce tasks per TaskTracker, several assumptions are made. The first assumption is that the map tasks will be threaded, and the individual jobs will choose a thread count that optimizes CPU utilization. This results in a setting of 1 for mapred.tasktracker.map.tasks.maximum.

```
<property>
    <name>mapred.tasktracker.map.tasks.maximum
    </name>
    <value>1</value>
    <description>The maximum number of map tasks that will be run
        simultaneously by a task tracker.
    </description>
</property>
```

The DataNode also will run on the same machine; therefore, you must budget for CPU and memory resources. In the best of all worlds, the DataNode would use a different set of disks for I/O than the TaskTracker. The setting for the `mapred.tasktracker.reduce.tasks.maximum` parameter is 6. This leaves CPU and I/O available for the DataNode. It is possible that the reduce tasks jobs are CPU-bound, but generally, the shuffle phase is CPU-bound and the reduce phase is I/O bound. In a high-performance cluster, these parameters will be carefully tuned for specific jobs.

```
<property>
    <name>mapred.tasktracker.reduce.tasks.maximum
    </name>
    <value>6</value>
    <description>The maximum number of reduce tasks that will be run
        simultaneously by a task tracker.
    </description>
</property>
```

Very few MapReduce jobs will run in the 200MB default heap size specified for the JVMs. To alter this to a more reasonable default, `mapred.child.java.opts` is set to `-Xmx512m -server`. The `-server` configures the JVM for the HotSpot JVM and provides other performance optimizations.

```
<property>
    <name>mapred.child.java.opts</name>
    <value>-Xmx512m -server</value>
    <description>Java opts for the task tracker child processes.
        The following symbol, if present, will be interpolated: @taskid@ is
        replaced by current TaskID. Any other occurrences of '@' will go
        unchanged.
        For example, to enable verbose gc logging to a file named
        for the taskid in /tmp and to set the heap maximum to be a gigabyte,
        pass a 'value' of:
        -Xmx1024m -verbose:gc -Xloggc:/tmp/@taskid@.gc
        The configuration variable mapred.child.ulimit can be used to control
        the maximum virtual memory of the child processes.
        Leave this unchanged from the default as io.sort.mb has been reduced for
        our test purposes.
     </description>
</property>
```

Finally, webinterface.private.actions is set to true, the recommended value for any cluster that doesn't require significant security.

```
    <property>
        <name>webinterface.private.actions</name>
        <value>true</value>
        <description> If set to true, the web interfaces of JT and NN may
            contain actions, such as kill job, delete file, etc., that should
            not be exposed to public. Enable this option if the interfaces
            are only reachable by those who have the right authorization.
            Enable this option if at all possible as it greatly simplifies
            debugging.
        </description>
    </property>

</configuration>
```

The slaves and masters Files

The slaves file contains five lines, each with one slave machine's hostname.

```
slave01
slave02
slave03
slave04
slave05
```

The masters file controls which machines run secondary NameNodes. The default configuration contains a single line containing localhost, which provides no real protection from machine or disk failure. It is a wise precaution to have a secondary NameNode on another machine. For this simple configuration example, the masters file has a single line with slave01.

```
slave01
```

The hadoop-metrics.properties File

The hadoop-metrics.properties file assumes that Ganglia is set up on the machine master01, and that some machine is set up with the Ganglia web interface and pulling data from master01. Installation and use of Ganglia are covered in Chapter 8.

```
# Configuration of the "dfs" context for null
#dfs.class=org.apache.hadoop.metrics.spi.NullContext

# Configuration of the "dfs" context for file
#dfs.class=org.apache.hadoop.metrics.file.FileContext
#dfs.period=10
#dfs.fileName=/tmp/dfsmetrics.log
```

```
# Configuration of the "dfs" context for ganglia
dfs.class=org.apache.hadoop.metrics.ganglia.GangliaContext
dfs.period=10
dfs.servers=master01:8649


# Configuration of the "mapred" context for null
# mapred.class=org.apache.hadoop.metrics.spi.NullContext

# Configuration of the "mapred" context for file
#mapred.class=org.apache.hadoop.metrics.file.FileContext
#mapred.period=10
#mapred.fileName=/tmp/mrmetrics.log

# Configuration of the "mapred" context for ganglia
mapred.class=org.apache.hadoop.metrics.ganglia.GangliaContext
mapred.period=10
mapred.servers=master01:8649


# Configuration of the "jvm" context for null
#jvm.class=org.apache.hadoop.metrics.spi.NullContext

# Configuration of the "jvm" context for file
#jvm.class=org.apache.hadoop.metrics.file.FileContext
#jvm.period=10
#jvm.fileName=/tmp/jvmmetrics.log

# Configuration of the "jvm" context for ganglia
jvm.class=org.apache.hadoop.metrics.ganglia.GangliaContext
jvm.period=10
jvm.servers=master01:8649
```

Distributing the Configuration

One of the reasons for requiring that all of the machines be essentially identical is that this greatly simplifies managing the cluster configuration. All of the core configuration files can be identical, which allows the use of the Unix `rsync` command to distribute the configuration files.

The command I like to use assumes that `HADOOP_HOME` is set correctly:

```
for a in `sort -u $HADOOP_HOME/conf/{slaves,masters}`; do ➥
rsync -e ssh -a --exclude 'logs/*' --exclude 'src/* ➥
--exclude 'docs/*' "${HADOOP_HOME}" ${a}:"${HADOOP_HOME}"; done
```

■ **Note** This command assumes that the current machine is not listed in the `slaves` or `masters` file. It's a very good test of the passwordless SSH configuration.

This command says that for each host that will run a server, distribute all of the Hadoop installation files except for the `src` (`--exclude 'src/*'`), logs (`--exclude 'logs/*'`), and `docs` (`--exclude 'docs/*'`). This ensures that all servers are running the same configuration files and the same Hadoop JARs. The `-e ssh` forces `rsync` to use SSH to establish the remote machine connections.

If your installation requires different configuration files on a per-machine basis, some other mechanism will be required to ensure consistency and correctness for the Hadoop installations and configuration files.

Verifying the Cluster Configuration

You should take a few steps to verify that that the cluster is installed and configured properly. At this point, you can assume that the Hadoop installation has been replicated to the same location across the cluster machines and that passwordless SSH is working. So, you should check the location of the JVM and make sure that `HADOOP_PID_DIR` can be written.

To verify that the JVM is in place and that `JAVA_HOME` is set correctly, run the following commands from the master machine:

```
master01% $for a in ➥
`sort -u "${HADOOP_HOME}"/conf/slaves ${HADOOP_HOME}"/conf/masters`; ➥
do echo -n $a " ";ssh $a -n ls -1 '"${JAVA_HOME}"/bin/java'; done
```

```
slave01 /usr/java/..../bin/java
slave02 /usr/java/..../bin/java
slave03 /usr/java/..../bin/java
slave04 /usr/java/..../bin/java
slave05 /usr/java/..../bin/java
```

```
master01% ls -1 "${JAVA_HOME}"/bin/java
```

```
/usr/java/..../bin/java
```

Every slave machine, as well as the local machine, should have an output line. If Java is not available on a machine in the expected location, install the JVM on that machine and set the `JAVA_HOME` environment variable to reflect the JVM installation directory.

The next item to verify is that the various required paths exist with the proper permissions or that the proper paths can be created. There are two directories to check: the directory specified as the `hadoop.tmp.dir`, in this case `/hadoop`, and the directory specified in the `conf/hadoop.env.sh` script for `HADOOP_PID_DIR`, in this case `/var/hadoop/pids`.

■ **Caution** All of the commands in this section must be run from the master machine and as the user that will own the cluster servers. Failure to do this will invalidate the verification process and may cause the cluster startup to fail in complex ways. The shell environment is also expected to be set up, as detailed in Chapter 2, such that both `java` and the `bin` directory of the Hadoop installation are the first two components of the `PATH` environment variable. The user that owns the cluster processes will be referred to in some of the following text as `CLUSTER_USER`.

Execute the following command to verify that `HADOOP_PID_DIR` and `hadmp.tmp.dir` are writable.

```
master01% for a in ➥
`sort -u "${HADOOP_HOME}/conf/slaves" "${HADOOP_HOME}/conf/masters"` `hostname`; ➥
do echo "${a} "; touch /var/hadoop/pids/dummy; touch /hadoop/dummy; done
```

```
slave01
slave02
slave03
slave04
slave05
master01
```

The command must not have any errors. Any error message about being unable to create the file `dummy` must be corrected before the next step is attempted.

At this point, you have checked that the cluster installation and configuration are correct, and that passwordless SSH is enabled. It is time to format the HDFS file system and to start the HDFS and MapReduce services.

Formatting HDFS

The command to format HDFS is very simple. If there has been an existing HDFS file system with the same `hadoop.tmp.dir`, it is best to remove all traces of it with the Unix `rm` command before formatting a new file system.

```
hadoop namenode -format
```

```
08/12/21 19:45:32 INFO dfs.NameNode: STARTUP_MSG:
/************************************************************
STARTUP_MSG: Starting NameNode
STARTUP_MSG:   host = master01/127.0.0.1
STARTUP_MSG:   args = [-format]
STARTUP_MSG:   version = 0.18.2-dev
STARTUP_MSG:   build =  -r ; compiled by 'jason' on Sun Nov 16 20:16:42 PST 2008
************************************************************/
```

```
fs.FSNamesystem: fsOwner=jason,jason,lp
fs.FSNamesystem: supergroup=supergroup
fs.FSNamesystem: isPermissionEnabled=true
dfs.Storage: Image file of size 79 saved in 0 seconds.
dfs.Storage: Storage directory /hadoop/dfs/name has been successfully formatted.
dfs.NameNode: SHUTDOWN_MSG:
/************************************************************
SHUTDOWN_MSG: Shutting down NameNode at at/127.0.0.1
************************************************************/
```

Starting HDFS

Starting HDFS is also quite simple. It may be started with the MapReduce portion of the cluster via `start-all.sh`. Here, I'll show you how to start HDFS separately to demonstrate the commands, as it is common to separate the HDFS master and configuration from the MapReduce master and configuration.

```
master01% bin/start-dfs.sh
```

```
starting namenode, logging to ➥
/home/training/hadoop-0.19.0/bin/../logs/HT-namenode-master01.out
slave01: starting datanode, logging to ➥
/home/training/hadoop-0.19.0/bin/../logs/HT-datanode-slave01.out
slave02: starting datanode, logging to ➥
/home/training/hadoop-0.19.0/bin/../logs/HT-datanode-slave02.out
slave05: starting datanode, logging to ➥
/home/training/hadoop-0.19.0/bin/../logs/HT-datanode-slave05.out
slave03: starting datanode, logging to ➥
/home/training/hadoop-0.19.0/bin/../logs/HT-datanode-slave03.out
slave04: starting datanode, logging to ➥
/home/training/hadoop-0.19.0/bin/../logs/HT-datanode-slave04.out
slave01: starting secondarynamenode, logging to ➥
➥/home/training/hadoop-0.19.0/bin/../logs/HT-secondarynamenode-slave01.out
```

Your output should look similar to the preceding sample output. Some common reasons for failures are listed in the next section.

After roughly one minute, allowing the DataNodes time to start and connect to the NameNode, issue the command that reports on the status of the DataNodes:

```
hadoop dfsadmin -report
```

```
Total raw bytes: 5368709120 (5 GB)
Remaining raw bytes: 5368709120 (5 GB)
Used raw bytes: 0 (0 GB)
% used: 0.00%
```

```
Total effective bytes: 0 (0 KB)
Effective replication multiplier: Infinity
-------------------------------------------------
Datanodes available: 5

Name: 192.168.0.10:50010
State          : In Service
Total raw bytes: 1073741824 (1 GB)
Remaining raw bytes: 1073741824(1.04 GB)
Used raw bytes: 0 (2.18 GB)
% used: 0.00%
Last contact: Sun Dec 21 19:30:14 PST 2008

Name: 192.168.0.11:50010
State          : In Service
Total raw bytes: 1073741824 (1 GB)
Remaining raw bytes: 1073741824(1.04 GB)
Used raw bytes: 0 (2.18 GB)
% used: 0.00%
Last contact: Sun Dec 21 19:30:14 PST 2008


Name: 192.168.0.12:50010
State          : In Service
Total raw bytes: 1073741824 (1 GB)
Remaining raw bytes: 1073741824(1.04 GB)
Used raw bytes: 0 (2.18 GB)
% used: 0.00%
Last contact: Sun Dec 21 19:30:14 PST 2008


Name: 192.168.0.13:50010
State          : In Service
Total raw bytes: 1073741824 (1 GB)
Remaining raw bytes: 1073741824(1.04 GB)
Used raw bytes: 0 (2.18 GB)
% used: 0.00%
Last contact: Sun Dec 21 19:30:14 PST 2008


Name: 192.168.0.14:50010
State          : In Service
Total raw bytes: 1073741824 (1 GB)
Remaining raw bytes: 1073741824(1.04 GB)
Used raw bytes: 0 (2.18 GB)
% used: 0.00%
Last contact: Sun Dec 21 19:30:14 PST 2008
```

In the sample cluster configuration, five DataNodes should report. If there are not five DataNodes, or the HDFS reports that it is in safe mode, something has gone wrong. Detailed log messages will be available in the logs directory of the Hadoop installation on the slave that is not reporting. The log file will be named hadoop-datanode-slave*XX*.log.

The final test is to attempt to copy a file into HDFS, as follows:

```
bin/hadoop dfs -touchz my_first_file
bin/hadoop dfs -ls my_first_file.
```

```
-rw-r--r--   3 training supergroup          0 2009-04-03 01:57 ➥
/user/training/my_first_file
```

A zero-length file named my_first_file will be created in the /user/*USERNAME* directory, where *USERNAME* is the username of the user running the touchz command.

Correcting Errors

The most common errors should not occur if the verification steps detailed in the preceding sections completed with no errors. Here are some errors you might come across:

- HADOOP_PID_DIR is not writable by the CLUSTER_USER.
- The directory specified for hadoop.tmp.dir does not exist with full access permissions for CLUSTER_USER, or could not be created.
- The JVM may be missing or installed in a different location.
- The environment set up on login for CLUSTER_USER may not set up JAVA_HOME and HADOOP_HOME correctly.
- HADOOP_HOME is not in an identical location on all of the cluster machines.
- Passwordless SSH to some subset of the machines may not be working, or the master machine cannot connect to the slave machines via SSH due to firewall or network topology reasons. This will be clear from the error response of the start command.
- The servers on the slave machines may not be able to connect to their respective master server due to firewall or network topology issues. If this is a problem, there will be a somewhat descriptive error message in the server (TaskTracker or DataNode) log file on the slave machine. Resolving network topology and firewall issues will require support from your local network administrator.

The most exotic error should not occur at this point, as your installation should be using the Hadoop Core default classpath. If there is an error message in a log file that indicates that Jetty could not start its web server, there is a nonvalidating XML parser in the classpath ahead of the validating XML parser that Hadoop Core supplies. This may be fixed by reordering the classpath or by explicitly setting the XML parser by setting a Java property. You can modify the HADOOP_OPTS environment variable to include this string:

```
-Djavax.xml.parsers.SAXParserFactory=org.apache.xerces.jaxp.SAXParserFactoryImpl
```

Alternatively, you can alter the setting of HADOOP_OPTS in conf/hadoop-env.sh.

Caution It is highly recommended that you verify the termination of all Hadoop Core server processes across the cluster before attempting a restart. It is also strongly recommended that the `hadoop.tmp.dir` have its contents wiped on all cluster machines between attempts to start a new HDFS, and then the file system be reformatted as described in this chapter.

The Web Interface to HDFS

The NameNode web interface will be available via HTTP on port 50070, `http://master01:50070/`. As shown in Figure 3-3, this interface shows information about the cluster. It also provides links to browse the file system, view the NameNode logs, and to drill down to specific node information.

NameNode '192.168.1.2:8020'

Started:	Thu Apr 23 12:42:29 GMT-08:00 2009
Version:	0.19.1-dev, r
Compiled:	Tue Mar 17 04:03:57 PDT 2009 by jason
Upgrades:	There are no upgrades in progress.

Browse the filesystem
Namenode Logs

Cluster Summary

10 files and directories, 0 blocks = 10 total. Heap Size is 7.93 MB / 992.31 MB (0%)

Configured Capacity	:	191.58 GB
DFS Used	:	76 KB
Non DFS Used	:	176.93 GB
DFS Remaining	:	14.65 GB
DFS Used%	:	0 %
DFS Remaining%	:	7.65 %
Live Nodes	:	2
Dead Nodes	:	0

Live Datanodes : 2

Node	Last Contact	Admin State	Configured Capacity (GB)	Used (GB)	Non DFS Used (GB)	Remaining (GB)	Used (%)	Used (%)	Remaining (%)	Blocks
192.168.1.119	1	In Service	165.6	0	151.92	13.67	0		8.26	0
192.168.1.2	0	In Service	25.98	0	25	0.98	0		3.77	0

Dead Datanodes : 0

Figure 3-3. *NameNode web interface*

Starting MapReduce

The MapReduce portion of the cluster will be started by the `start-mapred.sh` command.

If there are any errors when starting the TaskTrackers, the detailed error message will be in the `logs` directory on the specific slave node of the failed TaskTracker. The log file will be `hadoop-slaveXX-tasktracker.log`. The common reasons for failure are very similar to those for HDFS node startup failure.

For example, when I ran the example, on `slave01`, there was a process using the standard TaskTracker port that I was unaware of, and when the cluster was started, the TaskTracker did not start on `slave01`. I received the log message shown in Listing 3-1. This is a clear indication that some process is holding open a required port. The log file was in `${HADOOP_HOME}/logs/HT-tasktracker-slave01.log`.

Listing 3-1. *Tasktracker Error Log Message Due to TCP Port Unavailability*

```
2009-04-03 01:42:00,511 INFO org.apache.hadoop.mapred.TaskTracker: STARTUP_MSG:
/************************************************************
STARTUP_MSG: Starting TaskTracker
STARTUP_MSG:   host = slave01/192.168.1.121
STARTUP_MSG:   args = []
STARTUP_MSG:   version = 0.19.0
STARTUP_MSG:   build = ➥
https://svn.apache.org/repos/asf/hadoop/core/branches/branch-0.19 -r 713890; ➥
compiled by 'ndaley' on Fri Nov 14 03:12:29 UTC 2008
************************************************************/
INFO org.mortbay.http.HttpServer: Version Jetty/5.1.4
INFO org.mortbay.util.Credential: Checking Resource aliases
INFO org.mortbay.util.Container: Started ➥
org.mortbay.jetty.servlet.WebApplicationHandler@dc57db
INFO org.mortbay.util.Container: Started WebApplicationContext[/static,/static]
INFO org.mortbay.util.Container: Started ➥
org.mortbay.jetty.servlet.WebApplicationHandler@8e32e7
INFO org.mortbay.util.Container: Started WebApplicationContext[/logs,/logs]
INFO org.mortbay.util.Container: Started ➥
org.mortbay.jetty.servlet.WebApplicationHandler@15253d5
WARN org.mortbay.http.HttpContext: Can't reuse /tmp/Jetty__50060__, using ➥
/tmp/Jetty__50060___954083005982349324
INFO org.mortbay.util.Container: Started WebApplicationContext[/,/]
WARN org.mortbay.util.ThreadedServer: Failed to start: SocketListener0@0.0.0.0:50060
ERROR org.apache.hadoop.mapred.TaskTracker: Can not start task tracker because ➥
java.net.BindException: Address already in use
    at java.net.PlainSocketImpl.socketBind(Native Method)
    at java.net.PlainSocketImpl.bind(PlainSocketImpl.java:359)
    at java.net.ServerSocket.bind(ServerSocket.java:319)
    at java.net.ServerSocket.<init>(ServerSocket.java:185)
    at org.mortbay.util.ThreadedServer.newServerSocket(ThreadedServer.java:391)
    at org.mortbay.util.ThreadedServer.open(ThreadedServer.java:477)
    at org.mortbay.util.ThreadedServer.start(ThreadedServer.java:503)
    at org.mortbay.http.SocketListener.start(SocketListener.java:203)
    at org.mortbay.http.HttpServer.doStart(HttpServer.java:761)
    at org.mortbay.util.Container.start(Container.java:72)
    at org.apache.hadoop.http.HttpServer.start(HttpServer.java:321)
    at org.apache.hadoop.mapred.TaskTracker.<init>(TaskTracker.java:894)
    at org.apache.hadoop.mapred.TaskTracker.main(TaskTracker.java:2698)
```

Running a Test Job on the Cluster

To test the cluster configuration, let's run our old friend the Hadoop Core example pi (introduced in Chapter 1). Recall that this program takes two arguments: the number of maps and the number of samples. In this case, the cluster has five map slots, so you will set the number of maps to five. You have a couple of machines, so you can set the number of samples to a moderately large number—say 10,000. Your results should be similar to the following:

```
cd $HADOOP_HOME; hadoop jar hadoop-0.18.2-examples.jar 5 10000
```

```
Number of Maps = 5 Samples per Map = 10000
Wrote input for Map #0
Wrote input for Map #1
Wrote input for Map #2
Wrote input for Map #3
Wrote input for Map #4
Starting Job
jvm.JvmMetrics: Initializing JVM Metrics with processName=JobTracker, sessionId=
mapred.FileInputFormat: Total input paths to process : 5
mapred.FileInputFormat: Total input paths to process : 5
mapred.JobClient: Running job: job_local_0001
mapred.FileInputFormat: Total input paths to process : 5
mapred.FileInputFormat: Total input paths to process : 5
mapred.LocalJobRunner: Generated 9001 samples.
mapred.JobClient:  map 100% reduce 0%
mapred.LocalJobRunner: Generated 9001 samples.
mapred.LocalJobRunner: Generated 9001 samples.
mapred.LocalJobRunner: Generated 9001 samples.
mapred.LocalJobRunner: Generated 9001 samples.
mapred.ReduceTask: Initiating final on-disk merge with 5 files
mapred.Merger: Merging 5 sorted segments
mapred.Merger: Down to the last merge-pass, with 5 ➥
segments left of total size: 190 bytes
mapred.LocalJobRunner: reduce > reduce
mapred.JobClient:  map 100% reduce 100%
mapred.JobClient: Job complete: job_local_0001
mapred.JobClient: Counters: 11
mapred.JobClient:   File Systems
mapred.JobClient:     Local bytes read=632904
mapred.JobClient:     Local bytes written=731468
mapred.JobClient:   Map-Reduce Framework
mapred.JobClient:     Reduce input groups=2
mapred.JobClient:     Combine output records=0
mapred.JobClient:     Map input records=5
mapred.JobClient:     Reduce output records=0
mapred.JobClient:     Map output bytes=160
mapred.JobClient:     Map input bytes=120
```

```
mapred.JobClient:     Combine input records=0
mapred.JobClient:     Map output records=10
mapred.JobClient:     Reduce input records=10
Job Finished in 1.807 seconds
Estimated value of PI is 3.14816
```

Summary

This chapter provided a simple walk-through of configuring a small Hadoop Core cluster. It did not discuss the tuning parameters required for a larger or a high-performance cluster. These parameters will be covered in Chapters 4 and 6.

For a multimachine cluster to run, the configuration must include the following:

- List of slave machines (`conf/slaves`)
- Network location of the JobTracker server (`mapred.job.tracker`)
- Network location of the NameNode server (`fs.default.name`)
- Persistent location on the cluster machines to store the data for HDFS (`hadoop.tmp.dir`)

You now have an understanding of what a master node is, what the NameNode and JobTracker servers do, and what DataNode and TaskTracker servers are. You may have even set up a multimachine cluster and run jobs over it. Go forth and grid compute.

CHAPTER 4

■ ■ ■

HDFS Details for Multimachine Clusters

As you learned in the previous chapter, the defaults provided for multimachine clusters will work well for very small clusters, but they are not suitable for large clusters (the clusters will fail in unexpected and difficult-to-understand ways). This chapter covers HDFS installation for multimachine clusters that are not very small, as well as HDFS tuning factors, recovery procedures, and troubleshooting tips. But first, let's look at some of the configuration trade-offs faced by IT departments.

Configuration Trade-Offs

There appears to be an ongoing conflict between the optimal machine and network configurations for Hadoop Core and the configurations required by IT departments.

IT departments are generally looking for low-overhead ways of maintaining high availability for all equipment. The IT department model commonly requires RAID 1 and RAID 5 for disks to minimize machine downtime from disk failures. IT departments also prefer managed network switches, as this allows for reporting and virtual local area network (VLAN) configurations. These strategies reduce the risk of machine failure and provide network diagnostics, flexibility, and simplified administration. Operations staff also prefer high-availability solutions for production applications.

Hadoop Core does not need highly reliable storage on the DataNode or TaskTracker nodes. Hadoop Core greatly benefits from increased network bandwidth.

The highest performance Hadoop Core installations will have separate and possibly multiple disks or arrays for each stream of I/O. The DataNode storage will be spread over multiple disks or arrays to allow interleaved I/O, and the TaskTracker intermediate output will also go to a separate disk or array. This configuration reduces the contention for I/O on any given array or device, thus increasing the maximum disk I/O performance of the machine substantially. If the switch ports are inexpensive, using bonded network interface cards (NICs) to increase per machine network bandwidth will greatly increase the I/O performance of the cluster.

Hadoop Core provides high availability of DataNode and TaskTracker services without requiring special hardware, software, or configuration. However, there is no simple solution for high availability for the NameNode or JobTracker. High availability for the NameNode is an active area of development within the Hadoop community. The techniques described in this chapter allow for rapid recovery from NameNode failures. All of these techniques require special configuration and have some performance cost. With Hadoop 0.19.0, there is built-in recovery for JobTracker failures, and generally, high availability for the JobTracker is not considered critical.

AFFORDABLE DISK PERFORMANCE VS. NETWORK PERFORMANCE

Hadoop Core is designed to take advantage of commodity hardware, rather than more expensive special-purpose hardware.

In my experience, the bulk of custom-purchased Hadoop nodes appear to be 2U 8-way machines, with six internal drive bays, two Gigabit Ethernet (GigE) interfaces, and 8GB of RAM. Most of the disk drives that are being used for Hadoop are inexpensive SATA drives that generally have a sustained sequential transfer rate of about 70Mbps.

The RAID setup preferred by my IT department groups six of these drives in a RAID 5 array that provides roughly 250 Mbps sequential transfers to the application layer. Mixed read/write operations provide about 100 Mbps, as all I/O operations require seeks on the same set of drives. If the six drives were provided as individual drives or as three RAID 0 pairs, the mixed read/write I/O transfer rate would be higher, as seeks for each I/O operation could occur on different drives.

The common network infrastructure is GigE network cards on a GigE switch, providing roughly 100 Mbps I/O. I've used bonded pairs of GigE cards to provide 200 Mbps I/O for high-demand DataNodes to good effect.

IT departments seem to prefer large managed switches, resulting in a high per-port cost. For Hadoop nodes, providing dumb, unmanaged crossbar switches for the DataNodes is ideal.

HDFS Installation for Multimachine Clusters

Setting up an HDFS installation for a multimachine cluster involves the following steps:

- Build the configuration.
- Distribute your installation data to all of the machines that will host HDFS servers.
- Format your HDFS.
- Start your HDFS installation.
- Verify HDFS is running.

The following sections detail these steps.

Building the HDFS Configuration

As discussed in the previous chapter, building the HDFS configuration requires generating the `conf/hadoop-site.xml`, `conf/slaves`, and `conf/masters` files. You also need to customize the `conf/hadoop-env.sh` file.

Generating the conf/hadoop-site.xml File

In the conf/hadoop-site.xml file, you tell Hadoop where the data files reside on the file system. At the simplest level, this requires setting a value for hadoop.tmp.dir, and providing a value for fs.default.name to indicate the master node of the HDFS cluster, as shown in Listing 4-1.

Listing 4-1. *A Minimal hadoop-site.xml for an HFS Cluster (conf/hadoop-site.xml)*

```
<property>
  <name>fs.default.name</name>
  <value>hdfs://master:54310/</value>
  <description>The name of the default file system.  A URI whose
  scheme and authority determine the FileSystem implementation. The
  uri's scheme determines the config property (fs.SCHEME.impl) naming
  the FileSystem implementation class. The uri's authority is used to
  determine the host, port, etc. for a filesystem.</description>
</property>

<property>
  <name>hadoop.tmp.dir</name>
  <value>/hdfs</value>
  <description>A base for other all storage directories,
          temporary and persistent.</description>
</property>
```

This will configure a cluster with a NameNode on the host master, and all HDFS storage under the directory /hdfs.

Generating the conf/slaves and conf/masters Files

On the machine master, create the conf/slaves file, and populate it with the names of the hosts that will be DataNodes, one per line.

In the file conf/masters, add a single host to be the secondary NameNode. For safety, make it a separate machine, rather than localhost.

Customizing the conf/hadoop-env.sh File

The conf/hadoop-env.sh file provides system environment configuration information for all processes started by Hadoop, as well as all processes run by the user through the scripts in the bin directory of the installation. At a very minimum, this script must ensure that the correct JAVA_HOME environment variable is set. Table 4-1 provides a list of the required, commonly set, and optional environment variables.

Table 4-1. *Environment Variables for Hadoop Processes*

Variable	Description	Default
JAVA_HOME	This is required. It must be the root of the JDK installation, such that ${JAVA_HOME}/bin/java is the program to start a JVM.	System JDK
HADOOP_NAMENODE_OPTS	Additional command-line arguments for the NameNode server. The default enables local JMX access.	"-Dcom.sun.management.jmxremote $HADOOP_NAMENODE_OPTS"
HADOOP_SECONDARYNAMENODE_OPTS	Additional command-line arguments for the secondary NameNode server. The default enables local JMX access.	"-Dcom.sun.management.jmxremote $HADOOP_SECONDARYNAMENODE_OPTS"
HADOOP_DATANODE_OPTS	Additional command-line arguments for the DataNode servers. The default enables local JMX access.	"-Dcom.sun.management.jmxremote $HADOOP_DATANODE_OPTS"
HADOOP_BALANCER_OPTS	Additional command-line arguments for the Balancer service. The default enables local JMX access.	"-Dcom.sun.management.jmxremote $HADOOP_BALANCER_OPTS"
HADOOP_JOBTRACKER_OPTS	Additional command-line arguments for the JobTracker server. The default enables local JMX access.	"-Dcom.sun.management.jmxremote $HADOOP_JOBTRACKER_OPTS"
HADOOP_TASKTRACKER_OPTS	Additional command-line arguments for the TaskTracker servers.	
HADOOP_CLIENT_OPTS	Additional command-line arguments for all nonserver processes started by bin/hadoop. This is not applied to the server processes, such as the NameNode, JobTracker, TaskTracker, and DataNode.	
HADOOP_SSH_OPTS	Additional command-line arguments for any ssh process run by scripts in bin. This commented-out option in the stock hadoop-env.sh file sets the ssh connection timeout to 1 second and instructs ssh to forward the HADOOP_CONF_DIR environment variable to the remote shell.	"-o ConnectTimeout=1 -o SendEnv=HADOOP_CONF_DIR"
HADOOP_LOG_DIR	The root directory path that Hadoop logging files will be created under.	${HADOOP_HOME}/logs
HADOOP_SLAVES	The path of the file containing the list of hostnames to be used as DataNode and or TaskTracker servers.	${HADOOP_HOME}/conf/slaves
HADOOP_SLAVE_SLEEP	The amount of time to sleep between ssh commands when operating on all of the slave nodes.	0.1
HADOOP_PID_DIR	The directory that server process ID (PID) files are written to. Used by the service start and stop scripts to determine if a prior instance of a server is running. The default, /tmp, is not a good location, as the server PID files will be periodically removed by the system temp file cleaning service.	/tmp

Variable	Description	Default
HADOOP_IDENT_STRING	Used in constructing path names for cluster instance-specific file names, such as the file names of the log files for the server processes and the PID files for the server processes.	$USER
HADOOP_NICENESS	CPU scheduling nice factor to apply to server processes. 5 is recommended for DataNode servers and 10 for TaskTrackers. The suggested settings prioritize the DataNode over the TaskTracker to ensure that DataNode requests are more rapidly serviced. They also help ensure that the NameNode or JobTracker servers have priority if a DataNode or TaskTracker is colocated on the same machine. These suggested settings facilitate smooth cluster operation and enable easier monitoring.	
HADOOP_CLASSPATH	Extra entries for the classpath for all Hadoop Java processes. If your jobs always use specific JARs, and these JARs are available on all systems in the same location, adding the JARs here ensures that they are available to all tasks and reduces the overhead in setting up a job on the cluster.	
HADOOP_HEAPSIZE	The maximum process heap size for running tasks. 2000, indicating 2GB, is suggested. See this book's appendix for details on the JobConf object.	
HADOOP_OPTS	Additional command-line options for all processes started by bin/hadoop. This setting is normally present but commented out.	-server

Distributing Your Installation Data

Distribute your Hadoop installation to all the machines in conf/masters and conf/slaves, as well as the machine master. Ensure that the user that will own the servers can write to /hdfs and the directory set for HADOOP_PID_DIR in conf/hadoop-env.sh, on all of the machines in conf/masters, conf/slaves, and master.

Finally, ensure that passwordless SSH from master to all of the machines in conf/masters and conf/slaves works.

SETTING JAVA_HOME AND PERMISSIONS

The user is responsible for ensuring that the `JAVA_HOME` environment variable is configured. The framework will issue an `ssh` call to each slave machine:

```
execute ${HADOOP_HOME}/bin/hadoop-daemon.sh start SERVER
```

This sources `${HADOOP_HOME}/conf/hadoop-env.sh`.

The administrator is also responsible for ensuring that the `JAVA_HOME` environment variable is set correctly in the login scripts or in the `hadoop-env.sh` script.

Additionally, the administrator must ensure that the storage directories are writable by the Hadoop user. In general, this simply means constructing the `hadoop.tmp.dir` directory set in the `conf/hadoop-site.xml` file, and `chown`ing the directory to the user that the Hadoop servers will run as.

Formatting Your HDFS

At this point, you are ready to actually format your HDFS installation. Run the following to format a NameNode for the first time:

```
hadoop namenode -format
```

```
namenode.NameNode: STARTUP_MSG:
/************************************************************
STARTUP_MSG: Starting NameNode
STARTUP_MSG:   host = master/127.0.0.1
STARTUP_MSG:   args = [-format]
STARTUP_MSG:   version = 0.19.0
STARTUP_MSG:   build = …
************************************************************/
namenode.FSNamesystem: fsOwner=jason,jason,lp,wheel,matching
namenode.FSNamesystem: supergroup=supergroup
namenode.FSNamesystem: isPermissionEnabled=true
common.Storage: Image file of size 95 saved in 0 seconds.
common.Storage: Storage directory /hdfs/dfs/name has been successfully formatted.
namenode.NameNode: SHUTDOWN_MSG:
/************************************************************
SHUTDOWN_MSG: Shutting down NameNode at master/127.0.0.1
************************************************************/
```

Note The exact stack traces and line numbers will vary with your version of Hadoop Core.

If you have already formatted a NameNode with this data directory, you will see that the command will try to reformat the NameNode:

```
hadoop namenode -format
```

```
09/01/24 12:03:57 INFO namenode.NameNode: STARTUP_MSG:
/************************************************************
STARTUP_MSG: Starting NameNode
STARTUP_MSG:   host = master/127.0.0.1
STARTUP_MSG:   args = [-format]
STARTUP_MSG:   version = 0.19.0
STARTUP_MSG:   build = …
************************************************************/
Re-format filesystem in /hdfs/dfs/name ? (Y or N) y
Format aborted in /hdfs/dfs/name
09/01/24 12:04:01 INFO namenode.NameNode: SHUTDOWN_MSG:
/************************************************************
SHUTDOWN_MSG: Shutting down NameNode at master/127.0.0.1
************************************************************/
```

If the user doing the formatting does not have write permissions, the output will be as follows:

```
bin/hadoop namenode -format
```

```
INFO namenode.NameNode: STARTUP_MSG:
/************************************************************
STARTUP_MSG: Starting NameNode
STARTUP_MSG:   host = master/127.0.0.1
STARTUP_MSG:   args = [-format]
STARTUP_MSG:   version = 0.19.0
STARTUP_MSG:   build = …
************************************************************/
INFO namenode.FSNamesystem: fsOwner=jason,jason,lp,wheel,matching
INFO namenode.FSNamesystem: supergroup=supergroup
INFO namenode.FSNamesystem: isPermissionEnabled=true
ERROR namenode.NameNode: java.io.IOException:
    Cannot create directory /tmp/test1/dir/dfs/name/current
    at org.apache.hadoop.hdfs.server.common.Storage$StorageDirectory. ➥
    clearDirectory(Storage.java:295)
    at org.apache.hadoop.hdfs.server.namenode.FSImage.format(FSImage.java:1067)
    at org.apache.hadoop.hdfs.server.namenode.FSImage.format(FSImage.java:1091)
    at org.apache.hadoop.hdfs.server.namenode.NameNode.format(NameNode.java:767)
```

```
    at org.apache.hadoop.hdfs.server.namenode.NameNode. ➥
    createNameNode(NameNode.java:851)
    at org.apache.hadoop.hdfs.server.namenode.NameNode.main(NameNode.java:868)

09/01/25 19:14:37 INFO namenode.NameNode: SHUTDOWN_MSG:
/************************************************************
SHUTDOWN_MSG: Shutting down NameNode at master/127.0.0.1
```

If you are reformatting an HDFS installation, it is recommended that you wipe the `hadoop.tmp.dir` directories on all of the DataNode machines.

Starting Your HDFS Installation

After you've configured and formatted HDFS, it is time to actually start your multimachine HDFS cluster. You can use the `bin/start-dfs.sh` command for this, as follows:

```
bin/start-dfs.sh
```

```
starting namenode, logging to
     /home/jsn/src/hadoop-0.19.0/bin/../logs/hadoop-jsn-namenode-master.out
slave1: starting datanode, logging to
    /home/jsn/src/hadoop-0.19.0/bin/../logs/hadoop-jsn-datanode-at.out
slave1: starting secondarynamenode, logging to
    /home/jsn/src/hadoop-0.19.0/bin/../logs/hadoop-jsn-secondarynamenode-slave1.out
slave2: starting datanode, logging to
    /home/jsn/src/hadoop-0.19.0/bin/../logs/hadoop-jsn-datanode-at.out
slave2: starting secondarynamenode, logging to
    /home/jsn/src/hadoop-0.19.0/bin/../logs/hadoop-jsn-secondarynamenode-slave2.out
slave3: starting datanode, logging to
    /home/jsn/src/hadoop-0.19.0/bin/../logs/hadoop-jsn-datanode-at.out
slave3: starting secondarynamenode, logging to
    /home/jsn/src/hadoop-0.19.0/bin/../logs/hadoop-jsn-secondarynamenode-slave3.out
slave4: starting datanode, logging to
    /home/jsn/src/hadoop-0.19.0/bin/../logs/hadoop-jsn-datanode-at.out
slave4: starting secondarynamenode, logging to
    /home/jsn/src/hadoop-0.19.0/bin/../logs/hadoop-jsn-secondarynamenode-slave4.out
slave5: starting datanode, logging to
    /home/jsn/src/hadoop-0.19.0/bin/../logs/hadoop-jsn-datanode-at.out
slave5: starting secondarynamenode, logging to
    /home/jsn/src/hadoop-0.19.0/bin/../logs/hadoop-jsn-secondarynamenode-slave5.out
```

If you have any lines like the following in your output, then the script was unable to `ssh` to the slave node:

```
slave1: ssh: connect to host slave1 port 22: No route to host
```

If you have a block like the following, you have not distributed your Hadoop installation correctly to all of the slave nodes:

```
slave1: bash: line 0: cd: /home/jason/src/hadoop-0.19.0/bin/..: ➥
No such file or directory
slave1: bash: /home/jason/src/hadoop-0.19.0/bin/hadoop-daemon.sh: ➥
No such file or directory
slave1: bash: line 0: cd: /home/jason/src/hadoop-0.19.0/bin/..: ➥
No such file or directory
slave1: bash: /home/jason/src/hadoop-0.19.0/bin/hadoop-daemon.sh: ➥
No such file or directory
```

The following output indicates that the directory specified in `hadoop-env.sh` for the server PID files was not writable:

```
slave1: mkdir: cannot create directory `/var/bad-hadoop': Permission denied
slave1: starting datanode, logging to
 /home/jason/src/hadoop-0.19.0/bin/../logs/hadoop-jason-datanode-slave1.out
slave1: /home/jason/src/hadoop-0.19.0/bin/hadoop-daemon.sh: line 118:
  /var/bad-hadoop/pids/hadoop-jason-datanode.pid: No such file or directory
slave1: mkdir: cannot create directory `/var/bad-hadoop': Permission denied
slave1: starting secondarynamenode, logging to
 /home/jason/src/hadoop-0.19.0/bin/../ ➥
logs/hadoop-jason-secondarynamenode-slave1.out
slave1: /home/jason/src/hadoop-0.19.0/bin/hadoop-daemon.sh: line 118: ➥
/var/bad-hadoop/pids/hadoop-jason-secondarynamenode.pid: ➥
No such file or directory
```

When Hadoop Core is starting services on the cluster, the required directories for the server operation are created if needed. These include the PID directory and the working and temporary storage directories for the server. If the framework is unable to create a directory, there will be error messages to that effect logged to the error stream of the script being used to start the services. Any messages in the startup output about files or directories that are not writable, or directories that could not be created, must be addressed. In some cases, the server will start, and the cluster may appear to run, but there will be stability and reliability issues.

At this point, the next step is to verify that the DataNode servers started and that they were able to establish service with the NameNode.

Verifying HDFS Is Running

To verify that the server processes are in fact running, wait roughly 1 minute after the finish of `start-dfs.sh`, and then check that the NameNode and DataNode exist.

Checking the NameNodes

On the master machine, run the `jps` command as follows:

```
${JAVA_HOME}/bin/jps
```

```
1887 Jps
19379 NameNode
```

The first value in the output is the PID of the `java` process, and it will be different in your output.

If there is no NameNode, the initialization failed, and you will need to examine the log file to determine the problem. The log data will be in the file `logs/hadoop-${USER}-namenode.log`.

> ■ **Note** The NameNode actually performs the formatting operation, and the formatting results end up in the NameNode log.

The examples in Listings 4-2 and 4-3 are excerpts from a NameNode log, demonstrating different failure cases.

Listing 4-2. *Did the NameNode Format Fail Due to Insufficient Permissions?*

```
bin/hadoop namenode -format
```

```
INFO namenode.NameNode: STARTUP_MSG:
/************************************************************
STARTUP_MSG: Starting NameNode
STARTUP_MSG:   host = master/127.0.0.1
STARTUP_MSG:   args = [-format]
STARTUP_MSG:   version = 0.19.0
STARTUP_MSG:   build = ...
************************************************************/
INFO namenode.FSNamesystem: fsOwner=jason,jason,lp,wheel,matching
INFO namenode.FSNamesystem: supergroup=supergroup
INFO namenode.FSNamesystem: isPermissionEnabled=true
ERROR namenode.NameNode: java.io.IOException: ➥
Cannot create directory /tmp/test1/dir/dfs/name/current
    at org.apache.hadoop.hdfs.server.common.Storage$StorageDirectory.➥
clearDirectory(Storage.java:295)
    at org.apache.hadoop.hdfs.server.namenode.FSImage.format(FSImage.java:1067)
    at org.apache.hadoop.hdfs.server.namenode.FSImage.format(FSImage.java:1091)
```

```
    at org.apache.hadoop.hdfs.server.namenode.NameNode.format(NameNode.java:767)
    at org.apache.hadoop.hdfs.server.namenode.NameNode. ➥
    createNameNode(NameNode.java:851)
    at org.apache.hadoop.hdfs.server.namenode.NameNode.main(NameNode.java:868)

09/01/25 19:14:37 INFO namenode.NameNode: SHUTDOWN_MSG:
/************************************************************
SHUTDOWN_MSG: Shutting down NameNode at master/127.0.0.1
```

Listing 4-2 indicates that the NameNode process was unable to find a valid directory for HDFS metadata. When this occurs, the command `hadoop namenode -format` must be run, to determine the actual failure. If the `format` command completes successfully, the next `start-dfs.sh` run should complete successfully. The example in Listing 4-3 demonstrates the failed `format` command output. The actual directory listed will be different in actual usage, the directory path `/tmp/test1/dir/dfs/name/current` *was constructed just for this test.*

Listing 4-3. *A Failed Format Due to Directory Permissions*

```
bin/hadoop namenode -format
```

```
09/04/04 13:13:37 INFO namenode.NameNode: STARTUP_MSG:
/************************************************************
STARTUP_MSG: Starting NameNode
STARTUP_MSG:   host = at/192.168.1.119
STARTUP_MSG:   args = [-format]
STARTUP_MSG:   version = 0.19.1-dev
STARTUP_MSG:   build =  -r ; compiled by 'jason' on Tue Mar 17 04:03:57 PDT 2009
************************************************************/
INFO namenode.FSNamesystem: fsOwner=jason,jason,lp
INFO namenode.FSNamesystem: supergroup=supergroup
INFO namenode.FSNamesystem: isPermissionEnabled=true
ERROR namenode.NameNode: java.io.IOException: ➥
Cannot create directory /tmp/test1/dir/dfs/name/current
    at org.apache.hadoop.hdfs.server.common.Storage$StorageDirectory. ➥
clearDirectory(Storage.java:295)
    at org.apache.hadoop.hdfs.server.namenode.FSImage.format(FSImage.java:1067)
    at org.apache.hadoop.hdfs.server.namenode.FSImage.format(FSImage.java:1091)
    at org.apache.hadoop.hdfs.server.namenode.NameNode.format(NameNode.java:767)
    at org.apache.hadoop.hdfs.server.namenode.NameNode. ➥
createNameNode(NameNode.java:851)
    at org.apache.hadoop.hdfs.server.namenode.NameNode.main(NameNode.java:868)

INFO namenode.NameNode: SHUTDOWN_MSG:
/************************************************************
SHUTDOWN_MSG: Shutting down NameNode at at/192.168.1.119
************************************************************/
```

The web interface provided by the NameNode will show information about the status of the NameNode. By default, it will provide a service on `http://master:50070/`.

Checking the DataNodes

You also need to verify that there are DataNodes on each of the slave nodes. Use `jps` via the `bin/slaves.sh` command to look for DataNode processes:

```
bin/slaves.sh jps | grep Datanode | sort
```

```
slave1: 2564 DataNode
slave2: 2637 DataNode
slave3: 1532 DataNode
slave4: 7810 DataNode
slave5: 8730 DataNode
```

This example shows five DataNodes, one for each slave. If you do not have a DataNode on each of the slaves, something has failed. Each machine may have a different reason for failure, so you'll need to examine the log files on each machine.

The common reason for DataNode failure is that the `dfs.data.dir` was not writable, as shown in Listing 4-4.

Listing 4-4. *Excerpt from a DataNode Log File on Failure to Start Due to Permissions Problems*

```
2009-01-28 07:50:05,441 INFO org.apache.hadoop.hdfs.server.datanode. ➥
DataNode: STARTUP_MSG:
/************************************************************
STARTUP_MSG: Starting DataNode
STARTUP_MSG:   host = slave1/127.0.0.1
STARTUP_MSG:   args = []
STARTUP_MSG:   version = 0.19.1-dev
STARTUP_MSG:   build =  -r ; compiled by 'jason' on Wed Jan 21 18:10:58 PST 2009
************************************************************/
2009-01-28 07:50:05,653 WARN org.apache.hadoop.hdfs.server.datanode. ➥
DataNode: Invalid directory in dfs.data.dir: can not create directory: ➥
 /tmp/test1/dir/dfs/data
2009-01-28 07:50:05,653 ERROR org.apache.hadoop.hdfs.server.datanode. ➥
DataNode: All directories in dfs.data.dir are invalid.
2009-01-28 07:50:05,654 INFO org.apache.hadoop.hdfs.server.datanode. ➥
DataNode: SHUTDOWN_MSG:
/************************************************************
SHUTDOWN_MSG: Shutting down DataNode at at/127.0.0.1
************************************************************/
```

The DataNode may also be unable to contact the NameNode due to network connectivity or firewall issues. In fact, I had half of a new cluster fail to start, and it took some time to realize that the newly installed machines had a default firewall that blocked the HDFS port. Listing 4-5 shows an excerpt from a DataNode log for a DataNode that failed to start due to network connectivity problems.

Listing 4-5. *DataNode Log Excerpt, Failure to Connect to the NameNode*

```
INFO org.apache.hadoop.hdfs.server.datanode.DataNode: STARTUP_MSG:
/************************************************************
STARTUP_MSG: Starting DataNode
STARTUP_MSG:   host = slave1/127.0.0.1
STARTUP_MSG:   args = []
STARTUP_MSG:   version = 0.19.1-dev
STARTUP_MSG:   build =…
************************************************************/
INFO org.apache.hadoop.ipc.Client: Retrying connect to server:
    master/192.168.1.2:8020. Already tried 0 time(s).
INFO org.apache.hadoop.ipc.Client: Retrying connect to server:
    master/192.168.1.2:8020. Already tried 1 time(s).
INFO org.apache.hadoop.ipc.Client: Retrying connect to server:
    master/192.168.1.2:8020. Already tried 2 time(s).
INFO org.apache.hadoop.ipc.Client: Retrying connect to server:
    master/192.168.1.2:8020. Already tried 3 time(s).
INFO org.apache.hadoop.ipc.Client: Retrying connect to server:
    master/192.168.1.2:8020. Already tried 4 time(s).
INFO org.apache.hadoop.ipc.Client: Retrying connect to server:
    master/192.168.1.2:8020. Already tried 5 time(s).
INFO org.apache.hadoop.ipc.Client: Retrying connect to server:
    master/192.168.1.2:8020. Already tried 6 time(s).
INFO org.apache.hadoop.ipc.Client: Retrying connect to server:
    master/192.168.1.2:8020. Already tried 7 time(s).
INFO org.apache.hadoop.ipc.Client: Retrying connect to server:
    master/192.168.1.2:8020. Already tried 8 time(s).
INFO org.apache.hadoop.ipc.Client: Retrying connect to server:
    master/192.168.1.2:8020. Already tried 9 time(s).
ERROR org.apache.hadoop.hdfs.server.datanode.DataNode:
  java.io.IOException: Call to master/192.168.1.2:8020 failed on local exception:
    No route to host
    at org.apache.hadoop.ipc.Client.call(Client.java:699)
    at org.apache.hadoop.ipc.RPC$Invoker.invoke(RPC.java:216)
    at $Proxy4.getProtocolVersion(Unknown Source)
    at org.apache.hadoop.ipc.RPC.getProxy(RPC.java:319)
    at org.apache.hadoop.ipc.RPC.getProxy(RPC.java:306)
    at org.apache.hadoop.ipc.RPC.getProxy(RPC.java:343)
    at org.apache.hadoop.ipc.RPC.waitForProxy(RPC.java:288)
    at org.apache.hadoop.hdfs.server.datanode.DataNode. ➥
    startDataNode(DataNode.java:258)
```

```
    at org.apache.hadoop.hdfs.server.datanode.DataNode.<init>(DataNode.java:205)
    at org.apache.hadoop.hdfs.server.datanode.DataNode. ➥
    makeInstance(DataNode.java:1199)
    at org.apache.hadoop.hdfs.server.datanode.DataNode. ➥
    instantiateDataNode(DataNode.java:1154)
    at org.apache.hadoop.hdfs.server.datanode.DataNode. ➥
    createDataNode(DataNode.java:1162)
    at org.apache.hadoop.hdfs.server.datanode.DataNode.main(DataNode.java:1284)
Caused by: java.net.NoRouteToHostException: No route to host
    at sun.nio.ch.SocketChannelImpl.checkConnect(Native Method)
    at sun.nio.ch.SocketChannelImpl.finishConnect(SocketChannelImpl.java:574)
    at sun.nio.ch.SocketAdaptor.connect(SocketAdaptor.java:100)
    at org.apache.hadoop.ipc.Client$Connection.setupIOstreams(Client.java:299)
    at org.apache.hadoop.ipc.Client$Connection.access$1700(Client.java:176)
    at org.apache.hadoop.ipc.Client.getConnection(Client.java:772)
    at org.apache.hadoop.ipc.Client.call(Client.java:685)
    ... 12 more

INFO org.apache.hadoop.hdfs.server.datanode.DataNode: SHUTDOWN_MSG:
/************************************************************
SHUTDOWN_MSG: Shutting down DataNode at at/127.0.0.1
************************************************************/
```

■ **Note** It is important to note the IP address to which the DataNode is trying to connect. If the IP address is correct, verify that the NameNode is accepting connections on that port.

Hadoop also provides the `dfsadmin -report` command-line tool, which will provide a somewhat verbose listing of the DataNodes in a service. You can run this useful script from your system monitoring tools, so that alerts can be generated if DataNodes go offline. The following is an example run of this tool:

```
bin/hadoop dfsadmin -report
```

```
Configured Capacity: 190277042176 (177.21 GB)
Present Capacity: 128317767680 (119.51 GB)
DFS Remaining: 128317718528 (119.51 GB)
DFS Used: 49152 (48 KB)
DFS Used%: 0%
```

```
-------------------------------------------------
Datanodes available: 5 (5 total, 0 dead)

Name: 192.168.1.3:50010
Decommission Status : Normal
Configured Capacity: 177809035264 (165.6 GB)
DFS Used: 24576 (24 KB)
Non DFS Used: 50479497216 (47.01 GB)
DFS Remaining: 127329513472(118.58 GB)
DFS Used%: 0%
DFS Remaining%: 71.61%
Last contact: Wed Jan 28 08:27:20 GMT-08:00 2009

...
```

Tuning Factors

Here, we will look at tuning the cluster system and the HDFS parameters for performance and reliability.

Commonly, the two most important factors are network bandwidth and disk throughput. Memory use and CPU overhead for thread handling may also be issues.

HDFS uses a basic file system block size of 64MB, and the JobTracker also chunks task input into 64MB segments. Using large blocks helps reduce the cost of a disk seek compared with the read/write time, thereby increasing the aggregate I/O rate when multiple requests are active. The large input-split size reduces the ratio of task setup time to task run time, as there is work to be done to set up a task before the TaskTracker can start the mapper or reducer on the input split.

The various tuning factors available control the maximum number of requests in progress. In general, the more requests in progress, the more contention there is for storage operations and network bandwidth, with a corresponding increase in memory requirements and CPU overhead for handling all of the outstanding requests. If the number of requests allowed is too low, the cluster may not fully utilize the disk or network bandwidth, or cause requests to timeout.

Most of the tuning parameters for HDFS do not yet have exact science behind the settings. Currently, tuning is ad hoc and per cluster. In general, the selections are a compromise between various factors. Finding the sweet spot requires knowledge of the local system and experience.

Let's examine each of the tuning factors in turn.

File Descriptors

Hadoop Core uses large numbers of file descriptors for MapReduce, and the `DFSClient` uses a large number of file descriptors for communicating with the HDFS NameNode and DataNode server processes. The `DFSClient` code also presents a misleading error message when there has been a failure to allocate a file descriptor: `No live nodes contain current block.`

Note The class `org.apache.hadoop.hdfs.DFSClient` provides the actual file services for applications interacting with the HDFS file system. All file system operations and file operations performed by the application will be translated into method calls on the `DFSClient` object, which will in turn issue the appropriate Remote Procedure Call (RPC) calls to the NameNode and the DataNodes relevant for the operations. As of Hadoop 0.18, these RPC calls use the Java NIO services. Prior to Hadoop 0.18, blocking operations and fixed timeouts were used for the RPC calls.

Most sites immediately bump up the number of file descriptors to 64,000, and large sites, or sites that have MapReduce jobs that open many files, might go higher.

For a Linux machine, the simple change is to add a line to the `/etc/security/limits.conf` file of the following form:

```
* hard nofile 64000
```

This changes the per-user file descriptor limit to 64,000 file descriptors. If you will run a much larger number of file descriptors, you may need to alter the per-system limits via changes to `fs.file-max` in `/etc/sysctl.conf`. A line of the following form would set the system limit to 640,000 file descriptors:

```
fs.file-max = 640000
```

At this point, you may alter the `limits.conf` file line to this:

```
* hard nofile 640000
```

Changes to `limits.conf` take effect on the next login, and changes to `sysctl.conf` take place on the next reboot. You may run `sysctl` by hand (`sysctl -p`) to cause the `sysctl.conf` file to be reread and applied.

The web page at `http://support.zeus.com/zws/faqs/2005/09/19/filedescriptors` provides some instructions for several Unix operating systems. For Windows XP instructions, see `http://weblogs.asp.net/mikedopp/archive/2008/05/16/increasing-user-handle-and-gdi-handle-limits.aspx`.

Any user that runs processes that access HDFS should have a large limit on file descriptor access, and all applications that open files need careful checking to make sure that the files are explicitly closed. Trusting the JVM garbage collection to close your open files is a critical mistake.

Block Service Threads

Each DataNode maintains a pool of threads that handle block requests. The parameter `dfs.datanode.handler.count` controls the number of threads the DataNode will use for handling IPC requests. These are the threads that accept connections on `dfs.datanode.ipc.address`, the configuration value described in Chapter 3.

Currently, there does not appear to be significant research or tools on the tuning of this parameter. The overall concept is to balance JVM overhead due to the number of threads with disk and network I/O. The more requests active at a time, the more overlap for disk and network I/O there is. At some point, the overlap results in contention rather than increased performance.

The default value for the `dfs.datanode.handler.count` parameter is 3, which seems to be fine for small clusters. Medium-size clusters may use a value of 30, set as follows:

```
<property>
  <name>dfs.datanode.handler.count</name>
  <value>30</value>
  <description>The number of server threads for the datanode.</description>
</property>
```

NameNode Threads

Each HDFS file operation—create, open, close, read, write, stat, and unlink—requires a NameNode transaction. The NameNode maintains the metadata of the file system in memory. Any operation that changes the metadata, such as open, write, or unlink, results in the NameNode writing a transaction to the disks, and an asynchronous operation to the secondary NameNodes.

■Note Some installations add an NFS directory to the list of local locations to which the file system is written. This adds a substantial latency to any metadata altering transaction. Through Hadoop 0.19.0, the NameNode edit logs are forcibly flushed to disk storage, but space for the updates is preallocated before the update to reduce the overall latency.

The parameter `dfs.namenode.handler.count` controls the number of threads that will service NameNode requests. The default value is 10. Increasing this value will substantially increase the memory utilization of the NameNode and may result in reduced performance due to I/O contention for the metadata updates. However, if your map and reduce tasks create or remove large numbers of files, or execute many sync operations after writes, this number will need to be higher, or you will experience file system timeouts in your tasks.

In my setup, I have a cluster of 20 DataNodes that are very active. For this cluster, the `dfs.namenode.handler.count` parameter is set to 512:

```
<property>
  <name>dfs.namenode.handler.count</name>
  <value>512</value>
  <description>The number of server threads for the namenode.</description>
</property>
```

■Note The NameNode serializes the entire directory metadata and then sends it when a request for the status of files in a directory is made. For large directories, this can be a considerable amount of data. With many threads servicing requests, the amount of memory used by in transit serialized requests may be very large. I have worked with a cluster where this transitory memory requirement exceeded 1GB.

Server Pending Connections

Hadoop Core's `ipc.server.listen.queue.size` parameter sets the listen depth for accepting connections. This value is the number of outstanding, unserviced connection requests allowed by the operating system, before connections are refused. Its default value is `128`.

Many operating systems have hard and small limits for this value. If your thread counts are lower, this value will need to be higher, to prevent refused connections. If you find connection-refused errors in your log files, you may want to increase the value of this parameter.

Reserved Disk Space

Hadoop Core historically has not handled out-of-disk-space exceptions gracefully. As a method for preventing out-of-disk conditions, Hadoop Core provides four parameters: two for HDFS and two for MapReduce.

It is common, especially in smaller clusters, for a DataNode and a TaskTracker to reside on each computer node in the cluster. It is also common for the TaskTracker's temporary storage to be stored in the same file system as the HDFS data blocks. With HDFS, blocks written by an application that is running on the same machine as a DataNode will have one replica placed on that DataNode. It is very easy for a task to fill up the disk space on a partition and cause HDFS failures, or for HDFS to fill up a partition and result in task failures. Tuning the disk space reservation parameters will minimize these failures.

You can adjust the following parameters:

`mapred.local.dir.minspacestart`: This parameter controls how much space must be available in the temporary area used for map and reduce task output, before a task may be assigned to a TaskTracker. This is checked only before task assignment. If you have more than one TaskTracker on the node, you may need to multiply this minimum by the TaskTracker count to ensure sufficient space. This parameter has a default value of `0`, disabling the check.

`mapred.local.dir.minspacekill`: This parameter will cause the tasks in progress to be killed if the available MapReduce local space falls below this threshold. This parameter has a default value of `0`, disabling the check.

`dfs.datanode.du.reserved`: This parameter specifies how much of the available disk space to exclude from use for HDFS. If your file system has 10GB available and `dfs.datanode.du.reserved` is set to 10GB (`10737418240`), HDFS will not store new blocks on this DataNode. This parameter has a default value of `0`, disabling the check.

`dfs.datanode.du.pct`: This value controls what percentage of the physically available space is available for use by HDFS for new data blocks. For example, if there are 10GB available in the host file system partition, and the default value of `0.98` is in effect, only 10GB * 0.98 bytes (10,522,669,875 bytes) will actually be considered available. This parameter has been dropped from Hadoop 0.19.

■ **Caution** The `dfs.datanode.du.pct` parameter is undocumented and may not have the meaning described here in your version of Hadoop.

Storage Allocations

The Balancer service will slowly move data blocks between DataNodes to even out the storage allocations. This is very helpful when DataNodes have different storage capacities. The Balancer will also slowly re-replicate underreplicated blocks

The parameter `dfs.balance.bandwidthPerSec` controls the amount of bandwidth that may be used for balancing between DataNodes. The default value is 1MB (`1048576`).

The Balancer is run by the command `start-balancer.sh [-threshold VALUE]`. The argument `-threshold VALUE`, is optional, and the default threshold is 10%. The Balancer task will run until the free space percentage for block storage of each DataNode is approximately equal, with variances in free space percentages of up to the defined threshold allowed.

The `start-balancer.sh` script prints the name of a file to the standard output that will contain the progress reports for the Balancer. It is common for the Balancer task to be run automatically and periodically. The `start-balancer.sh` script will not allow multiple instances to be run, and it is safe to run this script repeatedly.

In general, the Balancer should be used if some DataNodes are close to their free space limits while other DataNodes have plenty of available space. This usually becomes an issue only when the DataNode storage capacity varies significantly or large datasets are written into HDFS from a machine that is also a DataNode.

The Balancer may not be able to complete successfully if the cluster is under heavy load, the threshold percentage is very small, or there is insufficient free space available. If you need to stop the Balancer at any time, you can use the command `stop-balancer.sh`.

Disk I/O

Hadoop Core is designed for jobs that have large input datasets and large output datasets. This I/O will be spread across multiple machines and will have different patterns depending on the purpose of the I/O, as follows:

- The NameNode handles storing the metadata for the file system. This includes the file paths, the blocks that make up the files, and the DataNodes that hold the blocks. The NameNode writes a journal entry to the edit log when a change is made. The NameNode keeps the entire dataset in memory to enable faster response time for requests that don't involve modification.
- The DataNodes store, retrieve, and delete data blocks. The basic block size is 64MB per block. Unless an archive (such as a `.zip`, `.tar`, `tgz`, `.tar.gz`, or `.har` file) is used, blocks will not be shared among multiple files.
- The TaskTracker will hold a copy of all of the unpacked elements of the distributed cache for a job, as well as store the partitioned and sorted map output, potentially the merge sort output, and the reduce input. There are also temporary data files and buffered HDFS data blocks.

The default for Hadoop is to place all of these storage areas under the directory defined by the parameter `hadoop.tmp.dir`. In installations where multiple partitions, each on a separate physical drive or RAID assembly, are available, you may specify directories and directory sets that are stored on different physical devices to minimize seek or transfer contention.

All of the HDFS data stores are hosted on native operating system file systems. These file systems have many tuning parameters. For the NameNode data stores, RAID 1 with hot spares is preferred for the low-level storage. For the DataNodes, no RAID is preferred, but RAID 5 is acceptable.

The file systems should be constructed with awareness of the stripe and stride of any RAID arrays and, where possible, should be mounted in such a way that access time information is not updated. Linux supports disabling the access time updates for several file systems with the `noatime` and `nodiratime` file system mount time options. I have found that this change alone has provided a 5% improvement in performance on DataNodes.

Journaled file systems are not needed for the NameNode, as the critical data is written synchronously. Journaled file systems are not recommended for DataNodes due to the increase in write loading. The downside of not having a journal is that crash recovery time becomes much larger.

Secondary NameNode Disk I/O Tuning

The secondary NameNode provides a replica of the file system metadata that is used to recover a failed NameNode. It is critically important that its storage directories be on separate physical devices from the NameNode.

There is some debate about locating the secondary NameNode itself on a separate physical machine. The merging of the edit logs into the file system image may be faster and have lower overhead when the retrieval of the edit logs and writing of the file system image are local. However, having the secondary NameNode on a separate machine provides rapid recovery to the previous checkpoint time in the event of a catastrophic failure of the NameNode server machine.

■ **Note** For maximum safety, the secondary NameNode and NameNode should run on separate physical machines, with physically separate storage.

The secondary NameNode uses the parameter `fs.checkpoint.dir` to determine which directory to use to maintain the file system image. The default value is `${hadoop.tmp.dir}/dfs/namesecondary`. This parameter may be given a comma-separated list of directories. The image is replicated across the set of comma-separated values.

The secondary NameNode uses the parameter `fs.checkpoint.edits.dir` to hold the edit log, essentially the journal for the file system. It, like `fs.checkpoint.dir`, may be a comma-separated list of items, and the data will be replicated across the set of values. The default value is the value of `fs.checkpoint.dir`. The data written to the `fs.checkpoint.edits.dir` tends to be many synchronous small writes. The update operations are allowed to run behind the NameNode's updates.

The secondary NameNode server will take a snapshot from the NameNode at defined time intervals. The interval is defined by the parameter `fs.checkpoint.period`, which is a time in seconds, with a default value of `3600`. If the NameNode edit log grows by more than `fs.checkpoint.size` bytes (the default value is `67108864`), a checkpoint is also triggered.

The secondary NameNode periodically (`fs.checkpoint.period`) requests a checkpoint from the NameNode. At that point, the NameNode will close the current edit log and start a new edit log. The file system image and the just closed edit log will be copied to the secondary NameNode. The secondary NameNode will apply the edit log to the file system image, and then transfer the up-to-date file system image back to the NameNode, which replaces the prior file system image with the merged copy.

The secondary NameNode configuration is not commonly altered, except in very high utilization situations. In these cases, multiple file systems on separate physical disks are used for the set of locations configured in the `fs.checkpoint.edits.dir` and `fs.checkpoint.dir` parameters.

NameNode Disk I/O Tuning

The NameNode is the most critical piece of equipment in your Hadoop Core cluster. The NameNode, like the secondary NameNode, maintains a journal and a file system image.

The parameter `dfs.name.dir` provides the directories in which the NameNode will store the file system image and is a comma-separated list (again, the image will be replicated across the set of values). The file system image is read and updated only at NameNode start time, as of Hadoop version 0.19. The default value for this parameter is `${hadoop.tmp.dir}/dfs/name`.

The parameter `dfs.name.edits.dir` contains the comma-separated list of directories to which the edit log or journal will be written. This is updated for every file system metadata-altering operation, synchronously. Your entire HDFS cluster will back up waiting for these updates to flush to the disk. If your cluster is experiencing a high rate of file create, rename, delete, or high-volume writes, there will be a high volume of writes to this set of directories. Any other I/O operations to file system partitions holding these directories will perform badly. The default value of this parameter is `${dfs.name.dir}`.

Some installations will place directories on remote mounted file systems in this list, to ensure that an exact copy of the data is available in the event of a NameNode disk failure.

RAID 1 or RAID 10 is recommended for the `dfs.name.edits.dir` set, rather than RAID 5, due to the increased write latency for small block writes.

Any data loss in the directories specified by `dfs.name.dir` and `dfs.name.edits.dir` will result in the loss of data in your HDFS. Caution, redundancy, and reliability are critical.

Caution The NameNode is a critical single point of failure. Any loss of data can wipe out your entire HDFS datastore.

DataNode Disk I/O Tuning

The DataNode provides two services: block storage, and retrieval of HDFS data and storage accounting information for the NameNode.

Through at least Hadoop 0.19.0, the storage accounting has significant problems if there are more than a few hundred thousand blocks per DataNode. This is because a linear scan of the blocks is performed on a frequent basis to provide accounting information.

The parameter `dfs.datanode.handler.count` (covered earlier in the "Block Service Threads" section) is the number of worker threads for storage requests. The following are some other DataNode parameters that may be adjusted:

`dfs.data.dir`: This parameter sets the location(s) for block storage. It is a comma-separated list of directories that will be used for block storage by the DataNode. Blocks will be distributed in a round-robin manner among the directories. If multiple directories are to be used, it is best if each directory resides on an independent storage device. This will allow concurrent I/O operations to be active on all of the devices. The default value for this parameter is `${hadoop.tmp.dir}/dfs/data`.

`dfs.replication`: This parameter controls the number of copies of each data block the cluster attempts to maintain. HDFS is resilient in the face of individual DataNode failures, because individual blocks are replicated to more than one machine. The HDFS cluster can withstand the failure of one less than the value of `dfs.replication` before there will be service degradation, in that some files may not be served as all of the blocks are not available. In small to medium-sized clusters, `3` is a good value. For large clusters, it should be a number that is at least two larger than the expected number of machine failures per day. If you have a 1,000-machine cluster and expect to have no more than 5 machines fail at a time, setting the replication factor to `7` is reasonable. The disk storage requirements and the write network bandwidth used are multiplied by this number.

Note When a client writes data to HDFS, each block is written to `dfs.replication` count DataNodes. Each node writes to the next node, to mitigate the network load on the individual machines. The nodes are selected more or less on a random basis, with some simple rules. If the origination machine is a DataNode for the HDFS cluster being written, one replica will go to that DataNode. HDFS has some concept of network topology, but this does not appear to be generally used as yet.

`dfs.block.size`: This parameter controls the basic file system block size for HDFS. HDFS is highly optimized for a medium number of very large files. The default value for an HDFS block is 64MB (67,108,864 bytes). Each file data block in the file system will be in a single file in the `dfs.data.dir` directory. A file in HDFS is composed of one or more data blocks. The last data block of a file may have less than the `dfs.block.size` bytes of data present. Any prior blocks will have `dfs.block.size` bytes of data. The block size may be specified on a per-file basis when the file is created. The individual blocks of the file are replicated onto `dfs.replication` DataNodes. The default input split size for a file being used as input in a MapReduce job is the `dfs.block.size` for the file. HDFS currently does not deal with very large numbers of files or very large numbers of blocks per DataNode.

Note Application developers are encouraged to use archives for storing multiple small files instead of creating them directly in HDFS. In one of my applications, there was a three-order-of-magnitude speed increase in an application when a set of zip files were constructed once per reduce task, instead of large numbers of small files in the reduce task.

Network I/O Tuning

Very little can be done to tune network I/O at the Hadoop Core level. The performance tuning must take place at the application layer and the hardware layer.

Hadoop Core does provide the ability to select the network to bind to for data services and the ability to specify an IP address to use for hostname resolution. This is ideal for clusters that have an internal network for data traffic and an external network for client communications. The following are the two relevant parameters:

`dfs.datanode.dns.interface`: This parameter may be set to an interface name, such as `eth0` or `en1`, and that interface will be used by the DataNode for HDFS communications.

`dfs.datanode.dns.nameserver`: This parameter may be set to provide an IP address to use for DNS-based hostname resolution, instead of the system default hostname resolution strategy.

If your installation has switch ports to spare and the switches support it, bonding your individual network connections can greatly increase the network bandwidth available to individual machines.

■ **Caution** The large-block HDFS traffic will dominate a network. It is best to isolate this traffic from your other traffic.

At the application layer, if your applications are data-intensive and your data is readily compressible, using block- or record-level compression may drastically reduce the I/O that the job requires. You can set this compression as follows:

```
FileOutput.setCompressOutput(jobConf,"true");
jobConf.setBoolean("mapred.output.compress",true);
```

The default compression codec will be used. Compressed input files that are not `SequenceFiles`, a Hadoop Core binary file format, will not be split, and a single task will handle a single file.

Recovery from Failure

Once you have configured your cluster and started running jobs, life will happen, and you may need to recover from a failure. Here, we will look at how HDFS protects from many individual failures and how to recover from other failures.

HDFS has two types of data, with different reliability requirements and recovery patterns.

- The NameNode stores the file system metadata and provides direct replication of the data and run-behind remote copies of the data.
- The DataNodes provide redundant storage by replicating data blocks to multiple DataNodes.

NameNode Recovery

The default Hadoop installation provides no protection from catastrophic NameNode server failures. The only default protection is that provided by a RAID 1 or RAID 5 storage device for the file system image and edit logs. You can avoid data loss and unrecoverable machine failures by running the secondary NameNode on an alternate machine. Storing the file system image and file system edit logs on multiple physical devices, or even multiple physical machines, also provides protection.

When a NameNode server fails, best practices require that all the JobTracker and TaskTrackers be restarted after the NameNode is restarted. All incomplete HDFS blocks will be lost, but there should be no file or data loss for existing files or for completed blocks in actively written files.

■ **Note** Hadoop 0.19.0 has initial support for a sync operator that allows the flushing of HDFS blocks that are not a full `dfs.block.size` value, but support for this is early and known to be unreliable. This feature has been disabled in Hadoop 0.19.1, but it may return in Hadoop 0.20.0.

The NameNode may be configured to write the metadata log to multiple locations on the host server's file system. In the event of data loss or corruption to one of these locations, the NameNode may be recovered by repairing or removing the failed location from the configuration, removing the data from that location, and restarting the NameNode. For rapid recovery, you may simply remove the failed location from the configuration and restart the NameNode.

If the NameNode needs to be recovered from a secondary NameNode, the procedure is somewhat more complex. Here are the steps:

1. Shut down the secondary NameNode.
2. Copy the contents of the `Secondary:fs.checkpoint.dir` to the `Namenode:dfs.name.dir`.
3. Copy the contents of the `Secondary: fs.checkpoint.edits.dir` to the `Namenode:dfs.name.edits.dir`.
4. When the copy completes, you may start the NameNode and restart the secondary NameNode.

All data written to HDFS after the last secondary NameNode checkpoint was taken will be removed and lost. The default frequency of the checkpoints is specified by the `fs.checkpoint.period` parameter

At present, there are no public forensic tools that will recover data from blocks on the DataNodes.

DataNode Recovery and Addition

The procedure for adding a new DataNode to a cluster and restarting a failed DataNode are identical and simple. The server process just needs to be started, assuming the configuration is correct, and the NameNode will integrate the new server or reintegrate a restarted server into the cluster.

Tip As long as your cluster does not have underreplicated files and no file's replication count is less than 3, it is generally safe to forcibly remove a DataNode from the cluster by killing the DataNode server process. The next section covers how to decommission a DataNode gracefully.

The command `hadoop-daemon.sh start datanode` will start a DataNode server on a machine, if one is not already running. The configuration in the `conf` directory associated with the script will be used to determine the NameNode address and other configuration parameters.

If more DataNodes than the `dfs.replication` value fail, some file blocks will be unavailable. Your cluster will still be able to write files and access the blocks of the files that remain available. It is advisable to stop your MapReduce jobs by invoking the `stop-mapred.sh` script, as most applications do not deal well with partial dataset availability. When sufficient DataNodes have been returned to service, you may resume MapReduce job processing by invoking the `start-mapred.sh` script.

When you add new nodes, or return a node to service after substantial data has been written to HDFS, the added node may start up with substantially less utilization than the rest of the DataNodes in the cluster. Running the Balancer via `start-balancer.sh` will rebalance the blocks.

DataNode Decommissioning

A running DataNode sometimes needs to be decommissioned. While you may just shut down the DataNode, and the cluster will recover, there is a procedure for gracefully decommissioning a running DataNode. This procedure becomes particularly important if your cluster has underreplicated blocks or you need to decommission more nodes than your `dfs.replication` value.

Caution You must not stop the NameNode during this process, or start this process while the NameNode is not running. The file specified by `dfs.hosts.exclude` has two purposes. One is to exclude the hosts from connecting to the NameNode, which takes effect if the parameter is set when the NameNode starts. The other starts the decommission process for the hosts, which takes place if the value is first seen after a Hadoop `dfsadmin -refreshNodes`.

The procedure is as follows:

1. Create a file on the NameNode machine with the hostnames or IP addresses of the DataNodes you wish to decommission, say `/tmp/nodes_to_decommission`. This file should contain one hostname or IP address per line, with standard Unix line endings.
2. Modify the `hadoop-site.xml` file by adding, or updating the following block:

```
<property>
  <name>dfs.hosts.exclude</name>
  <value>/tmp/nodes_to_decommission</value>
  <description>Names a file that contains a list of hosts that are
  not permitted to connect to the namenode. The full pathname of the
  file must be specified.  If the value is empty, no hosts are
  excluded.</description>
</property>
```

3. Run the following command to start the decommissioning process:

```
hadoop dfsadmin -refreshNodes
```

4. When the process is complete, you will see a line in the NameNode log file like the following for each entry in the file:

```
tmp/nodes_to_decommission.
```

```
Decommission complete for node IP:PORT
```

Deleted File Recovery

It is not uncommon for a user to accidentally delete large portions of the HDFS file system due to a program error or a command-line error. Unless your configuration has the delete-to-trash function enabled, via setting the parameter `fs.trash.interval` to a nonzero value, deletes are essentially immediate and forever.

If an erroneous large delete is in progress, your best bet is to terminate the NameNode and secondary NameNodes immediately, and then shut down the DataNodes. This will preserve as much data as possible. Use the procedures described earlier to recover from the secondary NameNode that has the edit log modification time closest to the time the deletion was started.

The `fs.trash.interval` determines how often the currently deleted files are moved to a date-stamped subdirectory of the deleting user's `.Trash` directory. The value is a time in minutes. Files that have not had a trash checkpoint will be under the `.Trash/current` directory in a path that is identical to their original path. Only one prior checkpoint is kept.

Troubleshooting HDFS Failures

The previous section dealt with the common and highly visible failure cases of a server process crashing or the machine hosting a server process failing. This section will cover how you can determine what has happened when the failure is less visible or why a server process is crashing.

There are a number of failures that can trip up an HDFS administrator or a MapReduce programmer. In the current state of development, it is not always clear from Hadoop's behavior or log messages what the failure or solution is. The usual first indication that something is in need of maintenance is a complaint from users that their jobs are performing poorly or failing, or a page from your installation monitoring tool such as Nagios.

NameNode Failures

The NameNode is the weak point in the highly available HDFS cluster. As noted earlier, currently there are no high-availability solutions for the NameNode. The NameNode has been designed to keep multiple copies of critical data on the local machine and close in time replicas on auxiliary machines. Let's examine how it can fail.

Out of Memory

The NameNode keeps all of the metadata for the file system in memory to speed request services. The NameNode also serializes directory listings before sending the result to requesting applications. The memory requirements grow with the number of files and the total number of blocks in the file system. If your file system has directories with many entries and applications are scanning the directory, this can cause a large transient increase in the memory requirements for the name server.

I once had a cluster that was using the Filesystem in Userspace (FUSE) `contrib` package to export HDFS as a read-only file system on a machine, which re-exported that file system via the Common Internet File System (CIFS) to a Windows server machine. The access patterns triggered repeated out-of-memory exceptions on the NameNode. (Using FUSE is discussed in Chapter 8.)

If the DataNodes are unreliable, and they are dropping out of service and then returning to service after a gap, the NameNode will build a large queue of blocks with invalid states. This may consume very large amounts of memory if large numbers of blocks become transitorily unavailable. For this problem, addressing the DataNode reliability is the only real solution, but increasing the memory size on the NameNode can help.

There are three solutions for NameNode out-of-memory problems:

- Increase the memory available to the NameNode, and ensure the machine has sufficient real memory to support this increase.
- Ensure that no directory has a large number of entries.
- Alter application access patterns so that there are not large numbers of directory listings.

For a 32-bit JVM, the maximum memory size is about 2.5GB, which will support a small HDFS cluster with a under a million files and blocks.

As a general rule, pin the full memory allocation by setting the starting heap size for the JVM to the maximum heap size.

Data Loss or Corruption

Ideally, the underlying disk storage used for the NameNode's file system image (`dfs.name.dir`) and edit log (`dfs.name.edits.dir`) should be a highly reliable storage system such as a RAID 1 or RAID 5 disk array with hot spares. Even so, catastrophic failures or user errors can result in data loss.

The NameNode configuration will accept a comma-separated list of directories and will maintain copies of the data in the full set of directories. This additional level of redundancy provides a current time backup of the file system metadata. The secondary NameNode provides a few-minutes-behind replica of the NameNode data.

If your configuration has multiple directories that contain the file system image or the edit log, and one of them is damaged, delete that directory's content and restart the NameNode. If the directory is unavailable, remove it from the list in the configuration and restart the NameNode.

If all of the `dfs.name.dir` directories are unavailable or suspect, do the following:

1. Archive the data if required.
2. Wipe all of the directories listed in `dfs.name.dir`.
3. Copy the contents of the `fs.checkpoint.dir` from the secondary NameNode to the `fs.checkpoint.dir` on the primary NameNode machine.
4. Run the following NameNode command:

```
hadoop namenode -importCheckpoint
```

If there is no good copy of the NameNode data, the secondary NameNode image may be imported. The secondary NameNode takes periodic snapshots, at `fs.checkpoint.period` intervals, so it is not as current as the NameNode data.

You may simply copy the file system image from the secondary NameNode to a file system image directory on the NameNode, and then restart.

As the imported data is older than the current state of the HDFS file system, the NameNode may spend significant time in safe mode as it brings the HDFS block store into consistency with restored snapshot.

No Live Node Contains Block Errors

Usually, if you see the `no live node contains block` error, it will be in the log files for your applications. It means that the client code in your application that interfaces with HDFS was unable to find a block of a requested file. For this error to occur, the client code received a list of DataNode and block ID pairs from the NameNode, and was unable to retrieve the block from any of the DataNodes.

This error commonly occurs when the application is unable to open a connection to any of the DataNodes. This may be because there are no more file descriptors available, there is a DNS resolution failure, there is a network problem, or all of the DataNodes in question are actually unavailable. The most common case is the out-of-file-descriptor situation. This may be corrected by increasing the number of file descriptors available, as described earlier in this chapter. An alternative is to minimize unclosed file descriptors in the applications.

I've seen DNS resolution failures transiently appear, and as a general rule, I now use IP addresses instead of hostnames in the configuration files. It is very helpful if the DNS reverse lookup returns the same name as the hostname of the machine.

Write Failed

If there are insufficient DataNodes available to allow full replication of a newly written block, the write will not be allowed to complete. This may result in a zero-length or incomplete file, which will need to be manually removed.

DataNode or NameNode Pauses

Through at least Hadoop 0.19.0, the DataNode has two periodic tasks that do a linear scan of all of the data blocks stored by the DataNode. If this process starts taking longer than a small number of minutes, the DataNode will be marked as disconnected by the NameNode.

When a DataNode is marked as disconnected, the NameNode queues all of the blocks that had a replica on that DataNode for replication. If the number of blocks is large, the NameNode may pause for a noticeable period while queuing the blocks.

The only solutions for this at present are to add enough DataNodes, so that no DataNode has more than a few hundred thousand data blocks, or to alter your application's I/O patterns to use Hadoop archives or zip files to pack many small HDFS subblock-size files into single HDFS files. The latter approach results in a reduction in the number of blocks stored in HDFS and the number of blocks per DataNode.

A simple way to work out how many blocks is too many is to run the following on a DataNode:

```
time find dfs.data.dir -ls > /dev/null
```

If it takes longer than a few hundred seconds, you are in the danger zone. If it takes longer than a few minutes, you are in the pain zone. Replace `dfs.data.dir` with an expanded value from your Hadoop configuration. The timeout interval is 600 seconds, and hard-coded. If your `ls` takes anywhere close to that, your cluster will be unstable.

■ **Note** The NameNode may pause if the one of the directories used for the `dfs.name.edits.dir` or `dfs.name.dir` is taking time to complete writes.

Summary

HDFS is a wonderful global file system for a medium number of very large files. With reasonable care and an understanding of HDFS's limitations it will serve you well.

HDFS is not a general-purpose file system to be used for large numbers of small files or for rapid creation and deletion of files.

Through HDFS 0.19.0, the HDFS NameNode is a single point of failure and needs careful handling to minimize the risk of data loss. Using a separate machine for your secondary NameNode, and having multiple devices for the file system image and edit logs, will go a long way toward providing a fail-safe, rapid recovery solution.

Monitoring your DataNode's block totals will go a long way toward avoiding congestion collapse issues in your HDFS. You can do this by simply running a `find` on the `dfs.data.dir` and making sure that it takes less than a couple of minutes.

Ensuring that your HDFS data traffic is segregated from your normal application traffic and crosses as few interswitch backhauls as possible will help to avoid network congestion and application misbehavior.

Ultimately, remember that HDFS is designed for a small to medium number of very large files, and not for transitory storage of large numbers of small files.

CHAPTER 5

MapReduce Details for Multimachine Clusters

Organizations run Hadoop Core to provide MapReduce services for their processing needs. They may have datasets that can't fit on a single machine, have time constraints that are impossible to satisfy with a small number of machines, or need to rapidly scale the computing power applied to a problem due to varying input set sizes. You will have your own unique reasons for running MapReduce applications.

To do your job effectively, you need to understand all of the moving parts of a MapReduce cluster and of the Hadoop Core MapReduce framework. This chapter will raise the hood and show you some schematics of the engine. This chapter will also provide examples that you can use as the basis for your own MapReduce applications.

Requirements for Successful MapReduce Jobs

For your MapReduce jobs to be successful, the mapper must be able to ingest the input and process the input record, sending forward the records that can be passed to the reduce task or to the final output directly, if no reduce step is required. The reducer must be able to accept the key and value groups that passed through the mapper, and generate the final output of this MapReduce step.

The job must be configured with the location and type of the input data, the mapper class to use, the number of reduce tasks required, and the reducer class and I/O types.

The TaskTracker service will actually run your map and reduce tasks, and the JobTracker service will distribute the tasks and their input split to the various trackers.

The cluster must be configured with the nodes that will run the TaskTrackers, and with the number of TaskTrackers to run per node. The TaskTrackers need to be configured with the JVM parameters, including the classpath for both the TaskTracker and the JVMs that will execute the individual tasks.

There are three levels of configuration to address to configure MapReduce on your cluster. From the bottom up, you need to configure the machines, the Hadoop MapReduce framework, and the jobs themselves.

We'll get started with these requirements by exploring how to launch your MapReduce jobs.

■ Tip A Hadoop job is usually part of a production application, which may have many steps, some of which are MapReduce jobs. Hadoop Core, as of version 0.19.0, provides a way of optimizing the data flows between a set of sequential MapReduce jobs. This framework for descriptively and efficiently running sequential MapReduce jobs together is called *chaining*, and uses the `ChainMapper` and the `ChainReducer`, as discussed in Chapter 8. An alternative is the cascading package, available from `http://www.cascading.org/`.

Launching MapReduce Jobs

Jobs within a MapReduce cluster can be launched by constructing a `JobConf` object (details on the `JobConf` object are provided in this book's appendix) and passing it to a `JobClient` object:

```
JobConf conf = new JobConf(MyClass.class);
/** Configuration setup deleted for clarity*/
/** Launch the Job by submitting it to the Framework. */
RunningJob job = JobClient.runJob(conf);
```

You can launch the preceding example from the command line as follows:

```
> bin/hadoop [-libjars jar1.jar,jar2.jar,jar3.jar] jar myjar.jar MyClass
```

The optional `-libjars jar1.jar...` specifications add JARs for your job. The assumption is that `MyClass` is in the `myjar.jar`.

For this to be successful requires a considerable amount of runtime environment setup. Hadoop Core provides a shell script, `bin/hadoop`, which manages the setup for a job. Using this script is the standard and recommended way to start a MapReduce job. This script sets up the process environment correctly for the installation, including inserting the Hadoop JARs and Hadoop configuration directory into the classpath, and launches your application. This behavior is triggered by providing the initial command-line argument `jar` to the `bin/hadoop` script.

Hadoop Core provides several mechanisms for setting the classpath for your application:

- You can set up a fixed base classpath by altering `hadoop-env.sh`, via the `HADOOP_CLASSPATH` environment variable (on all of your machines) or by setting that environment variable in the runtime environment for the user that starts the Hadoop servers.
- You may run your jobs via the `bin/hadoop jar` command and supply a `-libjars` argument with a list of JARs.
- The `DistributedCache` object provides a way to add files or archives to your runtime classpath.

Tip The mapred.child.java.opts variable may also be used to specify non-classpath parameters to the child JVMs. In particular, the java.library.path variable specifies the path for shared libraries if your application uses the Java Native Interface (JNI). If your application alters the job configuration parameter mapred.child.java.opts, it is important to ensure that the JVM memory settings are reset or still present, or your tasks may fail with out-of-memory exceptions.

The advantage of using the DistributedCache and -libjars is that resources, such as JAR files, do not have to already exist on the TaskTracker nodes. The disadvantages are that the resources must be unpacked on each machine and it is harder to verify which versions of the resources are used.

When launching an application, a number of command-line parameters may be provided. Table 5-1 lists some common command-line arguments. The class org.apache.hadoop.util.GenericOptionsParser actually handles the processing of Table 5-1 arguments.

Table 5-1. *Hadoop Standard Command-Line Arguments*

Flag	Description
-libjars	A comma-separated list of JAR files to add to the classpath to the job being launched and to the map and reduce tasks run by the TaskTrackers. These JAR files will be staged into HDFS if needed and made available as local files in a temporary job area on each of the TaskTracker nodes.
-archives	A comma-separated list of archive files to make available to the running tasks via the distributed cache. These archives will be staged into HDFS if needed.
-files	A comma-separated list of files to make available to the running tasks via the distributed cache. These files will be staged into HDFS if needed.
-fs	Override the configuration default file system with the supplied URL, the parameter fs.default.name.
-jt	Override the configuration default JobTracker with the supplied host port, the parameter mapred.job.tracker.
-conf	Use this configuration in place of the conf/hadoop-default.xml and conf/hadoop-site.xml files.
-D	Supply an additional job configuration property in *key=value* format. This argument may be provided multiple times. There must be whitespace between the -D and the *key=value*.

You can use hadoop jar to launch an application, as follows:

```
hadoop jar [-fs hdfs://host:port] [-jt host:port] [-conf hadoop-config.xml] ➥
 [-D prop1=value] [-D prop2=value…] [-libjars jar1[,jar2,jar3]] ➥
 [-files file1[,file2,file3]] [-archives archive1[,archive2,archive3]] ➥
applicationjar [main class if not supplied in jar] [arguments to main…]
```

When hadoop jar is used, the main method of org.apache.hadoop.mapred.JobShell is invoked by the JVM, with all of the remaining command-line arguments. The JobShell in turn uses the class `org.apache.hadoop.util.GenericOptionsParser` to process the arguments, as described in Table 5-1.

There are two distinct steps in the argument processing of jobs submitted by the `bin/hadoop` script. The first step is provided by the framework via the `JobShell`. The arguments after `jar` are processed by the `JobShell`, per Table 5-1. The first argument not in the set recognized by the `JobShell` must be the path to a JAR file, which is the job JAR file. If the job JAR file contains a main class specification in the manifest, that class will be the main class called after the first step of argument processing is complete. If the JAR file does not have a main class in the manifest, the next argument becomes required, and is used as main class name. Any remaining unprocessed arguments are passed to the main method of the main class as the arguments. The second step is the processing of the remaining command-line arguments by the user-specified main class.

Using Shared Libraries

Jobs sometime require specific shared libraries. For example, one of my jobs required a shared library that handled job-specific image processing. You can handle this in two ways:

- Pass the shared library via the `DistributedCache` object. For example, using the command-line options `-file libMyStuff.so` would make `libMyStuff.so` available in the current working directory of each task. (The `DistributedCache` object is discussed shortly, in the "Using the Distributed Cache" section.)
- Install the shared library on every TaskTracker machine, and have the JVM library loader path `java.library.path` include the installation directory. The task JVM working directory is part of the `java.library.path` for a task, and any file that is symbolic-linked may be loaded by the JVM.

Caution If you are manually loading shared libraries, the library name passed to `System.loadLibrary()` must not have the trailing `.so`. `System.loadLibrary()` first calls `System.mapLibraryName()` and attempts to load the results. This can result in library load failures that are hard to diagnose.

MapReduce-Specific Configuration for Each Machine in a Cluster

For simplicity and ease of ongoing maintenance, this section assumes identical Hadoop Core installations will be placed on each of the machines, in the same location. The cluster-level configuration is covered in Chapter 3.

The following are the MapReduce-specific configuration requirements for each machine in the cluster:

- You need to install any standard JARs that your application uses, such as Spring, Hibernate, HttpClient, Commons Lang, and so on.
- It is probable that your applications will have a runtime environment that is deployed from a configuration management application, which you will also need to deploy to each machine.
- The machines will need to have enough RAM for the Hadoop Core services plus the RAM required to run your tasks.
- The `conf/slaves` file should have the set of machines to serve as TaskTracker nodes. You may manually start individual TaskTrackers by running the command `bin/hadoop-daemon.sh start tasktracker`, but this is not a recommended practice for starting a cluster.

The `hadoop-env.sh` script has a section for providing custom JVM parameters for the different Hadoop Core servers, including the JobTracker and TaskTrackers. As of Hadoop 0.19.0, the classpath settings are global for all servers. The `hadoop-env.sh` file may be modified and distributed to the machines in the cluster, or the environment variable `HADOOP_JOBTRACKER_OPTS` may be set with JVM options before starting the cluster via the `bin/start-all.sh` command or `bin/start-mapred.sh` command. The environment variable `HADOOP_TASKTRACKER_OPTS` may be set to provide per TaskTracker JVM options. It is much better to modify the file, as the changes are persistent and stored in a single Hadoop-specific location.

When starting the TaskTrackers via the `start-*.sh` scripts, the environment variable `HADOOP_TASKTRACKER_OPTS` may be set in the `hadoop-env.sh` file in the MapReduce `conf` directory on the TaskTracker nodes, or the value may be set in the login shell environment so that the value is present in the environment of commands started via `ssh`. The `start-*.sh` scripts will `ssh` to each target machine, and then run the `bin/hadoop-daemon.sh start tasktracker` command.

Using the Distributed Cache

The `DistributedCache` object provides a programmatic mechanism for specifying the resources needed by the mapper and reducer. The job is actually already using the `DistributedCache` object to a limited degree, if the job creates the `JobConf` object with a class as an argument: `new JobConf(MyMapper.class)`. You may also invoke your MapReduce program using the `bin/hadoop` script and provide arguments for `-libjars`, `-files`, or `-archives`.

The downloadable code for this book (available from this book's details page on the Apress web site, `http://www.apress.com`) includes several source files for the `DistributedCache` examples: `Utils.java`, `DistributedCacheExample.java`, and `DistributedCacheMapper.java`.

■Caution The paths and URIs for `DistributedCache` items are stored as comma-separated lists of strings in the configuration. Any comma characters in the paths will result in unpredictable and incorrect behavior.

Adding Resources to the Task Classpath

Four methods add elements to the Java classpath for the map and reduce tasks. The first three in the following list add archives to the classpath. The archives are unpacked in the job local directory of the task. You can use the following methods to add resources to the task classpath:

`JobConf.setJar(String jar)`: Sets the user JAR for the MapReduce job. It is on the `JobConf` object, but it manipulates the same configuration keys as the `DistributedCache`. The file `jar` will be found, and if necessary, copied into the shared file system, and the full path name on the shared file system stored under the configuration key `mapred.jar`.

`JobConf.setJarByClass(Class cls)`: Determines the JAR that contains the class `cls` and calls `JobConf.setJar(jar)` with that JAR.

`DistributedCache.addArchiveToClassPath(Path archive, Configuration conf)`: Adds an archive path to the current set of classpath entries. This is a static method, and the archive (a zip or JAR file) will be made available to the running tasks via the classpath of the JVM. The archive is also added to the list of cached archives. The contents will be unpacked in the local job directory on each TaskTracker node. The archive path is stored in the configuration under the key `mapred.job.classpath.archives`, and the URI constructed from `archive.makeQualified(conf).toUri()` is stored under the key `mapred.job.classpath.archives`. If the path component of the URI does not exactly equal `archive`, `archive` will not be placed in the classpath of the task correctly.

■ **Caution** The `archive` path must be on the JobTracker shared file system, and must be an absolute path. Only the path `/user/hadoop/myjar.jar` is correct; `hdfs://host:8020/user/hadoop/myjar.jar` will fail, as will `hadoop/myjar.jar` or `myjar.jar`.

`DistributedCache.addFileToClassPath(Path file, Configuration conf)`: Adds a file path to the current set of classpath entries. It adds the file to the cache as well. This is a static method that makes the file available to the running tasks via the classpath of the JVM. The file path is stored under the configuration key `mapred.job.classpath.files`, and the URI constructed from `file.makeQualified(conf).toUri()` is stored under the key `mapred.cache.files`. If `file` is not exactly equal to the path portion of the constructed URI, `file` will not be added to the classpath of the task correctly.

■ **Caution** The `file` path added must be an absolute path on the JobTracker shared file system, and be only a path. `/user/hadoop/myfile` is correct; `hdfs://host:8020/user/hadoop/myfile` will fail, as will `hadoop/myfile` or `myfile`.

Distributing Archives and Files to Tasks

In addition to items that become available via the classpath, two methods distribute archives and individual files: `DistributedCache.addCacheArchive(URI uri, Configuration conf)` and `DistributedCache.addCacheFile(URI uri, Configuration conf)`. Local file system copies of these items are made on all of the TaskTracker machines, in the work area set aside for this job.

Distributing Archives

The `DistributedCache.addCacheArchive(URI uri, Configuration conf)` method will add an archive to the list of archives to be distributed to the jobs. The URI must have an absolute path and be on the JobTracker shared file system.

If the URI has a fragment, a symbolic link to the archive will be placed in the task working directory as the fragment. The URI `hdfs://host:8020/user/hadoop/myfile#mylink` will result in a symbolic link `mylink` in the task working directory that points to the local file system location that `myfile` was unpacked into at task start. The archive will be unpacked into the local working directory of the task.

The URI will be stored in the configuration under the key `mapred.cache.archives`.

Distributing Files

This `DistributedCache.addCacheFile(URI` *`uri`*`, Configuration` *`conf`*`)` method will make a copy of the file *`uri`* available to all of the tasks, as a local file system file. The URI must be on the JobTracker shared file system.

If the URI has a fragment, a symbolic link to the URI fragment will be created in the JVM working directory that points to the location on the local file system where the *`uri`* was unpacked into at task start. The directory where `DistributedCache` stores the local copies of the passed items is not the current working directory of the task JVM. This allows the items to be referenced by names that do not have any path components. In particular, executable items may be referenced as `./name`.

To pass a script via the distributed cache, use `DistributedCache.addCacheFile( new URI ("hdfs://host:8020/user/hadoop/myscript.pl"), conf);`. To pass a script so that it may be invoked via `./script`, use `DistributedCache.addCacheFile( new URI("hdfs://host:8020/ user/hadoop/myscript.pl#`*`script`*`"), conf);`.

The URI is stored in the configuration key `mapred.cache.files`.

Accessing the DistributedCache Data

Three methods find the locations of the items that were passed to the task via the `DistributedCache` object: `URI JobConf.getResource(name)`, `public static Path[]getLocalCacheArchives (Configuration conf)`, and `public static Path[] getLocalCacheFiles(Configuration conf)`.

Looking Up Names

The `URI JobConf.getResource(name)` method will look up `name` in the classpath. If `name` has a leading slash, this method will search for it in each location in the classpath, and return the URI.

If the job passed a file into `DistributedCache` via the `-files` command or the `DistributedCache.addFileToClassPath(Path file, conf)` method, a `getResource()` call of the file name component, with a leading slash, will return the URI.

Note The standard search rules for resources apply. The cache items will be the last items in the class-path. This does not appear to work for files that are added via `DistributedCache.addFileToClassPath`. The full path is available via the set of paths returned by `DistributedCache.getFileClassPaths()`.

The `DistributedCache.addArchiveToClassPath(jarFileForClassPath, job)` method actually stores the JAR information into the configuration. In the following example, `Utils.setupArchiveFile` builds a JAR file with ten files in it, in the default file system (HDFS in this case). `Utils.makeAbsolute` returns the absolute path.

```
Path jarFileForClassPath = Utils.makeAbsolute(Utils.setupArchiveFile(job, ➥
10, true),job);
DistributedCache.addArchiveToClassPath(jarFileForClassPath, job);
```

Any file that is in the JAR may be accessed via the `getResource()` method of the configuration object. If there were a file `myfile` in the JAR, the call `conf.getResource("/myfile");` would return the URL of the resource. The call `conf. getConfResourceAsInputStream("/myfile");` would return an `InputStream` that, when read, would provide the contents of `myfile` from the JAR.

Looking Up Archives and Files

The `public static Path[]getLocalCacheArchives (Configuration conf)` method returns a list of the archives that were passed via `DistributedCache`. The paths will be in the task local area of the local file system. Any archive passed via the command-line `-libjars` and `-archives` options, or the methods `DistributedCache.addCacheArchive()` and `DistributedCache.addArchiveToClassPath()` and the `JobConf.setJar` line, will have its path returned by this call.

It is possible that the file name portion of your archive will be changed slightly. `DistributedCache` provides the following method to help with this situation:

```
public static String makeRelative(URI cache, Configuration conf)
```

This takes an original archive path and returns the possibly altered file name component.

The `public static Path[] getLocalCacheFiles(Configuration conf)` method returns the set of localized paths for files that are passed via `DistributedCache.addCacheFile` and `DistributedCache.addFileToClassPath` and the command-line option `-files`. The file name portions of the paths may be different from the original file name.

Finding a File or Archive in the Localized Cache

The `DistributedCache` object may change the file name portion of the files and archives it distributes. This is usually not a problem for classpath items, but it may be a problem for non-classpath items. The `Utils.makeRelativeName()` method, described in Table 5-2 provides a way to determine what the file name portion of the passed item was changed to. In addition to the file name portion, the items will be stored in a location relative to the working area for the task on each TaskTracker. Table 5-2 lists the methods provided in the downloadable code that make working with the DistributedCache object simpler. These methods are designed to be used in the mapper and reducer methods.

Table 5-2. *Utility Methods Provided in the Examples for Working with the DistributedCache Object*

Method	Description
Utils.makeRelativeName(name, conf)	Returns the actual name DistributedCache will use for the passed-in name.
Utils.findClassPathArchive(name, conf)	Returns the actual path on the current machine of the archive name that was passed via DistributedCache.addArchiveToClassPath.
Utils.findClassPathFile(name, conf)	Returns the actual path on the current machine of the file name that was passed via DistributeCacheAddFileToClasspath.
Utils.findNonClassPathArchive(name, conf)	Returns the actual path on the current machine of the archive name that was passed via DistributedCache.addCacheArchive.
Utils.findNonClassPathFile(name, conf)	Returns the actual path on the current machine of the file name that was passed via DistributedCache.addCacheFile.

Configuring the Hadoop Core Cluster Information

The JobConf object provides two basic and critical ways for specifying the default file system: the URI to use for all shared file system paths, and the connection information for the Job-Tracker server. These two items are normally specified in the conf/hadoop-site.xml file, but they may be specified on the command line or by setting the values on the JobConf object.

Setting the Default File System URI

The default file system URI is normally specified with the fs.default.name setting in the hadoop-site.xml file, as it is cluster-specific. The value will be hdfs://NamenodeHostname:PORT. The PORT portion is optional and defaults to 8020, as of Hadoop 0.18

Note The default value for the file system URI is file:///, which stores all files on the local file system. The file system that is used must be a file system that is shared among all of the nodes in the cluster.

```
<property>
  <name>fs.default.name</name>
  <value>hdfs://NamenodeHostname:PORT</value>
</property>
```

The Hadoop tools, examples, and any application that uses the GenericOptionParser class to handle command-line arguments will accept a -fs hdfs://NamenodeHostname:PORT command-line argument pair to explicitly set the fs.default.name value in the configuration. This will override the value specified in the hadoop-site.xml file.

Here's a sample command line for listing files on an explicitly specified HDFS file system:

```
bin/hadoop dfs -fs hdfs://AlternateClusterNamenodeHostname:8020 -ls /
```

You can also use the JobConf object to set the default file system:

```
conf.set( "fs.default.name", "hdfs://NamenodeHostname:PORT");
```

Setting the JobTracker Location

The JobTracker location is normally specified with the mapred.job.tracker setting in the hadoop-site.xml file, as it is cluster-specific. The value will be JobTrackerHostname:PORT. Through Hadoop 0.19, there is not a standard for the PORT. Many installations use a port one higher that the HDFS port.

■ **Note** The default value for the JobTracker location is local, which will result in the job being executed by the JVM that submits it. The value local is ideal for testing and debugging new MapReduce jobs. It is important to ensure that any required Hadoop configuration files are in the classpath of the test jobs.

```
<property>
  <name>mapred.job.tracker</name>
  <value>JobtrackerHostname:PORT</value>
 </property>
```

Here's a sample command line explicitly setting the JobTracker for job control for listing jobs:

```
bin/hadoop job -jt AlternateClusterJobtrackerHostname:8021 -list
```

And here's how to use the JobConf object to set the JobTracker information:

```
conf.set( "mapred.job.tracker", "JobtrackerHostname:PORT");
```

The Mapper Dissected

All Hadoop jobs start with a mapper. The reducer is optional. The class providing the map function must implement the org.apache.hadoop.mapred.Mapper interface, which in turn requires the interfaces org.apache.hadoop.mapred.JobConfigurable and org.apache.hadoop.io.Closeable. The Hadoop framework provides org.apache.hadoop.mapred.MapReduceBase from which to derive mapper and reducer classes. The JobConfigurable and Closable implementations are empty methods. In the utilities supplied with this book's downloadable code is `com.apress.hadoopbook.utils.MapReduceBase`, which provides more useful implementations.

Note The interface `org.apache.hadoop.io.Closeable` will be replaced with `java.io.Closeable` in a later release.

This section examines the sample mapper class `SampleMapperRunner.java`, which is available with the rest of the downloadable code for this book. When run as a Java application, this example accepts all of the standard Hadoop arguments and may be run with custom bean context and definitions:

```
bin/hadoop jar hadoopprobook.jar ➥
com.apress.hadoopbook.examples.ch5.SampleMapperRunner -D ➥
mapper.bean.context=mycontext.xml -D mapper.bean.name=mybean -files ➥
mycontext.xml -deleteOutput
```

where:

- `bin/hadoop jar` is the standard Hadoop program invocation.
- `hadoopprobook.jar com.apress.hadoopbook.examples.ch5.SampleMapperRunner` specifies the JAR file to use and the main class to run.
- `-D mapper.bean.context=mycontext.xml` and `-D mapper.bean.name=mybean` specify that the string `mycontext.xml` is stored in the configuration under the key `mapper.bean.context`, and that the string `mybean` is stored in the configuration under the key `mapper.bean.name`.
- `-files mycontext.xml` causes the file `mycontext.xml` to be copied into HDFS, and then unpacked and made available in the working directory of each task run by the job. The working directory is in the task classpath. `mycontext.xml` may have a directory path component, and not be just a stand-alone file name. The path and file name provided must be a path that can be opened from the current working directory.

Note If you are using the value `local` as the value of the `mapred.task.tracker` configuration key, using the `DistributedCache` object is less effective, as the task cannot change working directories.

- `--deleteOutput`, which must be the last argument, causes the output directory to be deleted before the job is started. This is convenient when running the job multiple times.

Mapper Methods

For the mapper, the framework will call three methods:

- `configure()` method, defined in the `Configurable` interface
- `map()` method, defined in the `Mapper` interface
- `close()` method, defined in the `Closable` interface

The following sections discuss these methods in detail.

■Note The framework uses the static method `org.apache.hadoop.util.ReflectionUtils.<T>newInstance(Class<T> theClass, Configuration conf)` to create instances of objects that need a copy of the configuration. This will create the instance using the no-argument constructor. If the class is an instance of `Configurable`, `newInstance` will call the `setConf` method with the supplied configuration. If the class is an instance of `JobConfiguration`, `newInstance` will call the `configure` method. Any exceptions that are thrown during the construction or initialization of the instance are rethrown as `RuntimeExceptions`.

The configure() Method

The `void JobConfigurable.configure(JobConf job)` method, defined in `org.apache.hadoop.conf.Configurable`, is called exactly one time per map task as part of the initialization of the `Mapper` instance. If an exception is thrown, this task will fail. The framework may attempt to retry this task on another host if the allowable number of failures for the task has not been exceeded. The methods `JobConf.getMaxMapAttempts()` and `JobConf.setMaxMapAttempts(int n)` control the number of times a map task will be retried if the task fails. The default is four times.

It is considered good practice for any `Mapper` implementation to declare a member variable that the `configure()` method uses to store a reference to the passed-in `JobConf` object. The `configure()` method is also used for loading any Spring application context or initializing resources that are passed via `DistributedCache`.

Listing 5-1 shows the `configure()` method used in `SampleMapperRunner.java` (the example available with the downloadable code for this chapter).

Listing 5-1. *configure Method from SampleMapperRunner.java*

```
/** Sample Configure method for a map/reduce class.
 * This method assumes the class derives from {@link MapReduceBase}
 * and saves a copy of the JobConf object, the taskName
 * and the taskId into member variables.
 * and makes an instance of the output key and output value
 * objects as member variables for the
 * map or reduce to use.
 *
```

```
 * If this method fails the Tasktracker will abort this task.
 * @param job The Localized JobConf object for this task
 */
public void configure(JobConf job) {
    super.configure(job);
    LOG.info("Map Task Configure");
    this.conf = job;
    try {
        taskName = conf.getJobName();
        taskId = TaskAttemptID.forName(conf.get("mapred.task.id"));
        if (taskName == null || taskName.length() == 0) {
            /** if the job name is essentially unset make something up. */
            taskName = taskId.isMap() ? "map." : "reduce."
                    + this.getClass().getName();
        }

        /**
         * These casts are safe as they are checked by the framework
         * earlier in the process.
         */
        outputKey = (K2) conf.getMapOutputKeyClass().newInstance();
        outputValue = (V2) conf.getMapOutputValueClass().newInstance();
    } catch (RuntimeException e) {
        LOG.error("Map Task Failed to initialize", e);
        throw e;
    } catch (InstantiationException e) {
        LOG.error(
                "Failed to instantiate the key or output value class",
                e);
        throw new RuntimeException(e);
    } catch (IllegalAccessException e) {
        LOG
                .error(
 "Failed to run no argument constructor for key or output value objects",
                        e);
        throw new RuntimeException(e);
    }
    LOG.info(taskId.isMap() ? "Map" : "Reduce" + " Task Configure complete");

}
```

In this example, K2 is the map output key type, which defaults to the reduce output key type, which defaults to LongWritable. V2 is the map output value key type, which defaults to the reduce output value type, which defaults to Text.

This configure() method saves a copy of the JobConf object taskId and taskName into member variables. This method also instantiates a local instance of the key and value classes, to be used during the map() method calls. By using the isMap method on the taskId, you can take different actions for map and reduce tasks in the configure() and close() methods. This becomes very useful when a single class provides both a map method and a reduce method.

The map( ) Method

A call to the void map(K1 key, V1 value, OutputCollector<K2,V2> output, Reporter reporter) throws IOException method, defined in org.apache.hadoop.mapred.Mapper, will be made for every record in the job input. No calls will be made to the map() method in an instance before the configure() method completes.

If the job is configured for running multithreaded map tasks, as follows, there may be multiple simultaneous calls to the map() method.

```
jobConf.setMapRunnerClass(MultithreadedMapRunner.class);
jobConf.setInt("mapred.map.multithreadedrunner.threads", 10);
```

When running multithreaded, each map() call will have a different key and value object. The output and reporter objects are shared. The default number of threads for a multithreaded map task is ten.

The contents of the key object and the contents of the value object are valid only during the map() method call. The framework will reset the object contents with the next key/value pair prior to the next call to map().

The class converting the input into records is responsible for defining the types of K1 and V1. The standard textual input format, KeyValueTextInput, defines K1 and V1 to be of type Text.

K2 and V2 are defined by the JobConf.setMapOutputKeyClass(clazz) and JobConf.setMapOutputValueClass(clazz) methods. The types of K2 and V2 default to the classes set for the reduce key and value output classes. The reduce key and value output classes are set by JobConf.setOutputKeyClass(clazz) and JobConf.setOutputValueClass(clazz). The defaults for K2 and V2 are LongWritable and Text, respectively. You can explicitly configure the map output key and value classes, as follows:

```
jobConf.setMapOutputKeyClass(MyMapOutputKey.class);
jobConf.setMapOutputValueClass(MyMapOutputValue.class)
```

If a map output class is set, the corresponding reduce input class is also set to the class. If the map output key class is changed to BytesWritable, the Reducer.reduce's key type will be BytesWritable.

The close() Method

The void close() method, defined in java.io.Closable, is called one time after the last call to the map() method is made by the framework. This method is the place to close any open files or perform any status checking. Unless your configure() method has saved a copy of the JobConf object, there is little interaction that can be done with the framework. The close() method example in Listing 5-2 checks the task status based on the ratio of exceptions to input keys.

Listing 5-2. *close Method from SampleMapperRunner.java*

```
/** Sample close method that sets the task status based on how
 * many map exceptions there were.
 * This assumes that the reporter object passed into the map method was saved and
 *  that the JobConf object passed into the configure method was saved.
 */
public void close() throws IOException {
    super.close();
    LOG.info("Map task close");
    if (reporter != null) {
        /**
         * If we have a reporter we can perform simple checks on the
         * completion status and set a status message for this task.
         */
        Counter mapExceptionCounter = reporter.getCounter(taskName,
                "Total Map Failures");
        Counter mapTotalKeys = reporter.getCounter(taskName,
                "Total Map Keys");
        if (mapExceptionCounter.getCounter() == mapTotalKeys
                .getCounter()) {
            reporter.setStatus("Total Failure");
        } else if (mapExceptionCounter.getCounter() != 0) {
            reporter.setStatus("Partial Success");
        } else {
            /** Use the Spring set bean to show we did get the values. */
 reporter.incrCounter( taskName, getSpringSetString(), getSpringSetInt());
            reporter.setStatus("Complete Success");
        }
    }
    /**
     * Ensure any HDFS files are closed here, to force them to be
     * committed to HDFS.
     */
}
```

The `close()` method in Listing 5-2 will report success or failure status back to the framework, based on an examination of the job counters. It assumes that the `map()` method reported an exception under the counter, `Total Map Failure`, in the counter group `taskName`, and the number of keys received is in the counter, `Total Map Keys`, in the counter group `taskName`.

If there are no exceptions, the method will report the task status as "Complete Success." If there are some exceptions, the status is set to "Partial Success," If the exception count equals the key count, the status is set to "Total Failure."

This example also logs to counters with the values received from the Spring initialization. I found the Spring value-based counters useful while working out how to initialize map class member variables via the Spring Framework, as described after the discussion of the mapper class declaration and member fields.

Mapper Class Declaration and Member Fields

It is a best practice to capture the JobConf object passed in the configure() method into a member variable. It is also a good practice to instantiate member variables, or thread local variables, for any key or value that would otherwise be created in the body of the map() method. Having the TaskAttemptId available is also useful, as it is easy to determine if this is the map phase or the reduce phase of a job.

It is convenient to capture the output collector and the reporter into member fields so that they may be used in the close() method. This has a downside in that they can be captured only in the map() method, requiring extra code in that inner loop.

Listing 5-3 shows an example that declares a number of local variables, which are initialized by the configure() method for use by the map() and close() methods.

Listing 5-3. *Class and Member Variable Declarations from SampleMapperRunner.java*

```
/**
 * Sample Mapper shell showing various practices
 *
 * K1 and V1 will be defined by the InputFormat. K2 and V2 will be the
 * {@link JobConf#getOutputKeyClass()} and
 * {@link JobConf#getOutputValueClass()}, which by default are LongWritable
 * and Text. K1 and V1 may be explicitly set via
 * {@link JobConf#setMapOutputKeyClass(Class)} and
 * {@link JobConf#setMapOutputValueClass(Class)}. If K1 and V1 are
 * explicitly set, they become the K1 and V1 for the Reducer.
 *
 * @author Jason
 *
 */
public static class SampleMapper<K1, V1, K2, V2> extends MapReduceBase
        implements Mapper<K1, V1, K2, V2> {

    /**
     * Create a logging object or you will never know what happened in your
     * task.
     */

    /** Used in metrics reporting. */
    String taskName = null;
    /**
     * Always save one of these away. They are so handy for almost any
     * interaction with the framework.
     */
    JobConf conf = null;
    /**
     * These are nice to save, but require a test or a set each pass through
     * the map method.
     */
```

```
    Reporter reporter = null;
    /** Take this early, it is handy to have. */
    TaskAttemptID taskId = null;

    /**
     * If we are constructing new keys or values for the output, it is a
     * best practice to generate the key and value object once, and reset
     * them each time. Remember that the map method is an inner loop that
     * may be called millions of times. These really can't be used without
     * knowing an actual type
     */
    K2 outputKey = null;
    V2 outputValue = null;
```

Initializing the Mapper with Spring

Many installations use the Spring Framework to manage the services employed by their applications. One of the more interesting issues is how to use Spring in environments where Spring does not have full control over the creation of class instances. Spring likes to be in full control of the application and manage the creation of all of the Spring bean objects. In the Hadoop case, the Hadoop framework is in charge and will create the object instances. The examples in this section demonstrate how to use Spring to initialize member variables in the mapper class. The same techniques apply to the reducer class.

Listing 5-4 shows the bean file used in the Spring example. The file `mapper.bean.context.xml` in the downloadable examples `src/config` directory is the actual file used.

Listing 5-4. *Simple Bean Resource File for the Spring-Initialized Task*

```
<?xml version="1.0" encoding="UTF-8"?>
<beans xmlns="http://www.springframework.org/schema/beans"
    xmlns:xsi="http://www.w3.org/2001/XMLSchema-instance"
    xmlns:context="http://www.springframework.org/schema/context"
    xsi:schemaLocation="http://www.springframework.org/schema/beans ➥
        http://www.springframework.org/schema/beans/spring-beans.xsd
        http://www.springframework.org/schema/context ➥
        http://www.springframework.org/schema/context/spring-context.xsd">

    <bean id="SampleMapperJob.mapper.bean.name"
       class="com.apress.hadoopbook.examples.ch5.SampleMapperRunner.SampleMapper"
       lazy-init="true"
       scope="singleton">
       <description> Simple bean definition to provide an example for
       using Spring to initialize context in a Mapper class.</description>
       <property name="springSetString"><value>SetFromDefaultFile</value></property>
       <property name="springSetInt"><value>37</value></property>
  </bean>
</beans>
```

Creating the Spring Application Context

To create an application context, you need to provide Spring with a resource set from which to load bean definitions. Being very JobConf-oriented, I prefer to pass the names of these resources, and possibly the resources themselves, to my tasks via the JobConf and DistributedCache objects.

The example in Listing 5-5 extracts the set of resource names from the JobConf object, and if not found, will supply a default set of resource names. This follows the Hadoop style of using comma-separated elements to store multiple elements in the configuration. The set of resources names are unpacked and passed to the Spring Framework. Each of these resources must be in the classpath, which includes the task working directory.

At the very simplest, the user may specify the specific Spring configuration files on the command line via the -files argument, when the GenericOptionsParser is in use. The mapper class will need to determine the name of the file passed in via the command line. For the example, set up the Spring initialization parameters on the application command line as follows:

```
hadoop jar appJar main-class -files spring1.xml,spring2,xml,spring3.xml ➥
-D mapper.bean.context=spring1.xml
```

■Note In the command-line specification, the -D mapper.bean.context=value argument must come after the main class reference to be stored in the job configuration. If it comes before the jar argument, it will become a Java system property.

The example in Listing 5-5 copies spring1.xml, spring2.xml, and spring3.xml from the local file system into HDFS, and then copies them to the task local directory and creates symbolic links from the local copy to the task working directory. The configuration parameter mapper.bean.context tells the map task which bean file to load. In the example, SampleMapperRunner looks up the configuration entry mapper.bean.context to determine which bean files to use when creating the application context.

Listing 5-5. *Extracting the Resource File Names from the JobConf Object and Initializing the Spring Application Context (from utils.Utils.java)*

```
/**
 * Initialize the Spring environment. This is of course completely
 * optional.
 *
 * This method picks up the application context from a file, that is in
 * the classpath. If the file items are passed through the
 * {@link DistributedCache} and symlinked
 * they will be in the classpath.
 *
```

```
 * @param conf The JobConf object to look for the Spring config file names.
 * If this is null, the default value is used.
 * @param contextConfigName The token to look under in the config for the names
 * @param defaultConfigString A default value
 * @return TODO
 *
 */
public static ApplicationContext initSpring(JobConf conf, String contextConfigName,
        String defaultConfigString) {
    /**
     * If you are a Spring user, you would initialize your application
     * context here.
     */
    /** Look up the context config files in the JobConf, provide a default value. */
    String applicationContextFileNameSet =
        conf == null ? defaultConfigString :
            conf.get( contextConfigName, defaultConfigString);
    LOG.info("Map Application Context File "
            + applicationContextFileNameSet);

    /** If no config information was found, bail out. */
    if (applicationContextFileNameSet==null) {
        LOG.error( "Unable to initialize Spring configuration using "
         + applicationContextFileNameSet );
        return null;
    }
    /** Attempt to split it into components using the config
      * standard method of comma separators. */
    String[] components = StringUtils.split(applicationContextFileNameSet, ",");

    /** Load the configuration. */
    ApplicationContext applicationContext =
        new ClassPathXmlApplicationContext( components);

    return applicationContext;
}
```

Using Spring to Autowire the Mapper Class

Once the Spring application context has been created, the task may instantiate beans. The confusing issue is that the mapper class has already been instantiated, so how can Spring be forced to initialize/autowire that class?

Accomplishing this autowiring requires two things. The first is that the bean definition to be used must specify lazy initialization, to prevent Spring from creating an instance of the bean when the application context is created. The second is to know the bean name/ID of the mapper class.

The example in Listing 5-6 makes some assumptions about how application contexts and task beans are named, and can be easily modified for your application.

Listing 5-6. *Example of a Spring Task Initialization Method*

```
/** Handle Spring configuration for the mapper.
 * The bean definition has to be <code>lazy-init="true"</code>
 * as this object must be initialized.
 * This will fail if Spring weaves a wrapper class for AOP around
 * the configure bean.
 *
 * The bean name is extracted from the configuration as
 * mapper.bean.name or reducer.bean.name
 * or defaults to taskName.XXXX.bean.name
 *
 * The application context is loaded from mapper.bean.context
 * or reducer.bean.context and may be a set of files
 * The default is jobName.XXX.bean.context
 *
 * @param job The JobConf object to look up application context ➥
   files and bean names in
 * @param RuntimeException if the application context can not be ➥
   loaded or the initializtion requires delegation of the task object.
 */
 void springAutoWire(JobConf job) {
    String springBaseName = taskId.isMap()? "mapper.bean": "reducer.bean";

    /** Construct a bean name for this class using the configuration
      * or a default name. */
    String beanName = conf.get(springBaseName + ".name",
        taskName + "." + springBaseName + ".name" );
    LOG.info("Bean name is " + beanName);
    applicationContext = Utils.initSpring(job, springBaseName
     + ".context", springBaseName + ".context.xml");
    if (applicationContext==null) {
        throw new RuntimeException(
 "Unable to initialize spring configuration for " + springBaseName);
    }
    AutowireCapableBeanFactory autowire =
  applicationContext.getAutowireCapableBeanFactory();
    Object mayBeWrapped = autowire.configureBean( this, beanName);
    if (mayBeWrapped != this) {
        throw new RuntimeException( "Spring wrapped our class for " + beanName);
    }
}
```

In Listing 5-6, a base name is constructed for looking up information in the configuration via the following:

```
String springBaseName = taskId.isMap()? "mapper.bean": "reducer.bean";
```

The example builds a context file name key, which will be `mapper.bean.context` in the case of a map, to look up the application context information in the configuration. If a value is found, it is treated as a comma-separated list of bean resource files to load. The application context is loaded and saved in the member variable `applicationContext`:

```
applicationContext = SpringUtils.initSpring(job, springBaseName
   + ".context", springBaseName + ".context.xml");
```

A default bean file is used if no value is found. In this example, the file is `mapper.bean.context.xml`.

A bean name key `mapper.bean.name`, with a default value of `mapper.bean.name`, is looked up in the configuration. This is the bean that will be used to configure the task. The following line constructs the bean name to use:

```
String beanName = conf.get(springBaseName + ".name", taskName + "."
   + springBaseName + ".name" );
```

An autowire-capable bean factory is extracted from the application context via the following:

```
AutowireCapableBeanFactory autowire =
  applicationContext.getAutowireCapableBeanFactory();
```

The following line actually causes Spring to initialize the task:

```
Object mayBeWrapped = autowire.configureBean( this, beanName);
```

The code must ensure that Spring did not return a delegator object when it was initializing the task from the bean definition:

```
if (mayBeWrapped != this) {
    throw new RuntimeException( "Spring wrapped our class for " + beanName);
}
```

■ **Note** This example does not handle the case where Spring returns a delegator object for the task. To handle this case, the `map()` method would need to be redirected through the delegated object.

Partitioners Dissected

A core part of the MapReduce concept requires that map outputs be split into multiple streams called *partitions*, and that each of these partitions is fed to a single reduce task. The reduce contract specifies that each reduce task will be given as input the fully sorted set of keys and their values in a particular partition. The entire partition is the input of the reduce task. For the framework to satisfy this contract, a number of things have to happen first. The outputs of each map task are partitioned and sorted. The partitioner is run in the context of the map task.

The Hadoop framework provides several partitioning classes and a mechanism to specify a class to use for partitioning. The actual class to be used must implement the `org.apache.hadoop.mapred.Partitioner` interface, as shown in Listing 5-7. The piece that provides a partition number is the `getPartition()` method:

```
int getPartition(K2 key, V2 value, int numPartitions)
```

Note that both the key and the value are available in making the partition choice.

Listing 5-7. *The Partitioner Interface in Hadoop 0.19.0*

```
/**
 * Partitions the key space.
 *
 * <p><code>Partitioner</code> controls the partitioning of the keys of the
 * intermediate map-outputs. The key (or a subset of the key) is used to derive
 * the partition, typically by a hash function. The total number of partitions
 * is the same as the number of reduce tasks for the job. Hence this controls
 * which of the <code>m</code> reduce tasks the intermediate key (and hence the
 * record) is sent for reduction.</p>
 *
 * @see Reducer
 */
public interface Partitioner<K2, V2> extends JobConfigurable {

  /**
   * Get the partition number for a given key (hence record) given the total
   * number of partitions i.e. number of reduce tasks for the job.
   *
   * <p>Typically a hash function on a all or a subset of the key.</p>
   *
   * @param key the key to be paritioned.
   * @param value the entry value.
   * @param numPartitions the total number of partitions.
   * @return the partition number for the <code>key</code>.
   */
  int getPartition(K2 key, V2 value, int numPartitions);
}
```

The key and value will be streamed into the partition number that this function returns. Each key/value pair output by the `map()` method has the partition number determined and is then written to that map local partition. Each of these map local partition files is sorted in key order by the class returned by the `JobConf.getOutputKeyComparator()` method.

For each reduce task, the framework will collect all the reduce task's partition pieces from each of the map tasks and merge-sort those pieces. The results of the merge-sort are then fed to the `reduce()` method. The merge-sort is also done by the class returned by the `JobConf.getOutputKeyComparator()` method.

The output of a reduce task will be written to the `part-`*XXXXX* file, where the *XXXXX* corresponds to the partition number.

The Hadoop framework provides the following partitioner classes:

- `HashPartitioner`, which is the default
- `TotalOrderPartitioner`, which provides a way to partition by range
- `KeyFieldBasedPartitioner`, which provides a way to partition by parts of the key

The following sections describe each of these partitioners.

The HashPartitioner Class

The default partitioner, `org.apache.hadoop.mapred.lib.HashPartitioner`, simply uses the hash code value of the key as the determining factor for partitioning. Listing 5-8 shows the actual code from the default partitioner used by Hadoop. The partition number is simply the hash value of the key modulus the number of partitions.

Listing 5-8. *The HashCode Partitioner from Hadoop 0.19.0*

```
/** Partition keys by their {@link Object#hashCode()}. */
public class HashPartitioner<K2, V2> implements Partitioner<K2, V2> {

  public void configure(JobConf job) {}

  /** Use {@link Object#hashCode()} to partition. */
  public int getPartition(K2 key, V2 value,
                          int numReduceTasks) {
    return (key.hashCode() & Integer.MAX_VALUE) % numReduceTasks;
  }

}
```

The hash value is converted to a positive value, `(key.hashCode() & Integer.MAX_VALUE)`, to ensure that the partition will be a positive integer. The resulting number has modulus the number of reduce tasks applied, `% numReduceTasks`, and the result returned. This produces a positive number between 0 and one less than the number of partitions.

The TotalOrderPartitioner Class

The `TotalOrderPartitioner`, `org.apache.hadoop.mapred.lib.TotalOrderPartitioner`, relies on a file that provides the class with range information. With this information, the partitioner is able to determine which range a key/value pair belongs in and route it to the relevant partition.

Note The TotalOrderParitioner grew out of the TetraSort example package. Jim Gray introduced a contest called the TeraByteSort, which was a benchmark to sort one terabyte of data and write the results to disk. In 2008, Yahoo! produced a Hadoop version of the test that completed in 209 seconds (http://developer.yahoo.net/blogs/hadoop/2008/07/apache_hadoop_wins_terabyte_sort_benchmark.html). The code is included with the Hadoop examples as bin/hadoop jar hadoop-*-examples.jar terasort in-dir out-dir. The class file is org.apache.hadoop.examples.terasort.TeraSort.

Building a Range Table

The org.apache.hadoop.mapred.lib.InputSampler class is used to generate a range partitioning file for arbitrary input sets. This class will sample the input to build an approximate range table.

This sampling strategy will take no more than the specified number of samples total from the input. The user may specify a maximum number of input splits to look in as well. The actual number of records read from each input split varies based on the number of splits and the number of records in the input split.

The Hadoop framework controls how the input is split based on the number of input files, the input format, the input file size, and the minimum split size and the HDFS block size. Let's look at a few examples of running InputSampler from the command line.

In the following example, the argument set -splitSample 1000 10 will sample a total of 1,000 input records out of no more than 10 input splits.

```
bin/hadoop jar hadoop-0.19.0-core.jar org.apache.hadoop.mapred.lib.InputSampler ➥
-inFormat org.apache.hadoop.mapred.KeyValueTextInputFormat ➥
-keyClass org.apache.hadoop.io.Text -r 15 -splitSample 1000 10 csvin csvout
```

If there are 10 or more input splits, each of which has more than 100 records, the first 100 records from each input split will be used for samples. The input is loaded from the directory csvin, and is parsed by the KeyValueTextInputFormat class. The range file is written to csvout, and the argument set -r 15 sets up the output for a job with 15 output partitions. The input splits are examined in the order in which they are returned by InputFormat.

The next example takes 1,000 samples from roughly 10 input splits. The input splits are sampled in a random order, and the records from each split read are sequentially.

```
bin/hadoop jar hadoop-0.19.0-core.jar org.apache.hadoop.mapred.lib.InputSampler ➥
-inFormat org.apache.hadoop.mapred.KeyValueTextInputFormat ➥
-keyClass org.apache.hadoop.io.Text -r 15 -splitRandom .1 1000 10 csvin csvout
```

Each record has a 0.1% chance of being selected. The -splitRandom .1 1000 10 argument set specifies the percentage, the total samples, and the maximum splits to sample. If the 1,000 samples are not selected after processing the recommended number of splits, more splits will be sampled. The index is set up for 15 reduce tasks, and the input comes from csvin. The index is written to csvout. The splits to examine are selected randomly.

In the final example, the argument set -splitInterval .01 10 will examine no more than 10 input splits and take one record in 100 from each split.

```
bin/hadoop jar hadoop-0.19.0-core.jar org.apache.hadoop.mapred.lib.InputSampler ➥
-inFormat org.apache.hadoop.mapred.KeyValueTextInputFormat ➥
-keyClass org.apache.hadoop.io.Text -r 15 -splitInterval .01 10 csvin csvout
```

The frequency parameter defines how many records will be sampled. For a frequency of 0.1, as in this example, one record in 10 will be used. For a frequency of 0.01, one record in 100 will be used. The index is set up for 15 reduce tasks. The input comes from csvin, and the index is written to csvout.

Using the TotalOrderPartitioner

Once an index is generated, a job may be set up to use the TotalOrderPartitioner and the index. Three configuration settings are required for this to work:

- The partitioner must be set to TotalOrderPartitioner in the JobConf object via conf.setPartitionerClass(TotalOrderPartitioner).
- The partitioning index must be specified via the configuration key total.order.partitioner.path:

  ```
  conf.set("total.order.partitioner.path", "csvout");
  ```

- The sort type for the keys must also be specified. If the binary representation of the keys is the correct sorting, the Boolean field total.order.partitioner.natural.order should be set to true in the configuration. If the binary representation of the keys is not the correct sort, the Boolean field total.order.partitioner.natural.order must be set to false. This Boolean field is set as follows:

  ```
  conf.setBoolean("total.order.partitioner.natural.order");
  ```

  If the binary representation of the key is the correct sort order, a binary trie (an ordered tree structure; see http://en.wikipedia.org/wiki/Trie) will be constructed and used for searching; otherwise, a binary search based on the output key comparator will be used.

Here's an example of how to put all this together:

```
TotalOrderPartitioner.setPartitionFile(conf,"csvout");
conf.setPartitionerClass(TotalOrderPartitioner.class);
conf.set("total.order.partitioner.natural.order",false);
conf.setNumReduceTasks (15);
```

In this example, csvin is the input file, and csvout is the index file. The csvout file was set up for 15 reduce tasks, and requires the comparator rather than binary comparison.

The KeyFieldBasedPartitioner Class

The KeyFieldBasedPartitioner, org.apache.hadoop.mapred.lib.KeyFieldBasedPartitioner, provides the job with a way of using only parts of the key for comparison purposes. The primary concept is that the keys may be split into pieces based on a piece separator string. Each piece is then numbered from 1 to *N*, and each character of each piece numbered from 1 to *M*.

The separator string is defined by the configuration key `map.output.key.field.separator` and defaults to the tab character. It may be set to another string, `str`, as follows:

```
conf.set(map.output.key.field.separator, str);
```

This is functionally equivalent to using the `String.split(Pattern.quote(str))` call on each key and treating the resulting array as if indexes were one-based instead of zero-based.

If the separator is `X` and the key is `oneXtwoXthree`, the pieces will be `1) one`, `2) two`, `3) three`.

Referencing individual characters within the pieces is also one-based rather than zero-based, with the index `0` being the index of the last character of the key part. For the first key piece in the preceding example, the string `one`, the characters will be `1) o`, `2) n`, `3) e`, `0) e`. Note that both `3` and `0` refer to `e`, which is the last character of the key piece.

■Note In addition to the one-based ordinal position within the key piece, the last character of the key piece may also be referenced by `0`.

The key pieces to compare are specified by setting the key field partition option, via the following:

```
conf. setKeyFieldPartitionerOptions(str).
```

The `str` format is very similar to the key field-based comparator.

The Javadoc from Hadoop 0.19.0 for `KeyFieldBasedPartitioner` provides the following definition:

> *Defines a way to partition keys based on certain key fields (also see KeyFieldBasedComparator). The key specification supported is of the form -k pos1[,pos2], where, pos is of the form f[.c][opts], where f is the number of the key field to use, and c is the number of the first character from the beginning of the field. Fields and character posns are numbered starting with 1; a character position of zero in pos2 indicates the field's last character. If '.c' is omitted from pos1, it defaults to 1 (the beginning of the field); if omitted from pos2, it defaults to 0 (the end of the field).*

In plain English, `-k#` selects piece # for the comparison, and `-k#1,#2` selects the pieces from `#1` through `#2`. In the preceding example, `-k1` selects `oneX` as the portion of the key to use for comparison, and `-k1,1` selects `one` as the portion of the key to use for comparison.

There is also the facility to select a start and stop point within an individual key. The option `-k1.2,1` is equivalent to `-k1.2,1.0`, and selects `ne` from the `one` for comparison.

You may also span key pieces. `-k1.2,3.2` selects `eXtwoXth` as the comparison region from the sample key. It means to start with key piece 1, character 2 and end with key piece 3 character 2.

■Note If your key specification may touch the last key piece, it is important to terminate with the last character of the key. Otherwise, the current code (as of Hadoop 0.19.0) will generate an `ArrayIndexOutOfBoundsException` as it tries to use the missing separator string. In this section's example, `-k3,3` would work, but `-k3` would throw the exception.

The Reducer Dissected

The reducer has a very similar shape to the mapper. The class may provide `configure()` and `close()` methods. All of the mapper good practices of saving the `JobConf` object and making instances of the output key and output value objects apply to the reducer as well.

The key difference is in the `reduce()` method. Unlike the `map()` method, which is given a single key/value pair on each invocation, each `reduce()` method invocation is given a key and all of the values that share that key.

The reducer is an operator on groups. The default is to define a group as all values that share a key. Common uses for reduce tasks are to suppress duplicates in datasets or to segregate ranges and order output of large datasets.

In the example shown in Listing 5-9, notice that the signature of the `reduce()` method contains an `Iterator<V>`, an iterator over the values that share key. The identity reducer simply outputs each value in the iterator.

Listing 5-9. *The Identity Reducer from Hadoop Core 0.19.0*

```
/** Performs no reduction, writing all input values directly to the output. */
public class IdentityReducer<K, V>
    extends MapReduceBase implements Reducer<K, V, K, V> {

  /** Writes all keys and values directly to output. */
  public void reduce(K1 key, Iterator<V1> values,
                     OutputCollector<K2, V2> output, Reporter reporter)
    throws IOException {
    while (values.hasNext()) {
      output.collect(key, values.next());
    }
  }

}
```

The `configure()` and `close()` methods have the same requirements and suggested usage as the corresponding mapper methods.

It is generally recommended that you do not make a copy of all of the value objects, as there may be very many of these objects.

■ **Note** In one of my early applications, I assumed that there would never be more than a small number of values per key. The reduce tasks started experiencing out-of-memory exceptions. It turned out that there were often more than 150,000 values per key!

It is possible to simulate a secondary sort/grouping of the values by setting the output value grouping. To do this requires the cooperation of the `OutputComparator`, `OutputPartitioner`, and `OutputValueGroupingComparator`. See this book's appendix for more information.

By default, the input key and value types are the same as the output key and value types, and are set by the `conf.setOutputKeyClass(class)` and `conf.setOutputValueClass(class)` methods. The defaults are `LongWritable` and `Text`, respectively.

If the map output keys must be different, using `conf.setMapOutputKeyClass(class)` and `conf.setMapOutputValueClass(class)` will also change the expected input key and value for the reduce task.

A Simple Transforming Reducer

Listing 5-10 shows the simple transformational reducer, `SimpleReduceTransformingReducer.java`, used in this chapter's `SimpleReduce.java` example.

Listing 5-10. *Transformational Reducer in SimpleReduceTransformingReducer.java*

```
/** Demonstrate some aggregation in the reducer
 *
 * Produce output records that are the key, the average, the count,
 * the min, max and diff
 *
 * @author Jason
 *
 */
public class SimpleReduceTransformingReducer extends MapReduceBase implements
        Reducer<LongWritable, LongWritable, Text, Text> {

    /** Save object churn. */
    Text outputKey = new Text();
    Text outputValue = new Text();

    /** Used in building the textual representation of the output key and values. */
    StringBuilder sb = new StringBuilder();
    Formatter fmt = new Formatter(sb);
```

```
@Override
public void reduce(LongWritable key, Iterator<LongWritable> values,
        OutputCollector<Text, Text> output,
        Reporter reporter) throws IOException {
    /** This is a bad practice, the transformation of
      * the key should be done in the map. */
    reporter.incrCounter("Reduce Input Keys", "Total", 1);
    try {
        long total = 0;
        long count = 0;
        long min = Long.MAX_VALUE;
        long max = 0;

        /** Examine each of the values that grouped with this key. */
        while (values.hasNext()) {
            final long value = values.next().get();
            if (value>max) {
                max = value;
            }
            if (value<min) {
                min = value;
            }
            total += value;
            count++;
        }

        sb.setLength(0);
        fmt.format("%12d %3d %12d %12d %12d", total/count,
                    count, min, max, max-min);
        fmt.flush();
        outputValue.set(sb.toString());

        sb.setLength(0);
        fmt.format("%4d", key.get());
        outputKey.set(sb.toString());

        reporter.incrCounter("Reduce Output Keys", "Total", 1);
        output.collect(outputKey, outputValue);

    } catch( Throwable e) {
        reporter.incrCounter("Reduce Input Keys", "Exception", 1);
        if (e instanceof IOException) {
            throw (IOException) e;
        }
```

```
                if (e instanceof RuntimeException) {
                    throw (RuntimeException) e;
                }
                throw new IOException(e);
            }

    }
}
```

It begins by establishing several member variables that will be used in the `reduce()` method to save object generation:

```
/** Save object churn. */
    Text outputKey = new Text();
    Text outputValue = new Text();

    /** Used in building the textual representation of the output key and values. */
    StringBuilder sb = new StringBuilder();
    Formatter fmt = new Formatter(sb);
```

The working body of the `reduce()` method is within a `try` block that catches `Throwables`, and the input count, output count, and failure count are reported to the framework:

```
reporter.incrCounter("Reduce Input Keys", "Total", 1);
try {
    ....
    reporter.incrCounter("Reduce Output Keys", "Total", 1);
    output.collect(outputKey, outputValue);

} catch( Throwable e) {
    reporter.incrCounter("Reduce Input Keys", "Exception", 1);
```

In the body of the example in Listing 5-10, each value that is passed in is examined and aggregated:

```
/** Examine each of the values that grouped with this key. */
while (values.hasNext()) {
    final long value = values.next().get();
    if (value>max) {
        max = value;
    }
    if (value<min) {
        min = value;
    }
    total += value;
    count++;
}
```

Finally, the output key and value are constructed with the aggregated data:

```
sb.setLength(0);
fmt.format("%12d %3d %12d %12d %12d", total/count, count, min, max, max-min);
fmt.flush();
outputValue.set(sb.toString());

sb.setLength(0);
fmt.format("%4d", key.get());
outputKey.set(sb.toString());
```

The example that runs this reducer also uses an output grouping comparator that groups the records in sets of ten. The comparator `Utils.GroupByLongGroupingComparator.java` (supplied with the downloadable code for this chapter) handles grouping `LongWritable` values in sets of 10, 0–9, 10–19, and so on.

The following is the core code in `SimpleReduce` that sets up the job that runs `SimpleReduceTransformingReducer`:

```
job.setInputFormat(KeyValueTextInputFormat.class);
FileInputFormat.setInputPaths(job, inputDir);

job.setMapperClass(SimpleReduceTransformingMapper.class);
job.setMapOutputValueClass(LongWritable.class);
job.setMapOutputKeyClass(LongWritable.class);

/** Force the reduce to take text as the output value class,
  * instead of the default. */
job.setOutputValueClass(Text.class);
job.setOutputKeyClass(Text.class);
job.setReducerClass(SimpleReduceTransformingReducer.class);

/** Cause the keys to be grouped by 10s. */
job.setOutputValueGroupingComparator(GroupByLongGroupingComparator.class);
job.setNumReduceTasks(1);    /** Ensure that all keys go to 1 reduce so
   * the group by is stable. */
```

The following command will run the `SimpleReduce` job (your output will vary slightly):

```
% HADOOP_CLASSPATH=/misc/HadoopSource/commons-lang-2.4.jar ➥
bin/hadoop jar /misc/HadoopSource/hadoop-0.19.0/hadoopprobook.jar ➥
com.apress.hadoopbook.examples.ch5.SimpleReduce -libjars ➥
/misc/HadoopSource/commons-lang-2.4.jar
```

```
Total input paths to process : 5
Running job: job_200902221346_0079
 map 0% reduce 0%
 map 20% reduce 0%
 map 60% reduce 0%
 map 80% reduce 0%
 map 100% reduce 0%
 map 100% reduce 100%
Job complete: job_200902221346_0079
Counters: 20
  File Systems
    HDFS bytes read=7103
    HDFS bytes written=2135
    Local bytes read=9006
    Local bytes written=18176
  Job Counters
    Launched reduce tasks=1
    Launched map tasks=5
    Data-local map tasks=5
  Map Input Keys
    Total=500
  Reduce Output Keys
    Total=35
  Map Output Keys
    Total=500
  Reduce Input Keys
    Total=35
  Map-Reduce Framework
    Reduce input groups=35
    Combine output records=0
    Map input records=500
    Reduce output records=35
    Map output bytes=8000
    Map input bytes=7103
    Combine input records=0
    Map output records=500
    Reduce input records=500
The Job is  complete  and  successfull
```

Note how the output keys are multiples of tens. This is the result of the output value grouping. The actual output is the key, the average value, the number of values averaged, the minimum value, the maximum value, and the difference between the minimum and the maximum.

Now you can print the job output (key, average, count, min, max, difference), as follows:

```
% hadoop dfs -cat SampleReduce.ouput/part-00000
```

```
  0     1032312560  10     8929475  2037836662  2028907187
 10      909677971  10    40027932  2084424645  2044396713
 20     1264310186  10   109435752  2002508155  1893072403
 30      984307588  10   112010776  1912518297  1800507521
 40      925589754  10    38065333  1782409589  1744344256
 50      923048786  10   374030611  1725384504  1351353893
 60      908071213  10   255471236  2115349080  1859877844
 70     1068729097  10    63376590  1954205116  1890828526
 80     1216389986  10    40846046  2120059182  2079213136
 90     1119730476  10   289044657  2002422718  1713378061
100      638218214  10    10905001  1679731545  1668826544
110     1208679389  10   351936606  1701974468  1350037862
120      958900520  10   116429037  1686303707  1569874670
130      871313033  10    52844729  2019468622  1966623893
140     1328295033  10   111275382  2059113431  1947838049
150     1038185198  20    47621146  1976756537  1929135391
160      980833493  20    72608499  2029753820  1957145321
170      912381685  20    54099516  1961970644  1907871128
180     1247773207  20    30716232  2116148228  2085431996
190      875941698  20     6692770  1663091528  1656398758
200     1051606085  20    18948588  2123342351  2104393763
210     1207066231  20   160161337  1952936377  1792775040
220     1327655145  20    75910389  2078268756  2002358367
230     1148152274  20   273711624  2074598677  1800887053
240      735579301  20    43456136  2094659831  2051203695
250     1115493614  20   190919486  1988623879  1797704393
260     1026999134  20    59805730  2072846822  2013041092
270     1109366173  20   198612696  2077682368  1879069672
280      954780820  20    44018855  2107358734  2063339879
290      778472644  20    22502766  2063051919  2040549153
300     1032042843  10   292411084  2097164456  1804753372
310      822060835  10    90530214  2135412572  2044882358
320      857131707  10   138285402  1675393365  1537107963
330     1153129237  10   231919805  1799184626  1567264821
340      851254291  10   135630114  1965837214  1830207100
```

A Reducer That Uses Three Partitions

A variant of the `SimpleReduce.java` example, called `TotalOrderSimpleReduce.java` (available with the rest of this chapter's downloadable code), uses three partitions, rather than just one. This example demonstrates how to use the `InputSampler` class and the `TotalOrderPartitioner` class, as well as some of the interesting errors that will occur if the partitioner and the OutputValueGroupingComparator do not coordinate fully.

In this example, the grouping operator groups by multiples of ten in the key space. The TotalOrderParititioner selects a random sample of the keys and creates three groups that are roughly even in size given the sample of keys. There is no guarantee that an entire group of keys will not be split into multiple partitions.

This application also requires a custom InputFormat, LongLongTextInputFormat, as the input key and the reduce key must be of the same type for the InputSampler. In the previous version, the map input keys are Text and the reduce input keys are LongWritable. Listing 5-11 shows the core of the LongLongTextInputFormat, the RecordReader.next method.

Listing 5-11. *The RecordReader.next Method of the LongLongTextInputFormat*

```
/** Delegated next, read the textual values from the the data source
  * and convert them into LongWritables.
 * @param key The key object to fill with the next record's key
 * @param value The value object to fill with the next record's value
 * @return true if a record was read or false if at EOF
 * @throws IOException
 * @see org.apache.hadoop.mapred.RecordReader#next(java.lang.Object, ➥
java.lang.Object)
 */
public boolean next(LongWritable key, LongWritable value) throws IOException {
    /** Perform the real read. */
    final boolean res = realReader.next(this.key, this.value);
    if (!res) { /** If at eof, we are done. */
        return false;
    }
    /** Attempt to convert the two text values read into LongWritables.
     * If there is an error, throw an IOException.
     */
    try {
        key.set(Long.valueOf(this.key.toString()));
        value.set(Long.valueOf(this.value.toString()));
        return true;
    } catch( NumberFormatException e) {
        throw new IOException("Invalid key, value " + key + ", " + value);
    }
}
```

The code in Listing 5-12 sets up the JobConf object for the TotalOrderParitioner. Note that natural ordering is set to true. As the keys are long values, they are binary comparable. The call to runInputSampler computes the partitioning index and stores it in the file TotalOrderSimpleReduce.index.

Listing 5-12. *TotalOrderPartition Setup, from TotalOrderSimpleReduce.java*

```
job.setInputFormat(LongLongTextInputFormat.class);
FileInputFormat.setInputPaths(job, inputDir);

job.setMapOutputValueClass(LongWritable.class);
job.setMapOutputKeyClass(LongWritable.class);

/** Setup for a total order partitioning. */
job.setPartitionerClass(TotalOrderPartitioner.class);
job.setBoolean("total.order.partitioner.natural.order", true);

/** Force reduce to take text as the output value class, instead of the default. */
job.setOutputValueClass(Text.class);
job.setOutputKeyClass(Text.class);
job.setReducerClass(SimpleReduceTransformingReducer.class);

/** Cause the keys to be grouped by 10s. */
job.setOutputValueGroupingComparator(GroupByLongGroupingComparator.class);
/** Ensure that all keys go to 3 reduce to demonstrate order based partitioning. */
job.setNumReduceTasks(3);
runInputSampler(job, inputDir.suffix(".index"));
```

The code in Listing 5-13 runs the InputSampler to compute and store the index in indexFile. The assumption here is that the JobConf object conf is already correctly set up with the InputPaths and InputReader. The sampling strategy is to randomly sample the records with a 0.1% chance that any record is chosen. No more than 100 samples and a suggested 10 input splits are to be read.

Listing 5-13. *Running the InputSampler*

```
/** Generate the TotalOrderPartitioner index file for our key space
 *
 * This will sample the input paths set in conf, using the input format reader.
 * The index file location is written to conf.
 *
 * @param conf The Configuration object to use
 * @param indexFile The index file to generate
 * @throws IOException
 */
public void runInputSampler(final JobConf conf, Path indexFile) throws IOException {
    TotalOrderPartitioner.setPartitionFile(conf, indexFile);
    RandomSampler<LongWritable, LongWritable> sampler = new
    InputSampler.RandomSampler<LongWritable,LongWritable>(0.1, 100, 10);
    InputSampler.<LongWritable,LongWritable>writePartitionFile(conf, sampler);
}
```

The following results show that the input for group 150 is split between partition 0 and partition 1, and that the group 220 is split between partition 2 and partition 3. Your results will differ, as random data generation and selection are occurring.

```
HADOOP_CLASSPATH=/misc/HadoopSource/commons-lang-2.4.jar hadoop jar /misc ➥
/HadoopSource/hadoop-0.19.0/hadoopprobook.jar com.apress.hadoopbook.examples.ch5 ➥
.TotalOrderSimpleReduce -libjars /misc/HadoopSource/commons-lang-2.4.jar
```

```
The Job is  complete  and  successfull
Counter Group: File Systems
    HDFS bytes read 8060
    HDFS bytes written  2257
    Local bytes read    9018
    Local bytes written 18488
Counter Group: Job Counters
    Launched reduce tasks   3
    Launched map tasks  5
    Data-local map tasks    5
Counter Group: Reduce Output Keys
    Total   37
Counter Group: Reduce Input Keys
    Total   37
Counter Group: Map-Reduce Framework
    Reduce input groups 37
    Combine output records  0
    Map input records   500
    Reduce output records   37
    Map output bytes    8000
    Map input bytes 7135
    Combine input records   0
    Map output records  500
    Reduce input records    500
```

Let's examine the reduce output data:

```
for a in 0 1 2; do echo part-0000$a; hadoop dfs -cat TotalOrderSimpleReduce ➥
.ouput/part-0000$a; done
```

```
part-00000
  0       1120696448  10    114767562   2024812642  1910045080
 10       1262245737  10    147134609   2118565837  1971431228
 20       1355678543  10    221719466   2058534489  1836815023
 30       1011945955  10     32549345   1964050949  1931501604
 40       1141622277  10     14444296   2091872332  2077428036
 50       1033598416  10    128237459   1923443602  1795206143
 60       1110802460  10    259693362   1904661969  1644968607
```

```
 70    1241399906 10    41832977 2059443669 2017610692
 80    1230683390 10   103825808 2063631220 1959805412
 90    1128499980 10   107614131 2028701766 1921087635
100    1088361665 10   376207299 1832969382 1456762083
110    1332495922 10   332169914 2049937661 1717767747
120     991086606 10    18158041 1954291526 1936133485
130    1020804065 10   117011726 2094067623 1977055897
140     967879564 10    78769539 2041673853 1962904314
150    1236638804  8   401939855 2012038507 1610098652
part-00001
154    1139330738 12    51795064 1954863887 1903068823
160     993478558 20    54628468 2078982662 2024354194
170    1036438744 20   156951559 1983508735 1826557176
180    1101282242 20    42570729 2097760736 2055190007
190    1193146388 20   113670430 2111312959 1997642529
200    1015890669 20   130204162 2104346838 1974142676
210    1234536770 20   105147150 2045372284 1940225134
220    1464315969  8   479100103 2046550989 1567450886
part-00002
224     954658466 12    96604844 1853232282 1756627438
230     964917299 20   116190161 2115557112 1999366951
240    1207841113 20   352735303 2136588979 1783853676
250    1047422883 20   158450293 2047289337 1888839044
260     884844748 20    54670426 1920120397 1865449971
270    1143486218 20   240046014 2139315373 1899269359
280    1345299024 20   267642220 2099770746 1832128526
290     997769299 20    53033105 2114447296 2061414191
300     566836001 10     3288468 1688928276 1685639808
310     871057357 10     2573252 2059752419 2057179167
320     827237669 10   120300136 2091904736 1971604600
330    1034732041 10    72330772 2053586973 1981256201
340     938330142 10    49826875 2145892833 2096065958
```

Combiners

A combiner is a mini-reducer. The purpose of a combiner is to reduce the volume of data that must be passed to the reducer from a map task by summarizing output records that share the same key. A combiner must implement the `Reducer` interface, and the `reduce()` method of the combiner will be called with each output key and all of the output values that share that key. The output of the combiner is what will be sent over the network to the actual reduce task for the job or written to the final output directory, if there is no reduce task configured. The combiner class `reduce()` method must have the same input and output key/value types as the reducer class.

For each call to `output.collect` made by the `map()` method, the framework will route the key/value pair to the applicable partition, based on the result of the `Partitioner.getPartition` call. When all of the map task input has been processed, these partitions are sorted, and each one is passed as input to the combiner. The combiner's `reduce()` method will be called once for each unique key in the partition, and the values will be the set of values that share that key. The output of the combiner will replace that set of original map outputs, ideally with fewer records or smaller records. This is suitable for jobs that are producing summary information from a large dataset.

Caution The combiner must not change the key values, as the map outputs are not re-sorted after the combiner runs. The reduce phase requires the map outputs to be sorted by key.

It is common for the same class that is used in the reduce task to be used for the combiner. However, this practice often leads to difficult-to-diagnose problems. The combiner must only aggregate values, in a manner that is suitable for processing by the actual reducer. The actual reducer has the larger job of producing the final job output. Problems occur when the reducer is modified to provide some change in the job output, and the person doing the modification is unaware that the reducer is also used as a combiner. It is very important that the combiner class not have side effects, and that the actual reducer be able to properly process the results of the combiner.

Tip It not always simple to build a correct combiner. If a job output has problems, try running the job without the combiner to see if the problem persists. If your actual `reduce()` method is nontrivial, do not also use it as a combiner; instead, write a separate object to combine the map outputs.

The classic example of using a combiner is the `org.apache.hadoop.examples.WordCount` example. This MapReduce job reads a set of text input files and counts the frequency of occurrence of each word in the input files. The map phase outputs each word in the file as a key, with the count of 1. There will be one output record for each word in the file. The combiner will aggregate these into a set that contains one output record per unique word in the input, and the value is the number of times the word appeared in the input. Unless the writer has such a large vocabulary that no word is used more than once, the combiner will greatly reduce the number of records to be processed by the reduce phase.

Listings 5-14, 5-15, and 5-16 show the `JobConf` setup and the `map()` and `reduce()` methods from the `WordCount.java` example, The default `InputFormat` is `TextInputFormat`, which returns a `LongWritable` key, the input line number, and a `Text` value, which is the full line from the input file. The `map()` method tokenizes the line and emits a record for each word of the input record, a `Text` and the value `1`, an `IntWritable`. The `reduce()` method simply sums the values and outputs the word as `Text` and the sum of values, an `IntWritable`. By using the `reduce()` method as a combiner, there is a large reduction in the size of each map task output.

Listing 5-14. *The JobConf Setup, from WordCount.java's run Method*

```
conf.setJobName("wordcount");

// the keys are words (strings)
conf.setOutputKeyClass(Text.class);
// the values are counts (ints)
conf.setOutputValueClass(IntWritable.class);

conf.setMapperClass(MapClass.class);
conf.setCombinerClass(Reduce.class);
conf.setReducerClass(Reduce.class);
```

Listing 5-15. *The Core of the map Method, from WordCount.java*

```
public void map(LongWritable key, Text value,
              OutputCollector<Text, IntWritable> output,
              Reporter reporter) throws IOException {
String line = value.toString();
StringTokenizer itr = new StringTokenizer(line);
while (itr.hasMoreTokens()) {
  word.set(itr.nextToken());
  output.collect(word, one);
}
```

Listing 5-16. *The Core of the reduce Method, from WordCount.java*

```
public void reduce(Text key, Iterator<IntWritable> values,
                 OutputCollector<Text, IntWritable> output,
                 Reporter reporter) throws IOException {
int sum = 0;
while (values.hasNext()) {
  sum += values.next().get();
}
output.collect(key, new IntWritable(sum));
```

When the map task has completed and the partitions are sorted, the combiner may run over the partitions and aggregate values, reducing the total number of key/value pairs that must go over the network to the reduce task.

For example, suppose the map partition dataset originally contained the following:

Key	Value
A	1
A	1
The	1
The	1
The	1
Xylophone	1

After the combiner has completed, the map partition dataset would contain these keys and values:

Key	Value
A	2
The	3
Xylophone	1

It is fairly simply to shoot yourself in the foot with a combiner. The combiner must not cause the loss of any information that is needed by the actual reducer. The classic example of this is a reducer that computes the average of the values for each key. If that reducer is also used as a combiner, the information on the number of records involved computing the average will be lost, and the reduce tasks will see only the average values for each key; the final result will be the average of the averages, instead of the actual average. Combiners also must be idempotent, as they may be run an arbitrary number of times by the Hadoop framework over a given map task's output.

File Types for MapReduce Jobs

The Hadoop framework supports text files, binary (sequence) files, and map files, which are actually a pair of sequence files. Let's take a closer look at each of these file types.

Text Files

The Hadoop framework supports a number of textual input files and output files. The input formats support transparent decompression of input files if an input file name ends in one of the recognized compression format suffixes (`.gz`, `.deflate`, `.lzo_deflate`, `.lzo`, and `.bz2`).

The following formats are available for text files:

`TextInputFormat`: This class reads each line of the input split and returns a record composed of the line number as a `LongWritable` key, and the line itself as a `Text` value. The workhorse class that actually produces the key/value pairs is `org.apache.hadoop.mapred.LineRecordReader`. There is only one tunable parameter: the configuration key, `mapred.linerecordreader.maxlength`, which sets the maximum number of characters allowed in a line. The default value is `Integer.MAX_VALUE`, essentially unlimited. The parameter may be adjusted using the `conf.setInt()` method. For example, `conf.setInt("mapred.linerecordreader.maxlength", 1024)` limits the line length to 1,204 characters.

`KeyValueTextInputFormat`: This class reads each line and splits the line into a key/value pair on a tab character. The workhorse class is `org.apache.hadoop.mapred.KeyValueLineRecordReader`. The separator may be configured by setting the configuration key `key.value.separator.in.input.line`. The key and value are both `Text`. If there is no separator found, the value will be an empty string.

`NLineInputFormat`: This format is ideal for using the input data as control information. It guarantees that each input split will be *N* lines long, with one split being the remaining lines. The configuration key `mapred.line.input.format.linespermap` controls the number of lines of input per map task. The default value is `1`. This may be changed using the `conf.setInt()` method. For example, `conf.setInt("mapred.line.input.format.linespermap", 10)` sets the value to `10`. Under the covers, this uses `org.apache.hadoop.mapred.LineRecordReader` to read the input data and produce `LongWritable`, `Text` key/value pairs.

`MultiFileInputFormat`: This is an abstract class that provides a way for a single task to receive multiple input files as the task's input split. This is commonly done for performance tuning. There is substantial time involved in setting up and starting a task, as well as collecting the results. If the input split is small, a substantial portion of the job runtime may be in the setup and teardown of tasks. The developer is responsible for implementing the `getRecordReader()` method. The `org.apache.hadoop.examples MultiFileWordCount` provides an example of a `RecordReader` that handles reading from multiple files.

`TextOutputFormat`: This is the standard textual output format. It basically calls the `toString` method on each key and value, producing a single-line key `SEPARATOR` value ASCII newline for each output record. The `SEPARATOR` is specified by the value of the configuration key `apred.textoutputformat.separator`, which defaults to `TAB`. If the value is `null`, no `SEPARATOR` and no value will be emitted. If key is `null`, `SEPARATOR value` is emitted. The end-of-record character is hard-coded as an ASCII newline character. Compression is supported if configured.

`MultipleTextOutputFormat`: This format allows you to write output records to different files based on the key and value. The test case `org.apache.hadoop.mapred.TestMultipleTextOutputFormat` provides a sample implementation. The Java source to this class is located in `src/test/org/apache/hadoop/mapred/TestMultipleTextOutputFormat.java` in your Hadoop distribution. Using `MultipleTextOutputFormat`, the user has the option of interceding in the selection of an output file for each output key/value pair in several different ways by overriding different methods.

- For map-only jobs, a portion of the input file path may be included in the output path, by setting the value of the configuration key `mapred.outputformat.numOfTrailingLegs`, to a positive integer. The default is no components of the input file path are used. The value +1 worth of components from the right side of the input file are inserted in the output file path before the file name. This happens after the call to `generateFileNameForKeyValue()`. The actual key and value parameters may be modified by overriding the `getActualKey()` and `getActualValue()` methods.
- You can change the final file name or leaf name via the `String generateLeafFileName(String name)` method. The parameter `name` is the original leaf name. The leaf name is normally the `part-`*XXXXX*, where the *XXXXX* corresponds to the reduce ordinal number, or the map ordinal number if this is a map-only job. (Changing the leaf name is not commonly done.)
- You can change the path to the output file via the `String generateFileNameForKeyValue(K key, V value, String name)` method. The `name` parameter is the result of `generateLeafFileName`. You can construct arbitrary paths out of the key, value, and name. This is the method commonly overridden by developers. The example in Listing 5-17 produces an output file name of the first letter of the key, a dash, and the partition number. If the key were `akey`, and the `name` were `part-00000`, this key/value pair would go to the file `a-part-00000`.

Listing 5-17. *Simple MultipleTextOutputFormat Output File Name Generator*

```
static class KeyBasedMultipleTextOutputFormat extends
     MultipleTextOutputFormat<Text, Text> {
  protected String generateFileNameForKeyValue(Text key, Text v, String name) {

    return key.toString().substring(0, 1) + "-" + name;
  }
}
```

Caution It is critically important to minimize the number of HDFS files that are opened. HDFS, through at least Hadoop 0.19.0, is designed for small numbers of very large files. Opening many small files will bring your cluster to its knees, and may result in catastrophic failure of your job, as well as your HDFS. It is very easy to open hundreds of thousands of files with `MultipleOutputFormats`.

Sequence Files

Sequence files are a binary format for storing sets of serialized key/value pairs. Sequence files support compression, encapsulate the key and value types, and provide validity checksums. They are an ideal format to use for data that is expensive or complex to parse.

The following formats are available for sequence files:

`SequenceFileInputFormat`: The basic workhorse, this format supports splitting and provides the key and value types. If the input file is a map file (described in the next section), the data file is read.

`SequenceFileAsBinaryInputFormat`: This format returns the raw key and value bytes. It returns `BytesWritable` keys and values.

`SequenceFileAsTextInputFormat`: This format returns the key and value as text. It calls the `toString` method on the key and value classes and returns the key/value pair as `Text,Text`.

`SequenceFileInputFilter`: This format returns only specific records from the sequence file. It provides the `static void setFilterClass(Configuration conf, Class filterClass)` method, which supplies a class that is used to determine which records are returned by the `next(key,value)` method on the reader. The `FilterClass` must implement the `SequenceFileInputFilter.Filter` interface and provide a method `boolean accept(Object key)`. Three filters are provided:

- `RegexFilter.setPattern(Configuration conf, String regex)` provides the regular expression to filter keys.
- `PercentFilter.setFrequency(Configuration conf, int frequency)` provides the way of accepting one record in `frequency` records.
- `MD5Filter.setFrequency(Configuration conf, int frequency)` provides a way of selecting only those records that have an MD5 hash that is evenly divisible by `frequency`.

`SequenceFileOutputFormat`: This format writes the serialized key/value records as output. This is the standard sequence file output. The key and value types must be specified via the `conf.setOutputKeyClass()` and `conf.setOutputValueClass()` methods.

`SequenceFileAsBinaryOutputFormat`: This format writes the raw bytes. The key and value types must be `BytesWritable`, and these raw bytes are written as the records.

Map Files

Map files are a pair of sorted sequence files. If a map file named `mymap` is created, there will be a directory `mymap` in HDFS, and two files in `mymap`: `index` and `data`. The `data` sequence file contains the key/value pairs as records, where the records are sorted in key order. The `index` sequence file is key/location information, where location is the location in `data` where the first record containing a key is located.

Map files provide a way to find a particular key, or region of a sorted file, without having to read the entire file. The HBase project (`http://hadoop.apache.org/hbase`) provides a persistent distributed hash table stored in HDFS, using map files as the underlying storage.

When a map file is specified as a job input, the `data` file is used as the actual input. There is not a `MapFileInputFormat` class; the `SequenceFileInputFormat` class is used. The path specified is the path to the directory containing the `index` and `data` files. `SequenceFileInputFormat` will use the `data` file as the input source.

Tip For best performance, it is strongly suggested that all key lookups be performed in the sort order of the underlying map file. HDFS is highly optimized for streaming files sequentially, and does a very poor job of providing low-latency access to random locations within a file.

For the `MapFileOutputFormat`, the value of the configuration key `io.map.index.interval` determines how many records are written to the `data` sequence file between writes to the `index` sequence file. The default is one index entry for every 128 records.

Map files provide the following methods for looking up key/value pairs.

- `void reset()`: Resets the read position to the beginning of the file.
- `WritableComparable midKey()`: Returns the key roughly in the middle of the file.
- `void finalKey(WritableComparable key)`: Reads the final key.
- `boolean seek(WritableComparable key)`: Seeks to the key, or to the first key after it, if it does not exist.
- `boolean next(WritableComparable key, Writable val)`: Reads the next key/value pair.
- `Writable get(WritableComparable key, Writable val)`: Gets the value for key.
- `WritableComparable getClosest(WritableComparable key, Writable val)`: Gets the closest match to the key, searching as `seek`.
- `WritableComparable getClosest(WritableComparable key, Writable val, final boolean before)`: Works like the previously described `getClosest()` method, unless `before` is `true`—in which case the key before is returned.

Compression

The Hadoop framework supports several types of compression and several compression formats. The framework supports the gzip, zip, sometimes LZO, and bzip2 compression codecs. Native libraries are supplied for Linux i386 and x86_64 for gzip, zip, and LZO for some releases. The framework will transparently compress and uncompress most input and output files. Input files are uncompressed when the input file name has a suffix that maps to one of the known codecs, as shown in Table 5-3.

■ **Note** LZO is licensed under the GPL. It is incompatible with the Apache license and has been removed from some distributions. I sincerely wish that this will be resolved and that native LZO becomes a standard part of the Hadoop distribution.

Table 5-3. *Compression Codecs and Mapped File Name Suffixes*

Codec	Suffix
GzipCodec	.gz
DefaultCodec	.deflate
LzoCodec	.lzo_deflate
LzopCodec	.lzo
Bzip2Codec	.bz2

Codec Specification

The Hadoop framework supports a number of codecs, with native implementations for a smaller number. `GzipCodec`, `LzoCodec`, and the `DefaultCodec` (zip) have native implementations. `Bzip2Codec` has a pure Java implementation. `LzoCodec` may not be available in some releases due to licensing issues. `Bzip2Codec` is available as of Hadoop 0.19.0.

The list of codecs is stored in the configuration under the key `io.compression.codecs`. In Hadoop 0.19, it has the following value:

```
org.apache.hadoop.io.compress.DefaultCodec,org.apache.hadoop.io.compress. ➥
GzipCodec,org.apache.hadoop.io.compress.BZip2Codec
```

If your environment requires additional codecs, the glue interface is `org.apache.hadoop.io.compress.CompressionCodec`. You would then add the class name to the list of codecs in the `io.compression.codecs` value. The selection of a compression codec is a choice between speed and compression rate. LZO is the fastest by far, and produces files about double the size of gzip. The bzip2 compression is the slowest—substantially slower than gzip—and produces files about one half the size of gzip.

Sequence File Compression

Sequence files are binary record-oriented files, where each record has a serialized key and serialized value. The Hadoop framework supports compressing and decompressing sequence files transparently.

Sequence files may be, and generally should be, compressed. The framework will transparently compress at the record level or the block level. The key `io.seqfile.compression.type` controls the record- or block-level compression for sequence files. A value of `BLOCK` requests block-level compression. A value of `RECORD`, the default, specifies record-level compression. A value of `NONE` disables compression.

In general, block-level compression is recommended, because it provides greater data reduction (at the expense of individual key access). The compression overhead is less, and the compression ratio is much greater. For sequence files that are being used as input to a map or reduce phase, block-level compression is ideal. Sequence files that were written using transparent compression may be divided into multiple input splits by the framework.

Many sites will set the default to `BLOCK` in their `hadoop-site.xml` file, as follows:

```
<property>
  <name> io.seqfile.compression.type</name>
  <value>BLOCK</value>
  <description>Force the default sequence file compression to
           be block compression for efficiency reasons
</value>
</property>
```

Map Task Output

The intermediate map task outputs are a set of sequence files, one per reduce task. As these files must be transferred across the network, a low-overhead compression type, such as gzip or LZO, can provide a substantial reduction in network traffic for little CPU cost. The blog entry at `http://blog.oskarsson.nu/2009/03/hadoop-feat-lzo-save-disk-space-and.html` has some interesting information about compression CPU and size reductions for different Hadoop codecs. Table 5-4 summarizes the compression speed results. For pretty decent compression `LzoCodec` provides high throughput.

■ **Note** I have spent some time running the same job with different compression codecs and `RECORD` or `BLOCK` set for compression, to determine which combination gave the overall performance for the job. At present, this must be done manually.

Table 5-4. *Compression Timings for Hadoop Compression Codecs*

Compressor	Original Size	Compressed Size	Compression Speed	Decompression Speed
bzip2	8.3GB	1.1GB	2.4MB/s	9.5MB/s
gzip	8.3GB	1.8GB	17.5MB/s	58MB/s
LZO—best	8.3GB	2GB	4MB/s	60.6MB/s
LZO	8.3GB	2.9GB	49.3MB/s	74.6MB/s

Map output block-level compression may be specified by the job or in the site configuration. If compressed, map output, destined for a reduce task, is always `BLOCK` compressed. Listing 5-18 provides an XML block suitable for inclusion in the `conf/hadoop-site.xml` file to make LZO compression the default for the map task outputs.

Listing 5-18. *A hadoop-site.xml Specification for Map Output Level Compression with LZO*

```
<property>
  <name>mapred.compress.map.output</name>
  <value>true</value>
  <description>Should the outputs of the maps be compressed before being
              sent across the network. Uses SequenceFile compression.
  </description>
</property>

<property>
  <name>mapred.map.output.compression.codec</name>
  <value>org.apache.hadoop.io.compress.LzoCodec</value>
  <description>If the map outputs are compressed, how should they be
              compressed? Use Lzo fast even though not as good compression.
  </description>
</property>
```

Listing 5-19 demonstrates configuring a cluster to always use compression for final output files, and if the final output file is a sequence file, to use `BLOCK` compression.

Listing 5-19. *A hadoop-site.xml Specification for Final Output Files to be Compressed with LZO, and If Sequence Files, BLOCK-Compressed*

```
<property>
  <name>mapred.output.compress</name>
  <value>true</value>
  <description>Should the job outputs be compressed?
  </description>
</property>
```

```
<property>
  <name>mapred.output.compression.type</name>
  <value>BLOCK</value>
  <description>The type of compression to use for final
   output sequence files. May be BLOCK, RECORD or None.
  </description>
</property>

<property>
  <name>mapred.output.compression.codec</name>
  <value>org.apache.hadoop.io.compress.LzoCodec</value>
  <description>If the job outputs are compressed, how should they be
              compressed? Use Lzo fast even though not as good compression.
  </description>
</property>
```

Listings 5-20 and 5-21 demonstrate specifying the compression codec and type via settings on the `JobConf` object.

Listing 5-20. *Setting Intermediate Map Output Compression via the JobConf*

```
conf. setCompressMapOutput(true);
conf. setMapOutputCompressorClass(LzoCodec.class);
```

Listing 5-21. *Setting Final Output Compression via the JobConf*

```
FileOutputFormat.setOutputCompress (conf, true);
FileOutputFormat.setOutputCompressorClass(LzoCodec.class);
SequenceFileOutputFormat.setOutputCompressionType(conf,CompressionType.BLOCK);
```

JAR, Zip, and Tar Files

The Hadoop framework knows how to unpack JAR, zip, and tar files, but this is only automatically done for archives passed via the `DistributedCache` object The class `org.apache.hadoop.fs.FileUtil` provides two static methods that may be used to unpack these files: `unTar()` for tar files and `unzip()` for zip files. The archives may be unpacked only onto the native file system, not into HDFS.

Summary

The Hadoop Core framework provides a rich set of tools to support a variety of use cases. As with most powerful tools, using them effectively requires training and experience. This chapter has provided a solid foundation for configuring jobs to run successfully and building classes that will actual perform the work for the job.

The effective use of counters in the map and reduce methods provides both the application writer and the organization with metrics for job performance. The DistributedCache object provides a way of distributing required data to all of the tasks, without needing to have the data already available on the TaskTracker nodes. You can choose from a variety of input and output formats. The use of compression can greatly reduce the wall clock runtime of a job, as can the use of a combiner. The KeyFieldBasedComparator and KeyFieldBasedPartitioner classes allow you to implement a secondary sort via the OutputValueGroupingComparator. Partitioning is a simple controllable process. You also know how to use MultipleTextOutputFormat, and the potential problems it can bring. It is now time to have fun writing MapReduce jobs!

CHAPTER 6

Tuning Your MapReduce Jobs

Once you have developed your MapReduce job, you need to be able to run it at scale on your cluster. A number of factors influence how your job scales. This chapter will cover how to recognize that your job is having a problem and how to tune the scaling parameters so that your job performs optimally.

First, we'll look at tunable items. The framework provides several parameters that let you tune how your job will run on the cluster. Most of these take effect at the job level, but a few work at the cluster level.

With large clusters of machines, it becomes important to have a simple monitoring framework that provides a visual indication of how the cluster is and has been performing. Having alerts delivered when a problem is developing or occurs is also essential. This chapter introduces several tools for monitoring Hadoop services.

Finally, you'll get some tips on what to do when your job isn't performing as it should. Your jobs may be failing or running slowly.

This chapter is focused on tuning jobs running on the cluster, rather than debugging the jobs themselves. Debugging is covered in the next chapter.

Tunable Items for Cluster and Jobs

Hadoop Core is designed for running jobs that have large input data sets and medium to large outputs, running on large sets of dissimilar machines. The framework has been heavily optimized for this use case.

Hadoop Core is optimized for clusters of heterogeneous machines that are not highly reliable. The HDFS file system is optimized for small numbers of very large files that are accessed sequentially. The optimal job is one that uses as input a dataset composed of a number of large input files, where each input file is at least 64MB in size and transforms this data via a MapReduce job into a small number of large files, again where each file is at least 64MB. The data stored in HDFS is generally not considered valuable or irreplaceable. The service level agreement (SLA) for jobs is long and can sustain recovery from machine failure.

Users commonly get into trouble when their jobs input large numbers of small files, output large numbers of small files, or require random access to files. Another problem is a need for rapid access to data or for rapid turnover of jobs.

HDFS installations get into trouble when large numbers of files are being created or exist on the DataNodes.

Hadoop Core does not provide high availability for HDFS or for job submission, and special care must be taken to ensure that required HDFS data can be recovered in the event of a critical failure of the NameNode.

Behind the Scenes: What the Framework Does

Each job has a number of steps in its execution: the setup, the map, the shuffle/sort, and the reduce. The framework sets up, manages, and tears down each step.

Note The following discussion assumes that no other job is running on the cluster and that on submission, the job is immediately started.

On Job Submission

The framework will first store any resources that must be distributed in HDFS. These are the resources provided via the `-files`, `-archives`, and `-libjars` command-line arguments, as well as the JAR file indicated as the job JAR file. This step is executed on the local machine sequentially. If there are a large number of resources, this may take some wall clock time. The XML version of the `JobConf` data is also stored in HDFS.

The replication factor on these resource items is set to the value stored in the configuration under the key `mapred.submit.replication`, with a default value of `10`. The framework will then examine the input data set, using the `InputFormat` class to determine which input files must be passed whole to a task and which input files may be split across multiple tasks.

The framework will use the parameters listed in Table 6-1 to determine how many map tasks must be executed. Input formats may override this; for instance, `NLineInputFormat` forces the splits to be made by line count.

Table 6-1. *Parameters Controlling the Number of Map Tasks for a Job*

Getter	Parameter	Description	Default
`JobConf.getNumMapTasks()`	`mapred.map.tasks`	The suggested number of map tasks for the job	1
No getter	`mapred.min.split.size`	The minimum size of a split	1
`FileInputFormat.getMinSplitSize()`		The minimum size to use for this input format (a protected method, currently used only by `SequenceFileInputFormat`)	`SequenceFileInput.SYNC_INTERVAL` or 1
`Path.getBlockSize()`	`dfs.block.size`	The file system block size, in bytes of the input file	67108864
`InputFormat.isSplitable()`	Not configurable	Whether this file may be split	Varies

The parameters in Table 6-1 are used to compute the actual split size for each input file. The input format for the input file is responsible for indicating if the underlying file may be split. The public method `FileInputFormat.getSplits()` returns the list of splits for the input files. For inputs that can be split, three things are computed before the actual split size is determined: the goal size, which is the total input size divided by `JobConf.getNumMapTask()`; the minimum split size, `Math.max(JobConf.getInt("mapred.min.split.size",1), FileInputFormat.getMinSplitSize())`; and the block size for the input file, `Path.getBlockSize()`. The protected method `FileInputFormat.computeSplitSize(goalSize, minSize,blockSize)` is called to produce the actual split size, and the calculation is `Math.max(minSize, Math.min(goalSize, blockSize))`. In summary, splits are determined as follows::

- If a file may not be split, `InputFormat.isSplitable()`, it will be queued as input to one map task.
- A split will be no smaller than the remaining data in the file or `minSize`.
- A split will be no larger than the lesser of the `goalSize` and the `blockSize`.

■ **Tip** Through at least Hadoop 0.19.1, compressed files may not be split. A number of patches enable splitting for various compression formats: bzip2 (`http://issues.apache.org/jira/browse/HADOOP-4012`), LZO (`http://issues.apache.org/jira/browse/HADOOP-4640`), and gzip (`http://issues.apache.org/jira/browse/HADOOP-4652`).

In general, a cluster will have the `mapred.map.tasks` parameter set to a value that approximates the number of map task slots available in the cluster or some multiple of that value. The ideal split size is one file system block size, as this allows the framework to attempt to provide data locally for the task that processes the split.

The end result of this process is a set of input splits that are each tagged with information about which machines have local copies of the split data. The splits are sorted in size order so that the largest splits are executed first. The split information and the job configuration information are passed to the JobTracker for execution via a job information file that is written to HDFS.

Some jobs require that the input files not be split. The simplest way to achieve this is to set the value of the configuration parameter `mapred.min.split.size` to `Long.MAX_VALUE`: `JobConf.setInt("mapred.min.split.size", Long.MAX_VALUE);`.

Map Task Submission and Execution

The JobTracker has a set of map task execution slots, *N* per machine. Each input split is sent to a task execution slot for execution. Sending tasks to a slot that is hosted on the machine that has a local copy of the input split data minimizes network I/O.

If there are spare execution slots, and map speculative execution is enabled, multiple instances of a map task may be scheduled. In this case, the results of the first map task to complete will be used, the other instances killed, and the output, including the counter values, removed.

When map speculative execution is not enabled, only one instance of a map task will be run at a time. The TaskTracker on the machine will receive the task information, and if necessary, unpack all of the `DistributedCache` data into the task local directory and localize the paths to that data in the `JobConf` object that is being constructed for the task. With speculative execution for map tasks disabled, the only time more than one instance of a map task will occur in the job will be if the task is retried after failing.

Caution The framework is able to back out only counter values and output files written to the task output directory. Any other side effects of killed speculative execution tasks or failed tasks must be handled by the application.

The TaskTracker picks a map runner class based on the content of the key `mapred.map.runner.class`. Its choices are the standard `MapRunner`, which runs a single thread; the `MultithreadedMapRunner`, which runs `mapred.map.multithreadedrunner.threads` (the default is ten threads); or the chain mapper.

A child JVM is allocated to run the mapper class, and the map task is started. The output data of the map task is partitioned and sorted by the output partitioner class and the output comparator class, and aggregated by the combiner class, if one is present. The result of this will be *N* sequence files on disk: one for each reduce task, or one file if there is no reduce task.

Each time the map method is called, an output record is emitted, or the reporter object is interacted with, a heartbeat timer is reset. The heartbeat timeout is stored in the configuration under the key `mapred.tasktracker.expiry.interval`, and has a default value of 600,000 milliseconds (msec), or 10 minutes. If this timeout expires, the map task is considered hung and terminated.

If a terminated task has not failed more than the allowed number of times, it is rescheduled to a different task execution slot. A failing task may have a debugging script invoked on it if the value of the configuration key `mapred.map.task.debug.script` is the path to an executable program. The script is invoked with the additional arguments of the paths to the stdout, stderr, and syslog output files for the task. See this book's appendix, which covers the `JobConf` object, for details on how to configure a debugging script for failing tasks.

When a task finishes, the output commit class is launched on the task output directory, to decide which files are to be discarded and which files are to be committed for the next step. The class name is stored in the configuration under the key `mapred.output.committer.class` and has the default class `FileOutputCommitter`.

If less than the required number of tasks succeed, the job is failed and the intermediate output is deleted. The TaskTracker will inform the JobTracker of the task's success and output locations.

Merge-Sorting

The JobTracker will queue the number of reduce tasks as specified by the `JobConf.setNumReduceTasks()` method and stored in the configuration under the key `mapred.reduce.tasks`. The JobTracker will queue these reduce tasks for execution among the available reduce slots.

The TaskTracker that receives a reduce task will set up the local task execution environment if needed, and then fetch each of the map outputs that are destined for this reduce task. HTTP is the protocol used to transfer the map outputs. These map outputs are merge-sorted. The number of pieces that are fetched at one time is configurable. The value stored in the configuration under the key `mapred.reduce.parallel.copies` determines how many fetches are done in parallel. The default is five fetches.

A number of parameters control how the merge-sorting is done, as shown in Table 6-2.

Table 6-2. *Merge-Sort Parameters*

Parameter	Description	Default
`io.sort.factor`	The number of map output partitions to merge at a time.	10
`io.sort.mb`	The amount of buffer space in megabytes to use when sorting streams. This parameter often causes jobs to run out of memory on small memory machines.	100
`io.sort.record.percent`	The amount of the sort buffer dedicated for collecting records. Actual buffer space is this value * `io.sort.mb` / 4.	0.05
`io.sort.spill.percent`	The amount of the sort buffer or collection buffer that may be used before the data is spilled to disk.	0.80
`io.file.buffer.size`	The buffer size for I/O operations on the disk files.	4096
`io.bytes.per.checksum`	The amount of data per checksum.	512
`io.skip.checksum.errors`	If `true`, a block with a checksum failure may be skipped.	`false`

The Reduce Phase

Once the data is sorted, the reduce method may be called with the key/value groups. The reduce output is written to the local file system. On successful completion, the output commit class is called to select which output files are staged to the output area in HDFS.

If more than the allowed number of reduce tasks fail, the job is failed. Once the reduce tasks have finished, the job is done.

Writing to HDFS

There are two cases for an HDFS write: the write originates on a machine that hosts a DataNode of the HDFS cluster for which the write is destined, or the write originates on a machine that does not host a DataNode of the cluster. In both cases, the framework buffers a file system block-size worth of data in memory, and when the file is closed or the block fills, an HDFS write is issued.

The write process requests a set of DataNodes that will be used to store the block. If the local host is a DataNode in the file system, the local host will be the first DataNode in the returned set. The set will contain as many DataNodes as the replication factor requires, up to the number of DataNodes in the cluster. The replication factor may be set via the configuration key `dfs.replication`, which defaults to a factor of three, and should never be less than three. The replication for a particular file may be set by the following:

```
FileSystem.setReplication(Path path, int replication);
```

The block is written to the first DataNode in the list, the local host if possible, with the list of DataNodes that are to be used. On receipt of a block, each DataNode is responsible for initiating the transfer of the block to the next DataNode in the list. This allows writes to HDFS on a machine that hosts a DataNode to be very fast for the application, as they do not require bulk network traffic.

Cluster-Level Tunable Parameters

The cluster-level tunable parameters require a cluster restart to take effect. Some of them may require a restart of the HDFS portion of the cluster; others may require a restart of the MapReduce portion of the cluster. These parameters take effect only when the relevant server starts.

Server-Level Parameters

The server-level parameters, shown in Table 6-3, affect basic behavior of the servers. In general, these affect the number of worker threads, which may improve general responsiveness of the servers with an increase in CPU and memory use.

The variables are generally configured by setting the values in the `conf/hadoop-site.xml` file. It is possible to set them via command-line options for the servers, either in the `conf/hadoop-env.sh` file or by setting environment variables (as is done in `conf/hadoop-env.sh`).

The `nofile` parameter is not a Hadoop configuration parameter. It is an operating system parameter. For users of the bash shell, it may be set or examined via the command `ulimit -n [value to set]`. Quite often, the operating system-imposed limit is too low, and the administrator must increase that value. The value `64000` is considered a safe minimum for medium-size busy clusters.

■ **Caution** A number of difficult-to-diagnose failures happen when an application or server is unable to allocate additional file descriptors. Java application writers are notorious for not closing I/O channels, resulting in massive consumption of file descriptors by the map and reduce tasks..

Table 6-3. *Server-Level Tunable Parameters*

Parameter	Description	Default
`dfs.datanode.handler.count`	The number of threads servicing DataNode block requests	3
`dfs.namenode.handler.count`	The number of threads servicing NameNode requests	10
`tasktracker.http.threads`	The number of threads for servicing map output files to reduce tasks	40
`ipc.server.listen.queue.size`	The number of network incoming connections that may queue for a server	128
`nofile`	The limit on the number of file descriptors a process can open (alter `/etc/security/limits.con` for Linux machines)	1024

Caution Hadoop Core uses large numbers of file descriptors in each server. Rarely is the system default of 1,024 sufficient for the Hadoop servers or Hadoop jobs. Most installations find that a minimum limit of 64,000 is required. If you see errors in your log files that say `Bad connect ack with firstBadLink`, `Could not obtain block`, or `No live nodes contain current block`, you must increase the file descriptor limit for your Hadoop servers and jobs. How to change the limit is covered in Chapter 4, in the "File Descriptors" section.

HDFS Tunable Parameters

The most commonly tuned parameter for HDFS is the file system block size. The default block size is 64MB, specified as `67108864` bytes in `dfs.block.size`. The larger this value, the fewer individual blocks will be stored on the DataNodes, and the larger the input splits will be.

The DataNodes through at least Hadoop 0.19.0 have a limit to the number of blocks that can be stored. This limit appears to be roughly 500,000 blocks. After this size, the DataNode will start to drop in and out of the cluster. If enough DataNodes are having this problem, the HDFS performance will tend toward full stop.

When computing the number of tasks for a job, a task is created per input split, and input splits are created one per block of each input file by default. There is a maximum rate at which the JobTracker can start tasks, at least through Hadoop 0.19.0. The more tasks to execute, the longer it will take the JobTracker to schedule them, and the longer it will take the TaskTrackers to set up and tear down the tasks.

The other reason for increasing the block size is that on modern machines, an I/O-bound task will read 64MB of data in a small number of seconds, resulting in the ratio of task overhead to task runtime being very large. A downside to increasing this value is that it sets the minimum amount of I/O that must be done to access a single record. If your access patterns are not linearly reading large chunks of data from the file, having a large block size will greatly increase the disk and network loading required to service your I/O.

The DataNode and NameNode parameters are presented in Table 6-4.

Table 6-4. *HDFS Tunable Parameters*

Parameter	Description	Default
`fs.default.name`	The URI of the shared file system. This should be `hdfs://NameNodeHostName:PORT`.	`file:///`
`fs.trash.interval`	The interval between trash checkpoints. If `0`, the trash feature is disabled. The trash is used only for deletions done via the `hadoop dfs -rm` series of commands.	`0`
`dfs.hosts`	The full path to a file containing the list of hostnames that are allowed to connect to the NameNode. If specified, only the hosts in this file are permitted to connect to the NameNode.	

Continued

Table 6-4. *Continued*

Parameter	Description	Default
`dfs.hosts.exclude`	A path to a file containing a list of hosts to blacklist from the NameNode. If the file does not exist, no hosts are blacklisted. If a set of DataNode hostnames are added to this file while the NameNode is running, and the command `hadoop dfsadmin -refreshNodes` is executed, the DataNodes listed will be decommissioned. Any blocks stored on them will be redistributed to other nodes on the cluster such that the default replication for the blocks is satisfied. It is best to have this point to an empty file that exists, so that DataNodes may be decommissioned as needed.	
`dfs.namenode.decommission.interval`	The interval in seconds that the NameNode checks to see if a DataNode decommission has finished.	`300`
`dfs.replication.interval`	The period in seconds that the NameNode computes the list of blocks needing replication.	`3`
`dfs.access.time.precision`	The precision in msec that access times are maintained. If this value is `0`, no access times are maintained. Setting this to `0` may increase performance on busy clusters where the bottleneck is the NameNode edit log write speed.	`3600000`
`dfs.max.objects`	The maximum number of files, directories, and blocks permitted.	`0`
`dfs.replication`	The number of replicas of each block stored in the cluster. Larger values allow more DataNodes to fail before blocks are unavailable but increase the amount of network I/O required to store data and the disk space requirements. Large values also increase the likelihood that a map task will have a local replica of the input split.	`3`
`dfs.block.size`	The basic block size for the file system. This may be too small or too large for your cluster, depending on your job data access patterns.	`67108864`
`dfs.datanode.handler.count`	The number of threads handling block requests. Increasing this may increase DataNode throughput, particularly if the DataNode uses multiple separate physical devices for block storage.	`3`
`dfs.replication.considerLoad`	Consider the DataNode loading when picking replication locations.	`true`
`dfs.datanode.du.reserved`	The amount of space that must be kept free in each location used for block storage.	`0.0`
`dfs.permissions`	Permission checking is enabled for file access.	`true`
`dfs.df.interval`	The interval between disk usage statistic collection in msec.	`60000`

Parameter	Description	Default
`dfs.blockreport.intervalMsec`	The amount of time between block reports. The block report does a scan of every block that is stored on the DataNode and reports this information to the NameNode. This report as of Hadoop 0.19.0 blocks the DataNode from servicing block reports and is the cause of the congestion collapse of HDFS when more than 500,000 blocks are stored on a DataNode.	`3600000`
`dfs.heartbeat.interval`	The heartbeat interval with the NameNode.	`3`
`dfs.namenode.handler.count`	The number of server threads for the NameNode. This is commonly greatly increased in busy and large clusters.	`10`
`dfs.name.dir`	The location where the NameNode metadata storage is kept. This may be a comma-separated list of directories. A copy will be kept in each location. Writes to the locations are synchronous. If this data is lost, your entire HDFS data set is lost. Keep multiple copies on multiple machines.	`${hadoop.tmp.dir}/dfs/name`, in `/tmp` by default
`dfs.name.edits.dir`	The location where metadata edits are synchronously written. This may be a comma-separated list of directories. Ideally, this should hold multiple locations on separate physical devices. If this is lost, your last few minutes of changes will be lost.	`${dfs.name.dir}`
`dfs.data.dir`	The comma-separated list of directories to use for block storage. This list will be used in a round-robin fashion for storing new data blocks. The locations should be on separate physical devices. Using multiple physical devices yields roughly 50% better performance than RAID 0 striping.	`${hadoop.tmp.dir}/dfs/data`
`dfs.safemode.threshold.pct`	The percentage of blocks that must be minimally replicated before the HDFS will start accepting write requests. This condition is examined only on HDFS startup.	`0.999f`
`dfs.balance.bandwidthPerSec`	The amount of bandwidth that may be used to rebalance block storage among DataNodes. This value is in bytes per second.	`1048576`

JobTracker and TaskTracker Tunable Parameters

The JobTracker is the server that handles the management of the queued and executing jobs. The TaskTrackers are the servers that actually execute the individual map and reduce tasks. Table 6-5 shows the tunable parameters for the JobTracker, and Table 6-6 shows those for TaskTrackers. The JobTracker parameters are global to the cluster. The TaskTracker parameters are for the individual TaskTrackers.

Table 6-5. *JobTracker Tunable Parameters*

Parameter	Description	Default
`mapred.job.tracker`	The host and port of the JobTracker server. A value of `local` means to run the job in the current JVM with no more than 1 reduce. If the configuration specifies local, no JobTracker server will be started. Per-job configurable.	`local`
`mapred.max.tracker.failures`	The number of task failures allowed on a TaskTracker before the TaskTracker is considered failed for the job with the failing tasks. Per-job configurable.	4
`mapred.system.dir`	An HDFS path used for storing job data. If multiple JobTracker servers will share an HDFS cluster, each must have a different `mapred.system.dir`, or the JobTrackers will delete each other's job files.	`${hadoop.tmp.dir}/mapred/system`
`mapred.temp.dir`	An HDFS path used for storing shared temporary data such as `DistributedCache` data. Per-cluster configurable.	`${hadoop.tmp.dir}/mapred/temp`
`mapred.job.tracker.handler.count`	The number of server threads for handling TaskTracker requests. The recommended value is 4% of the TaskTracker nodes. Per-cluster configurable.	10
`mapred.jobtracker.restart.recover`	If this value is `true`, a JobTracker will attempt to restart any queued or running jobs that were running before a crash/shutdown. Per-cluster configurable.	`false`
`mapred.jobtracker.job.history.block.size`	The basic block size used for writes to the history file. Keeping this relatively small ensures that the most data is persisted in the event of a crash. Per-cluster configurable.	3145728
`mapred.jobtracker.completeuserjobs.maximum`	The number of jobs to be kept in the JobTracker history. Per-cluster configurable.	100
`mapred.jobtracker.maxtasks.per.job`	The maximum number of tasks allowed for a single job. A value of `-1` means no limit. Per-cluster configurable.	`-1`
`mapred.jobtracker.taskScheduler.maxRunningTasksPerJob`	The maximum number of tasks a job can run before it may be preempted. Per-cluster configurable when the Capacity Scheduler services (discussed in Chapter 8) are enabled.	Unlimited
`mapred.job.tracker.persist.jobstatus.active`	Determines whether job status results are persisted to HDFS. Per-cluster configurable.	`false`
`mapred.job.tracker.persist.jobstatus.hours`	The number of hours that job status information is kept. Per cluster.	0
`mapred.job.tracker.persist.jobstatus.dir`	The directory where status information is kept. Per-cluster configurable.	`/jobtracker/jobsInfo`
`mapred.hosts`	The full path to a file of hostnames that are permitted to talk to the JobTracker. If specified, only the hosts in this file are permitted.	
`mapred.hosts.exclude`	The full path to a file of hostnames that are blacklisted from talking to the JobTracker.	

Table 6-6. *TaskTracker Tunable Parameters*

Parameter	Description	Default
mapred.local.dir	The set of directories to use for task local storage. If multiple directories are provided, the usage is spread over the multiple directories. The directories should be on separate physical devices. Per-TaskTracker configurable.	${hadoop.tmp.dir}/mapred/local
local.cache.size	The local cache directory limit. If more than this many bytes of data are in the task local DistributedCache directory, there will be an attempt to remove unreferenced files. Per-TaskTracker configurable.	10737418240 (10GB)
mapred.local.dir.minspacestart	If the space available in the directories specified by mapred.local.dir falls below this value, do not accept more tasks. This prevents tasks from failing due to lack of temp space. The 0 value should be changed to something reasonable for your jobs. Per-TaskTracker configurable.	0
mapred.local.dir.minspacekill	If the available space in the mapred.local.dir set of directories is below this, accept no more tasks (as if mapred.local.dir.minspace were set to this value) and start killing tasks, starting with reduce tasks, until there is this much space free. Per-TaskTracker configurable.	0
mapred.tasktracker.expiry.interval	The number of msec without a heartbeat that a TaskTracker may go without reporting, before being considered hung and being killed. Per-TaskTracker configurable.	600000
mapred.child.ulimit	Only valid on Unix machines. This is used for processes started by the org.apache.util.hadoop.Shell class. The framework uses this to launch external subprocesses, such as the pipes jobs and the external programs of streaming jobs. Per-TaskTracker configurable.	Unlimited
mapred.tasktracker.taskmemorymanager.monitoring-interval	The rate in msec that virtual memory use by tasks is monitored.	5000
mapred.tasktracker.tasks.maxmemory	The maximum amount of virtual memory a task and its children may use before the TaskTracker will kill the task. A value of t indicates no limit. Per-TaskTracker configurable.	-1
mapred.tasktracker.procfsbasedprocesstree.sleeptime-before-sigkill	A task over its memory limit is sent a SIGTERM. If the task has not exited within this time in msec, a SIGKILL is sent.	5000

Continued

Table 6-6. *Continued*

Parameter	Description	Default
`mapred.map.tasks.maximum`	The number of map tasks to run simultaneously on a TaskTracker. This should either be `1` (if there is only one CPU) or roughly one less than the number of CPUs on the machine. This parameter needs to be tuned for a particular job mix. Per-TaskTracker configurable.	2
`mapred.reduce.tasks.maximum`	The number of simultaneous reduce tasks to run. This value is really a function of the CPU and I/O bandwidth available to the machine. It needs to be tuned for the machines and job mix. Per-TaskTracker configurable.	2
`mapred.tasktracker.dns.interface`	For multihomed TaskTracker nodes, report this interface's IP address to the JobTracker. If not `default`, this value is the name of a network interface, such as `eth0`. Per-TaskTracker configurable.	`default`
`mapred.tasktracker.dns.nameserver`	For multihomed TaskTracker nodes, use this address for DNS hostname resolution when resolving the IP address of the network interface specified by `mapred.tasktracker.dns.interface`. The value `default` means use the system default.	`default`
`tasktracker.http.threads`	The number of threads serving HTTP requests for reduce tasks requesting map output. If your system has many reduce execution slots, the default may be too small.	40
`mapred.userlog.limit.kb`	The maximum amount of data that may be written to a task user log.	0
`mapred.userlog.retain.hours`	The number of hours that user logs are retained.	24

Per-Job Tunable Parameters

The framework provides rich control over the way individual jobs are executed on the cluster. You can tune file system and task-related parameters. Table 6-7 shows the tunable parameters for the file system.

Table 6-7. *File System Tunable Parameters*

Parameter	Description	Default
`fs.default.name`	This is the URI for the shared file system. Normally it will be set to `hdfs://NamenodeHostname:NameNodePort`.	`file:///`
`dfs.replication`	The job may configure this value.	3
`dfs.block.size`	The client may also configure this value.	67108864
`dfs.client.block.write.retries`	The number of write attempts before a write is considered failed. In general, if writes are being retried, there is a problem with the HDFS or machine configuration.	3

The task-tunable parameters directly control the behavior of tasks in the cluster. These are the heart of the MapReduce framework. A large number of parameters affect the job. Only those parameters that directly control core functions are listed in Table 6-8. Many of the parameters are detailed in this book's appendix, which discusses the `JobConf` object.

Table 6-8. *Core Job-Level Task Parameters*

Parameter	Description	Default
`mapred.map.tasks`	The suggested number of map tasks for a job.	2
`mapred.reduce.tasks`	The number of reduce tasks for the job.	1
`mapred.map.max.attempts`	The maximum number times a map task will be retried after an error, before it is considered failed.	4
`mapred.reduce.max.attempts`	The maximum number of times a reduce task will be retried after an error, before it is considered failed.	4
`mapred.reduce.parallel.copies`	The number of parallel fetches of map output data made via HTTP at a time.	5
`mapred.reduce.copy.backoff`	The maximum amount of time to try to fetch a map output partition, before abandoning that partition.	300
`mapred.task.timeout`	The amount of time in msec that a task may go without the map or reduce method finishing, or making a call on the reporter or output collector.	600000 (10 min)
`mapred.child.java.opts`	The options to use for initializing the task JVM. `@taskid@` is replaced with the current task ID.	`-Xmx200m`
`mapred.child.tmp`	The value passed to the JVM for `java.io.tmpdir`. If it is a relative path, it will be relative to the task's local working directory.	`/tmp`

Continued

Table 6-8. *Continued*

Parameter	Description	Default
mapred.map.tasks.speculative.execution	Whether idle map task slots will be used to set up execution races for executing identical map tasks. This will consume more cluster resources and may offer faster job throughput. This must be false if your map tasks have side effects that the framework cannot undo or have real costs.	true
mapred.reduce.tasks.speculative.execution	Enable the use of unused reduce task execution slots to try a task in multiple slots, to see if one slot may complete the task faster. This will consume more cluster resources and may offer faster job throughput. This *must* be false if your reduce tasks have side effects the framework cannot undo or have real costs.	true
mapred.job.reuse.jvm.num.tasks	The number of times a task JVM may be reused for additional tasks of the same type for the same job. A value of -1 indicates no limit.	1
mapred.submit.replication	The replication factor for per-job data. This needs to be tuned on a per-job basis.	10
keep.failed.task.files	Whether the local directories for failed tasks should be kept. This is for debugging. There is no automatic mechanism in the framework to clean these directories if this is set to false.	false
keep.task.files.pattern	If set, a java.util.Pattern will be applied to task names to determine if their local directories will be kept. This is normally not present.	Unset
mapred.output.compress	Use compression on the final output data files for the job. This is usually a significant win for jobs with large output.	false
mapred.output.compression.type	The type of compression to do for the job output files if they are SequenceFiles. BLOCK is generally considered better if random access to the output is not desired.	RECORD
mapred.output.compression.codec	The codec to use for compression.	org.apache.hadoop.io.compress.DefaultCodec
mapred.compress.map.output	If true, use compression on the map output that is destined for a reduce task. This is usually a significant win.	false

Parameter	Description	Default
mapred.map.output.compression.codec	The codec to use for intermediate map output files. The LzoCodec appears to be the current best choice if it is available.	org.apache.hadoop.io.compress.DefaultCodec
io.seqfile.compress.blocksize	The minimum block size to use for block-level compression of SequenceFiles.	1000000
io.seqfile.lazydecompress	Only decompress SequenceFile data when it is needed.	true
io.seqfile.sorter.recordlimit	The maximum number of records to attempt to keep in memory when sorting the records of a SequenceFile.	1000000
map.sort.class	The sort implementation to use when sorting keys using the OutputComparator.	org.apache.hadoop.util.QuickSort
jobclient.output.filter	The status of the tasks whose user log data is reported to the console of the JobClient that submitted the job. The values allowed are NONE, KILLED, FAILED, SUCCEEDED, and ALL.	FAILED
mapred.task.profile	If true, some tasks may be profiled.	false
mapred.task.profile.maps	The set of map tasks to profile. See this book's appendix for how this may be set.	0-2
mapred.task.profile.reduces	The set of reduce tasks to profile. See this book's appendix for how this may be set.	0-2
mapred.skip.attempts.to.start.skipping	The number of failures of a task before skip mode is engaged. This is covered in Chapter 8.	2
mapred.skip.map.auto.incr.proc.count	Automatically increment the counter ReduceProcessedGroups. This must be false for streaming jobs or jobs that buffer records before reducing.	true
mapred.skip.out.dir	If unset, skipped records are written to file _logs/skip in the output directory. If the value is exactly none, no records will be written. If set to anything else, it becomes the directory where skipped records are written.	Unset
mapred.skip.map.max.skip.records	The number of contiguous records, including the bad record that may be skipped. The framework will attempt to narrow down the region to skip to this size. If the value is 0, no skipping is allowed. If the value is Long.MAX_VALUE, the entire split will be skipped.	0
mapred.skip.reduce.max.skip.groups	The number of key/value set groups surrounding a bad record group that may be skipped by the reduce task. See Chapter 8 for details.	0

Monitoring Hadoop Core Services

To be able to detect incipient failures, or otherwise recognize that a problem is developing or has occurred, some mechanism must be available to monitor current status, and if possible provide historical status. The Hadoop framework provides several APIs for allowing external agents to provide monitoring services to the Hadoop Core services. Here, we will look at Java Management Extensions (JMX), Nagios, Ganglia, Chukwa, and FailMon.

JMX: Hadoop Core Server and Task State Monitor

Hadoop provides local JMX bean services for all services. This allows for the use of JMX-aware applications to collect information about the state of the servers. The default configuration provides for only local access to the managed beans (MBeans). To enable remote access, after determining a port for JMX use, alter the `conf/hadoop-env.sh` file (shown in Listing 6-1) and change the JMX properties being set on the servers.

Listing 6-1. *The Default hadoop-env.sh Settings for Hadoop Servers to Enable JMX*

```
export HADOOP_NAMENODE_OPTS="-Dcom.sun.management.jmxremote $HADOOP_NAMENODE_OPTS"
export HADOOP_SECONDARYNAMENODE_OPTS="-Dcom.sun.management.jmxremote ➥
$HADOOP_SECONDARYNAMENODE_OPTS"
export HADOOP_DATANODE_OPTS="-Dcom.sun.management.jmxremote $HADOOP_DATANODE_OPTS"
export HADOOP_BALANCER_OPTS="-Dcom.sun.management.jmxremote $HADOOP_BALANCER_OPTS"
export HADOOP_JOBTRACKER_OPTS="-Dcom.sun.management.jmxremote ➥
$HADOOP_JOBTRACKER_OPTS"
# export HADOOP_TASKTRACKER_OPTS=
```

The string `-Dcom.sun.management.jmxremote` enables the JMX management bean services in the servers. The string is a JVM argument and passed to the JVM at start time on the command line.

JMX supports several connection options. See the Sun-supplied documentation for configuring access control and remote access, at `http://java.sun.com/javase/6/docs/technotes/guides/jmx/index.html`.

Nagios: A Monitoring and Alert Generation Framework

Nagios (`http://www.nagios.org`) provides a flexible customizable framework for collecting data about the state of a complex system and triggering various levels of alerts based on the collected data. A service of this type is essential for your cluster administration and operations team.

The University of Nebraska has a web page (`http://t2.unl.edu/documentation/hadoop/monitoring-guide/`) that details how to use the Nagios `check_jmx` plug-in to monitor Hadoop servers. The information is reproduced here. This example assumes that you understand how to construct the JMX password file and access control file.

To enable JMX monitoring on Hadoop, add the following lines to hadoop-env.sh:

```
export HADOOP_NAMENODE_OPTS=" -Dcom.sun.management.jmxremote.authenticate=false ➥
    -Dcom.sun.management.jmxremote.ssl=false ➥
    -Dcom.sun.management.jmxremote.port=8004 ➥
    -Dcom.sun.management.jmxremote.password.file= ➥
    $HADOOP_HOME/conf/jmxremote.password ➥
    -Dcom.sun.management.jmxremote.access.file=$HADOOP_HOME/conf/jmxremote.access"
export HADOOP_DATANODE_OPTS=" -Dcom.sun.management.jmxremote.authenticate=false ➥
    -Dcom.sun.management.jmxremote.ssl=false ➥
    -Dcom.sun.management.jmxremote.port=8004 ➥
    -Dcom.sun.management.jmxremote.password.file= ➥
    $HADOOP_HOME/conf/jmxremote.password ➥
    -Dcom.sun.management.jmxremote.access.file=$HADOOP_HOME/conf/jmxremote.access"
```

The following lines add check_jmx to the Nagios deployment:

```
./check_jmx -U service:jmx:rmi:///jndi/rmi://node182:8004/jmxrmi ➥
     -O hadoop.dfs:service=DataNode,name=DataNodeStatistics ➥
     -A BlockReportsMaxTime -w 10 -c 150
./check_jmx -U service:jmx:rmi:///jndi/rmi://node182:8004/jmxrmi ➥
    -O java.lang:type=Memory -A HeapMemoryUsage -K used -C 10000000
```

Ganglia: A Visual Monitoring Tool with History

Hadoop has built-in support for Ganglia version 3.0 through Hadoop 0.19.0. Support for Ganglia 3.1 is expected for Hadoop 0.20. The Ganglia framework is available from http://ganglia.sourceforge.net.

Ganglia by itself is a highly scalable cluster monitoring tool, and provides visual information on the state of individual machines in a cluster or summary information for a cluster or sets of clusters. Ganglia provides the ability to view different time windows into the past, normally one hour, one day, one week, one month, and so on.

Caution Due to some limitations in the Ganglia support in Hadoop through at least Hadoop 0.19.1, the configuration requirements are not as simple as Ganglia configuration normally is.

Ganglia is composed of two servers: the gmetad server, which provides historical data and collects current data, and the gmond server, which collects and serves current statistics. The Ganglia web interface is generally installed on the host(s) running the gmetad servers, and in coordination with the host's httpd provides a graphical view of the cluster information. In general, each node will run gmond, but only one or a small number of nodes will also run gmetad.

For Hadoop reporting to work with Ganglia, the configuration changes shown in Table 6-9 must be made in the conf/hadoop-metrics.properties file. Each Hadoop cluster must be allocated a unique multicast address/port, and be considered a single reporting domain. Each cluster must also be allocated a unique cluster name. The cluster name is referred to as CLUSTER in this section. The UDP port for reporting is referred to as PORT, and for simplicity, the multicast port will be identical to PORT.

Table 6-9. *Required Parameters for Hadoop Ganglia Reporting Configuration*

Substitution String	Description
CLUSTER	The unique cluster name shared by all hosts within the cluster/reporting domain.
HOSTNAME	The hostname of the machine that will be the Ganglia reporting master for CLUSTER
PORT	The non-multicast UDP port that the gmond server on HOSTNAME will listen on. Also the multicast port unique to CLUSTER, which all gmond servers in CLUSTER will listen and transmit on.
MULTICAST	The multicast address that the gmond servers in the cluster will communicate over. The default of 239.2.11.71 is acceptable as long as each CLUSTER uses a unique PORT.

■ **Note** MULTICAST:PORT must be unique per CLUSTER, but generally MULTICAST is left at the default value, so PORT becomes the unique value per cluster. The gmond on HOSTNAME will need to be configured to listen on the non-multicast UDP port of PORT. Many enterprise-grade switches will need to have multicast enabled for each CLUSTER's MULTICAST:PORT.

All nodes in the cluster will have the gmond server configured with the cluster name parameter set with the cluster's unique name. One node in the cluster, traditionally the NameNode or a JobTracker node, is configured to also accept non-multicast reporting on a port, commonly the same port as the multicast reception port. This host will be considered the Ganglia cluster master, and its hostname is the value for HOSTNAME. This host is also the host used in the /etc/gmetad.conf file. The conf/hadoop-metrics file needs to be altered as shown in Listing 6-2. The HOSTNAME and PORT must be substituted for the actual values. This file must then be distributed to all of the Hadoop conf directories and all Hadoop servers restarted.

Listing 6-2. *The conf/hadoop-metrics.properties File for Ganglia Reporting*

```
# Configuration of the "dfs" context for ganglia
dfs.class=org.apache.hadoop.metrics.ganglia.GangliaContext
dfs.period=10
dfs.servers=HOSTNAME:PORT
```

```
# Configuration of the "mapred" context for ganglia
mapred.class=org.apache.hadoop.metrics.ganglia.GangliaContext
mapred.period=10
mapred.servers=HOSTNAME:PORT


# Configuration of the "jvm" context for ganglia
jvm.class=org.apache.hadoop.metrics.ganglia.GangliaContext
jvm.period=10
jvm.servers=HOSTNAME:PORT
```

All of the Hadoop servers will now deliver metric data to `HOSTNAME:PORT` via UDP, once every 10 seconds.

The `gmetad` server that will collect metric information for the cluster will need to be instructed to collect metric information about `CLUSTER` from the master node via a TCP connection to `HOSTNAME:PORT`. The following is the configuration line in the `gmetad.conf` file for `CLUSTER`:

```
data_source "CLUSTER" HOSTNAME:PORT
```

The Ganglia web interface will provide a graphical view of the clusters, as shown in Figure 6-1.

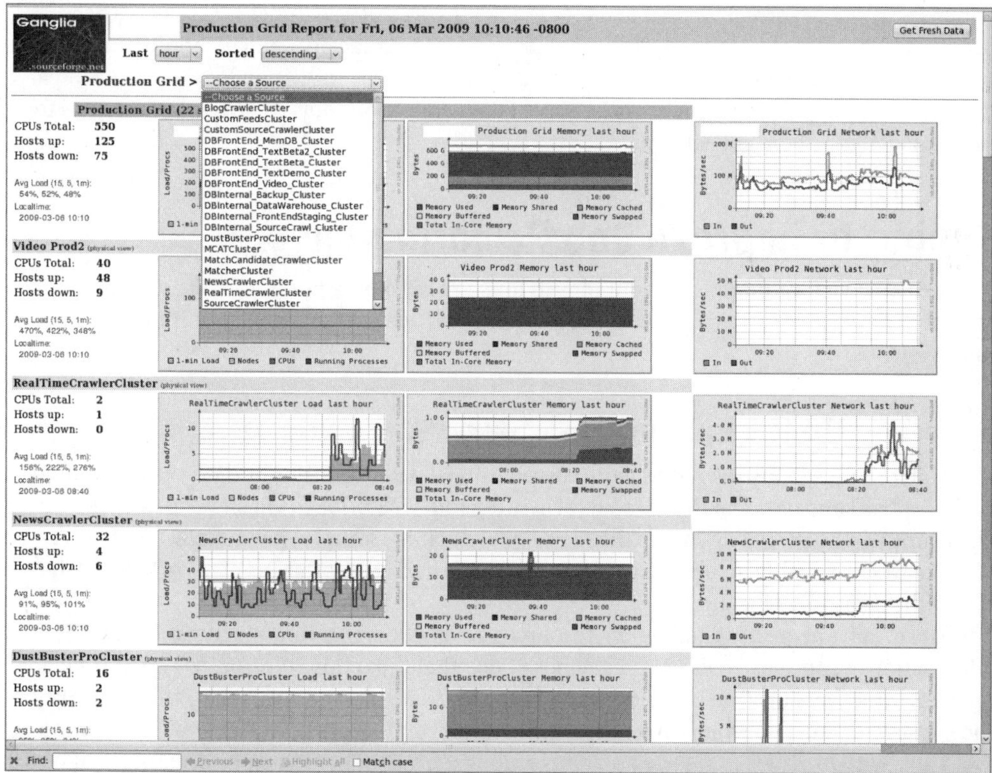

Figure 6-1. *The Ganglia web view of a running set of clusters*

When tuning jobs, Ganglia provides a wonderful interface to determine when your job is fully utilizing a cluster resource. Determining which resource is fully utilized, tuning the appropriate configuration parameters for that resource, and then rerunning the job will allow you to optimize your job's runtime on your cluster.

Chukwa: A Monitoring Service

Chukwa's goal is to provide extract, transform, and load (ETL) services for cluster logging data, thereby providing end users with a simple and efficient way to find the logging events that are actually important. Chukwa is new in Hadoop 0.19.0 and evolving rapidly.

Chukwa uses HDFS to collect data from various data providers, and MapReduce to analyze the collected data. The instance in Hadoop 0.19.0 appears to be currently optimized for the collection of data from log files, and then run a scheduled MapReduce job over the collected data. The Chukwa Quick Start is hosted on the Hadoop wiki, at `http://wiki.apache.org/hadoop/Chukwa_Quick_Start`.

FailMon: A Hardware Diagnostic Tool

The FailMon framework attempts to identify failures on large clusters by analyzing data collected from the Hadoop logs, the system logs, and other sources. The FailMon tools stem from a larger IBM effort to improve the operational reliability of large installations by predicting failures and taking corrective action before the failure occurs (see `https://issues.apache.org/jira/secure/attachment/12386597/failmon.pdf`). This is a very early technology and is expected to evolve rapidly.

The FailMon package consists primarily of data collection tools with MapReduce jobs to perform analysis of the collected data.

Tuning to Improve Job Performance

The general goal for tuning is for your jobs to finish as rapidly as possible using no more resources than necessary. This section covers best practices for achieving optimum performance of jobs.

Speeding Up the Job and Task Start

If the job requires many resources to be copied into HDFS for distribution via the distributed cache, or has large datasets that need to be written to HDFS prior to job start, substantial wall clock time can be spent copying in the files. For constant resources, it is simplest and quickest to make them available on all of the cluster machines and adjust the TaskTracker classpaths to reflect these resource locations.

The disadvantage of installing the resources on all of the machines is that it increases administrative complexity, as well as the possibility that the required resources are unavailable or an incorrect version. The advantage of this approach is that it reduces the amount of work the framework must do when setting up each task and may decrease the overall job runtime.

Table 6-10 provides a checklist of items to look for that affect write performance and what to do when the situations occur.

Table 6-10. *What to Monitor for Initial Bulk Transfer of Input Data*

Resource	What to Look For	What to Do
Source machine CPU utilization	The CPU is maxed out, the compression level is too high, or the compression algorithm is computationally too expensive	Change the compression or change the number of threads.
Source machine network	Saturation of the outbound network connection with traffic for HDFS	Increase the number of transfer threads or provide a higher-speed network connection.
Per DataNode network input	If it is not saturated, more writes could be delivered to this DataNode	Increase the number of simultaneous threads writing or reduce the number of files being created by increasing the individual file sizes.
DataNode I/O wait	I/O contention on a DataNode	Add more independent locations to `dfs.data.dir` or add more DataNodes.

If you have a large number of files that need to be stored in HDFS prior to the task start, such as might occur if your job needs to populate the job input directory, there are several things you may try, in varying combinations:

- It may be faster to copy the files from a machine that hosts a DataNode, as all of the writes will first go to the local DataNode, and the application will not have to wait for the data to traverse the network. The downside is that one replica of every block will end up on the local DataNode, greatly reducing the opportunity for data to be local to a map task. The DataNode may also get unbalanced with respect to storage, compared to other DataNodes. Ideally, bulk input of data to be used as input to a map task should be input from a host that does not also provide DataNode services, to ensure even distribution of the stored blocks across the DataNodes.
- It may be faster to run the copies in parallel. The limiting factor will be the network speed or the local DataNode disk speed in the event the copy host is also a DataNode.
- Use compression for data to be used once. LZO provides very good compression at little CPU overhead, provided that a native implementation is available.
- Create an archive of the input files, so that fewer files need to be created in HDFS. The downside is that zip and tar archives must be processed whole by a map task and may not be split into pieces (at least through Hadoop 0.19.0). Writing compressed sequence files, where the key/value pairs are of the type `BytesWritable`, will give you input that may be split and a reduction in file size.
- If you have large volumes of data, you may need to set up special machines with high-bandwidth network connections to the switching fabric that holds your DataNodes. Each block being written is sent directly to a DataNode. That DataNode will in turn send the block to the next DataNode in the chain and so on, until the required number of replicas are complete.
- If the origination machine has a higher bandwidth connection and is able to write multiple blocks in parallel (via multiple open files) while the bandwidth to each DataNode will be capped by the DataNode network speed, the origination machine will be able to write to HDFS at a higher rate.

There are very few tunable parameters at this point. You may change the `dfs.block.size` parameter to issue larger or smaller writes. You may decrease the `dfs.replication` parameter to reduce the overall HDFS write load, or increase it to increase the chance of local access by later MapReduce jobs. Compression generally helps but may cause issues later. Figure 6-2 illustrates how HDFS operations that your application issues are actually handled by the framework. Implicit in Figure 6-2 is that the replication count is three.

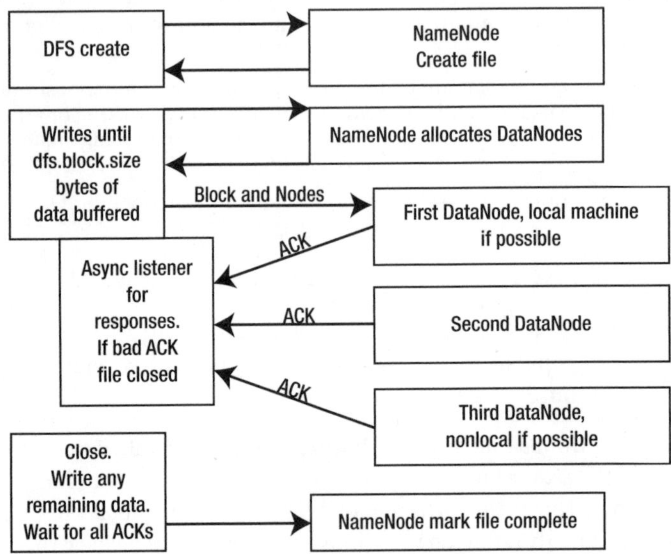

Figure 6-2. *Writing a block of data to HDFS*

From a monitoring perspective, you will want to monitor the network utilization on the upload machine and to a lesser extent on the DataNodes. If you are using compression, you will want to monitor CPU utilization on the machines doing the compression.

You may also wish to monitor the disk-write rate on the DataNodes, to verify that you are getting a good write rate. Since the incoming data rate is capped by the network rate, generally this is not a significant factor. If you see pauses in the network traffic or disk I/O, it implies that a Hadoop server may be unresponsive, and the client is timing out and will retry.

In general, increasing the server threads (`dfs.datanode.handler.count`) and the TCP listen queue depth (`ipc.server.listen.queue.size`) may help. It may be that the NameNode is not keeping up with requests, and in that case, increasing `dfs.namenode.handler.count` may help.

Optimizing a Job's Map Phase

The map phase involves dispatching a map task to a TaskTracker. Once the TaskTracker has a task to execute, it will prepare the task execution environment by building or refreshing the `DistributedCache` data. The TaskTracker maintains information about the `DistributedCache` for a particular job, and multiple tasks from the same job will share the same local execution environment. If you don't have an existing child JVM that has been used for this job's task and is within its reuse limit, start a new child JVM. The TaskTracker will then trigger the start of the map task in the child JVM.

The child JVM will start reading key/value pairs from its input, executing the map method for each pair. The output key/value pairs will be partitioned as needed and collected in the proper output format. If there is a reduce phase, the output format will be on the local disk in a sequence file. If there is not a reduce phase, the output will be in the job-specified output format and stored in HDFS.

Figure 6-3 shows a diagram of the job setup and map task execution. The left side follows the actions of the JobTracker from job submission through executing the map tasks on the available TaskTrackers. The right side follows the loop that a TaskTracker executes for map tasks. The diagram is read from top to bottom. The `Tasktracker$Child` is the class providing a `main()` method for the actual map task, which will be executed in a JVM launched and managed by the TaskTracker.

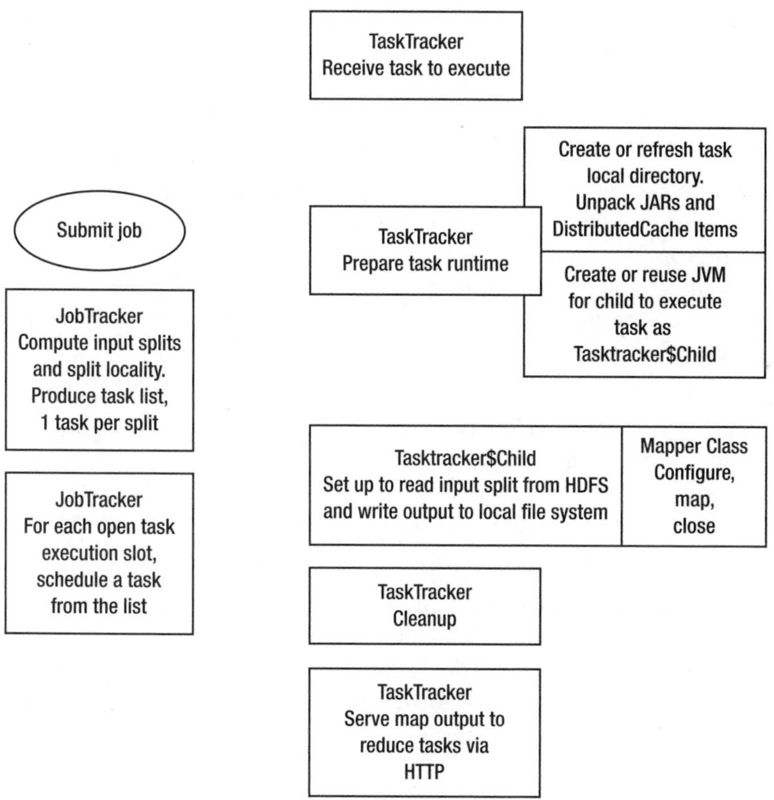

Figure 6-3. *Behind the scenes in a map task*

The following are some items you can tune for the map phase:

Map task run time: Each map task has some setup and teardown overhead. If the runtime of the map task is short, this overhead becomes the major time component. If the runtimes of tasks are too long, a single task may hold the cluster for a long period, or retrying a failed task becomes expensive. In general, less than a minute is usually too short, What is too long for a map task is job-specific. The primary tuning point for this is the `dfs.block.size` parameter. Increasing this parameter usually increases the split size and the task run time. On a per-job or per-cluster basis, you may also change `mapred.min.split.size`. It is better to use `dfs.block.size`, as the data is more likely to be local when the split size equals the HDFS file system block size.

TaskTracker node CPU utilization: If the map tasks are computationally intensive, a significant goal is to use all of the available CPU resources for that computation. There are two methods for controlling CPU utilization for map tasks:

- The job or cluster may configure the use of `MultithreadedMapRunner` for the `MapRunner` via `mapred.map.runner.class`, and specify the number of execution treads via `mapred.map.multithreadedrunner.threads`.
- The cluster may specify the number of map tasks to run simultaneously by a TaskTracker via `mapred.tasktracker.map.tasks.maximum`. This may be done on the command line for any job that uses the `GenericOptionsParser`.

Data location: If the map tasks are not receiving their input split from a local DataNode, the I/O performance will be limited to the network speed. This value is visible in the job counters of running and completed jobs, under the section titled "Data Local map tasks," the Total column gives the number of map tasks that ran with the input split served from a local DataNode. Other than increasing the replication factor and trying to ensure that the input split size is the file system block size, there is little tuning to be done.

Child garbage collection: If there is significant object churn in the `Mapper.map` method, and there is insufficient heap space allocated, the JVM hosting the task may spend a significant amount of wall clock time doing garbage collection. This is readily visible only via the Ganglia reporting framework or through the JMX MBean interface. The Ganglia reporting variable is `gcTimeMillis()` and is visible in the main reporting page for Ganglia, as shown in Figure 6-4.

Figure 6-4 shows an example of a Ganglia report for a two-host cluster, where one host is having problems. Note in the bottom-right graph, showing `gcTimeMilis`, how the host `cloud9` is spending roughly 2 to 400 msec per sample period doing garbage collection. This would imply that the child JVM has been configured with insufficient memory. At the current time, it is not possible to differentiate the garbage collection timing for the different server processes.

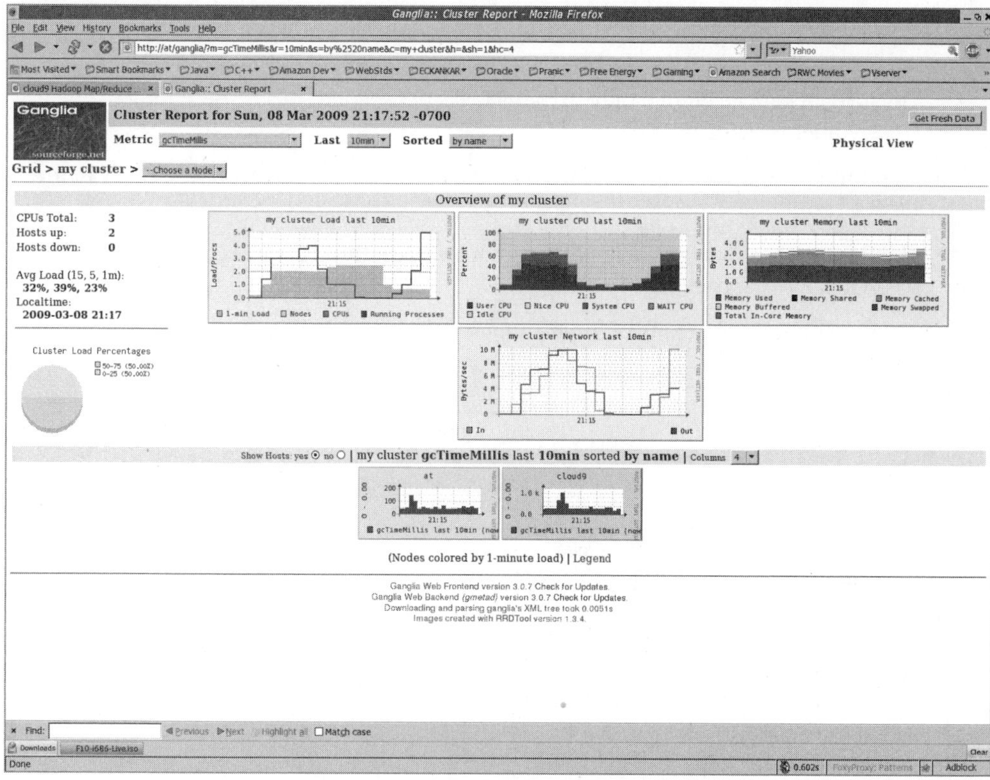

Figure 6-4. *Ganglia report showing gcTime for a two-host cluster, where one host is in trouble*

In this case, it's possible that increasing the child JVM memory limit, via `mapred.child.java.opts`, would be helpful. In this 10-minute window, the same task was run twice. The second time, it was run with twice as much memory per child JVM via `mapred.child.java.opts`. Note how much less time was taken in garbage collection on the right side of the graph for `cloud9` versus the left half of the graph.

Here are the command-line options to enable multithreaded map running with ten threads:

```
-D mapred.map.runner.class=org.apache.hadoop.mapred.lib.MultithreadedMapRunner ➥
    -D mapred.tasktracker.map.tasks.maximum=10
```

Tuning the Reduce Task Setup

The reduce task requires the same type of setup as the map task does with respect to the `DistributedCache` and the child JVM working environment. The two key differences relate to the input and the output. The reduce task input must be fetched from each of the TaskTrackers on which a map task has run, and these individual datasets need to be sorted. The reduce output is written to HDFS, unlike with the map task, which has output to the local file system.

As you can see from Figure 6-5, there are several steps for a reduce task, each of which has different constraints, as follows:

- The JobTracker can launch only so many tasks per second; this is something that will change after Hadoop 0.19.1.
- The tuning parameters for the map task with respect to job setup apply equally to the reduce task.
- The framework must fetch all of the map outputs for the reduce task, from the TaskTrackers that have them.
- The data to be fetched may be large enough that the network transfer speed becomes a bounding issue.

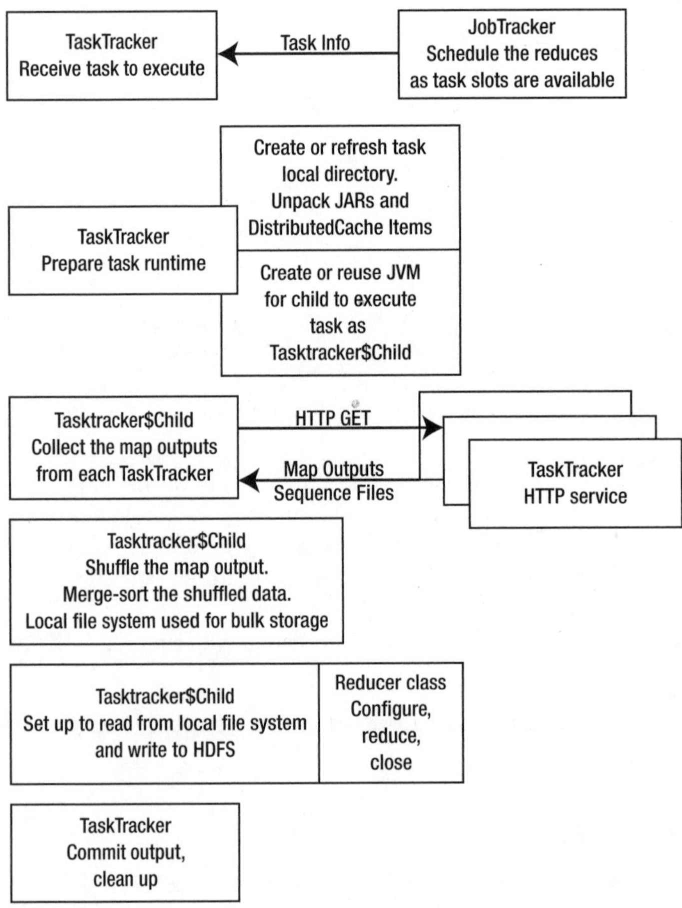

Figure 6-5. *Behind the scenes in the reduce task*

The following parameters affect the reduce task:

- `mapred.reduce.parallel.copies`: Controls how many fetches are run in parallel for each reduce task.
- `tasktracker.http.threads`: Controls the number of threads each TaskTracker runs to service these map output requests.
- `ipc.server.listen.queue.size`: At a lower layer, controls the number of requests that can queue before the client gets a connection-refused message.

There are several ways to reduce the size of the map output files, which can greatly speed up this phase. Some care needs to be used, as some of the compression options may slow down the shuffle and sort phase. The simplest thing is to specify a combiner class, which will act as a mini-reduce phase in each map task (as described in Chapter 5). This works very well for aggregation jobs, and not so well for jobs that need the full value space in the reduce task on which to operate. Many sites will enable map output file compression, via the Boolean value `mapred.compress.map.output`, in the `hadoop-site.mxl` file.

The choice of the compression algorithm is less clear. Native LZO is usually a good choice. The final trade-off is record-level versus block-level compression. The default is stored in `io.seqfile.compression.type`, and is `RECORD`. Conceptually, `RECORD` is better for the map output, as there will be a fair bit of reading through the files during the shuffle and sort phases. This is something that will have to be tried on a per-job basis. The other issue is that, at least through Hadoop 0.19.0, there is only one setting for this parameter, which affects all `SequenceFiles`.

■ **Note** There are a number of parameters that control the shuffle and merge. Tuning these parameters is expert work. There is a short discussion of the parameters in the Hadoop documentation, in the "Shuffle/Reduce Parameters" section (`http://hadoop.apache.org/core/docs/current/mapred_tutorial.html#Shuffle/Reduce+Parameters`).

Choosing the number of reduce tasks to run per machine and per cluster is the final level of tuning. A major determinant here is how the output data will be used, and that is application-specific. With reduce tasks, I/O, rather than CPU usage, is usually the bottleneck. If the DataNodes are coresident with the TaskTrackers, the reduce tasks will always have a local DataNode for the output. This will allow the initial writes to go at local speed, but the file closes will block until all the replicas are complete.

It is not uncommon for jobs to open many files in the reduce phase, which generally causes a huge slowdown, if not failure, in the HDFS cluster, so the job will take a significant amount of time to finish.

The following are some tuning points for the reduce phase:

Shuffle/sort time: The shuffle and sort cannot complete until all of the map output data is available. If this is an issue, you can try the following:

- Use a combiner class.
- Increase the number of `tasktracker.http.threads`.
- Increase the `ipc.server.listen.queue.size`.
- Set `mapred.compress.map.output` to `true`.
- Vary the compression codec stored in `mapred.map.output.compression.codec`.
- Experiment with `io.seqfile.compression.type` as `RECORD` or `BLOCK`.
- Change your algorithm so that less data needs to pass to the reduce phase. Try more reduce tasks, to reduce the volume of data that each reduce phase must sort.

Network saturation: The pull of the map outputs should just saturate your network. If the reduce tasks are timing out while trying to fetch outputs, increase the `tasktracker.http.threads`. If the network is saturated, enable compression, reduce the number of map tasks, improve the combiner class, or restructure the job to reduce the data passed to the reduce phase.

■ **Note** I once had a job where part of the value associated with each key was a large block of XML data that was unused by the reduce phase. Modifying the map to drop the XML data provided a tenfold improvement.

Actual reduce time: You may find that the time to actually reduce the data, after the shuffle and sort are done, is too long. If you are using `MultipleOutputFormat`, ensure that the number of files being created is small. If many small files must be created, write them as a zip archive. The Ganglia `gmetric` value `FilesCreated` will give you an idea of the rate of HDFS file creation.

Write time: The write time may be too long if the volume of data or the number of files are large. Enable output compression via setting `mapred.output.compress` to `true`. Experiment with codecs. Pack multiple files into zip files or other archive formats.

■ **Note** I had one job that needed to create many tens of thousands of small files. Writing the files as a zip archive in HDFS resulted in a hundredfold speed increase.

Overall reduce phase time: If the reduce phase is very long, you may want to tailor the number of reduce tasks per job and per machine. The job may specify the number of reduce tasks to run, but at least through Hadoop 0.19.1, the number of reduce tasks per TaskTracker is fixed at start time. If your cluster will run a specific set of jobs, with experimentation, you may find a reasonable number for the cluster-level parameter, and given that, identify a specific value for the number of reduce tasks for each job.

Addressing Job-Level Issues

One of the more interesting things to see is a cluster going almost idle while a job is running. This usually happens because a small number of tasks have not finished. This situation is called the *job tail*. This can happen with the map tasks or the reduce tasks. With multithreaded map tasks, the key or keys passed to one thread can sometimes take much longer to finish, and the task needs to wait for one or a small number of threads to complete, leaving the machine mostly idle. This is called the *task tail*.

Dealing with the Task Tail

I've had substantial experience with clusters set up with a single map task per TaskTracker, and have set the number of threads used by the `MultithreadedMapRunner` class to tune the task for full CPU utilization (roughly 80% to 90%). In one particular job, there was a large variance in the time it took to process a key: some keys took three hours, and others three seconds. If the long-running keys came late in an input split, the task would end up running one thread and idle six of the processors on the machine. The only solution for this was to reorder the input keys so that the long-running keys came first in the splits, or to abandon the long-running keys after a set elapsed run time, and reprocess all of the long-running keys in an additional job later.

Dealing with the Job Tail

The Hadoop standard is for very large jobs, spread over many machines, such that the time of one or two tasks is small compared to the run time of the job. This, in part, is where the 10-minute timeouts for server failures come from—a 10-minute period is considered short in the time of a job, so why not wait for that long? Many organizations have short timelines for jobs and limited budgets for hardware. These organizations must tune their jobs so that the clusters are well utilized.

The job tail really comes down to either a small number of reduce tasks taking much longer than others, either because the partitioning of the key space is very uneven or the duplicate keys fall unevenly in the partitions. The net result is that some reduce tasks have substantially more work to do. This is readily addressed only by turning the partitioning, via a custom partitioner class set via the `JobConf.setPartitionerClass(Class<? extends Partitioner> theClass)` method.

Tuning the number of reduce tasks so they fall evenly on your reduce slots may also help. Having one reduce task start after all the rest of the reduce tasks have finished can drastically increase the job runtime.

Summary

This chapter detailed how jobs are run by the Hadoop Framework and how MapReduce application writers and cluster administrators can tune both jobs and clusters for optimal performance.

The NameNode, JobTracker, DataNodes, and TaskTrackers have a number of start time parameters that directly affect how jobs are executed by the cluster and the overall run time of the jobs. The execution of a job is performed in several steps: setup, map, shuffle/sort, and reduce. It's possible to do some tuning to improve performance in each step.

This chapter discussed several tools for monitoring clusters and jobs. Ganglia is the tool I prefer for tuning and general dashboard-level awareness, and Nagios is the one I use for operational support. Using these tools enables rapid recognition of problems with jobs and clusters and provides insight into what parameters may need to be tuned, before the CEO calls you into the front office to explain why the mission-critical jobs haven't been running successfully. Ganglia also provides the informative, pretty graphs that higher-level management like so much.

CHAPTER 7

■ ■ ■

Unit Testing and Debugging

Two questions echo endlessly in the dark hours of small, enclosed areas lit by the dim glow of display screens: "Is it working?" and "How did *that* happen?" This chapter is about answering those questions.

"Is it working?" will be addressed by writing unit tests. The agile programming model suggests starting with unit tests and building the code afterwards. I generally try to follow this model, although under pressure, I have shifted to just writing the code, usually to my later regret. Testing the lower-level APIs in your applications is a known skill set. This chapter will cover unit tests that actually run MapReduce jobs at a small scale.

There are several ways to determine what is happening in a running program. The first level is through examining the log messages or other job output data. This requires detailed understanding of the output and may not provide sufficient information to isolate the problem. An alternative is to put custom code into the application to trigger different behavior or logging around the area of code that is in question. The most comprehensive method is to attach to the running application with a debugger and step through the execution of the code. This chapter will cover interactive debugging, using the Eclipse platform to provide the graphical interface.

Unit Testing MapReduce Jobs

MapReduce jobs, by their very nature, don't lend themselves to the traditional unit testing model. The common approach is to verify that all of the lower-level APIs are working correctly through their own unit tests. Next, you build small test datasets that have known outcomes and run them on a simulated MapReduce cluster, and then examine the output. These tests, when run on the simulated clusters, tend to be quite slow—on the order of minutes per test—due to the cluster setup and teardown times. If a real cluster is available for testing, the test run time will be shorter, but the tests must be coordinated among the cluster users.

The unit tests covered here are built on the Hadoop basic test case class `org.apache.hadoop.mapred.ClusterMapReduceTestCase`. This class provides a basic JUnit 3 test base that will start and stop a mini-HDFS with two DataNodes and a NameNode, and a mini-MapReduce cluster with two TaskTrackers and a JobTracker. This is a complete cluster, with web consoles for the JobTracker and the NameNode. All of the servers will run on the local machine, and the ports will be chosen from the free ports on the machine.

Because `ClusterMapReduceTestCase` is a JUnit 3-style test class, at least through Hadoop 0.19.0, it does not support annotations. The virtual cluster will be established and torn down for each test that any derived classes execute.

JUnit 4 supports the annotations `@BeforeClass` and `@AfterClass`, which allow for the cluster setup and teardown to happen one time per test class, providing each test with a clean cluster and saving significant wall clock time if many tests are run by the class. The JUnit 4-compliant delegate class demonstrated in this chapter allows `ClusterMapReduceTestCase` to be used with JUnit 4, so you can define when the virtual clusters are created and destroyed.

THE JAR FILES THAT COME WITH YOUR HADOOP DISTRIBUTION

In the archive that contains a Hadoop release are a number of prebuild release-specific JAR files. These JAR files are named in the form hadoop-*major release-minor release-component name*.jar.

The standard Hadoop JARs are found in the root directory of the installation. For Hadoop 0.19.0, the JAR files are `hadoop-0.19.0-core.jar`. This JAR is for the component `core`, for the major `release 0.19`, and the minor release `1`.

The root directory will contain JAR files for Ant, Core, examples, test, and tools. The `contrib.component` JARs also follow the same naming convention.

Each component may have an associated `lib` directory containing JAR files on which the component depends. By convention, the `lib` directory is located in the same directory as the component JAR.

In this chapter, I refer to the JARs as hadoop-rel-*component*.jar, where *component* is replaced with the actual component name, such as `hadoop-<rel>-core.jar` for the Hadoop Core JAR.

Requirements for Using ClusterMapReduceTestCase

The `ClusterMapReduceTestCase` class, like all Hadoop Core classes, makes strong assumptions about the runtime environment. For Hadoop Core unit tests, and for running standard jobs, the Hadoop Ant environment and the `bin/hadoop` script configure the runtime environment for the unit test or job, respectively. The unit tests that developers write to run in their workspace, or those created for build automation tools, do not generally have this luxury and must set up the runtime environment directly.

The `ClusterMapReduceTestCase` starts a virtual Hadoop cluster on which the tests are run. If the cluster does not start successfully, the test case will fail, without exercising the classes being tested. A startup failure is commonly due to a configuration issue, either with the runtime classpath or server or cluster configuration file. In particular, the NameNode and the JobTracker use Jetty to provide web servers for their web UIs. The error messages relating to the Jetty web server start failures do not provide sufficient information for the novice to resolve the configuration problem.

For developers running the tests from their IDE, it is not uncommon to load the Hadoop source code into the workspace, in place of using `hadoop-<rel>-core.jar`. Most of the Hadoop classes require configuration information that is provided in the distribution's `config/hadoop-default.xml` file. A copy of `hadoop-default.xml` is also bundled into the `hadoop-<rel>-core.jar`. When this configuration information is absent, the virtual cluster behavior is unpredictable.

The following are the requirements for using `ClusterMapReduceTestCase` in a unit test:

hadoop-<rel>-core jar: This is required for basic Hadoop classes. Include this JAR in the build path of your project and in the runtime classpath for the JUnit execution environment.

hadoop-<rel>- test.jar: This provides ClusterMapReduceTestCase and supporting classes. Include this JAR in the build path of your project and in the runtime classpath for the JUnit execution environment.

lib/*.jar: This provides the required services for the Hadoop Core classes. Include the needed JARs in the build path of your project and in the runtime classpath for the JUnit execution environment. It is often simpler to just include all of the JARs in the lib directory.

lib/jetty-ext/*.jar: This provides the additional classes that Jetty requires for the web consoles that will be run for the virtual cluster.

hadoop.log.dir: The framework requires this Java system property to be set to the path to an existing writable directory. This must be defined at test case start time, or by using the call System.setProperty("hadoop.log.dir", "path"), before the first call to ClusterMapReduceTestCase.setup(). The absence of a valid hadoop.log.dir system property results in a NullPointerExceptions or IOException being thrown by the test case during cluster setup.

javax.xml.parsers.SAXParserFactory: This must be Xerces, not SAX, or a validating parser error will be thrown. The correct value for this property is org.apache.xerces.jaxp.SAXParserFactoryImpl. It may be specified by the following argument on the JVM command line:

```
D javax.xml.parsers.SAXParserFactory= ➥
com.sun.org.apache.xerces.internal.jaxp.SAXParserFactoryImpl
```

Alternatively, you can call the following before the first call to ClusterMapReduceTestCase.setup():

```
System.setProperty("javax.xml.parsers.SAXParserFactory"," ➥
com.sun.org.apache.xerces.internal.jaxp.SAXParserFactoryImpl")
```

Let's look at some ways to check whether the requirements for using ClusterMapReduceTestCase have been met.

Troubles with Jetty, the HTTP Server for the Web UI

Jetty requires a validating parser that can handle its XML usage. Most parsers will do fine, but some will fail.

■ **Note** I was working on a large application that had a complex classpath. For unit testing, the entire classpath, including the Hadoop JARs, were folded together. All of a sudden, the unit tests started failing with an exception thrown by Jetty. The Saxon JAR was in the classpath before the Jetty JAR, so it was being used to deliver the XML parsers, and the parser was not validating.

A couple of system properties control the XML parser that applications will get:

`javax.xml.parsers.SAXParserFactory`: This contains the class name of the factory for constructing XML parsers. Set this to `com.sun.org.apache.xerces.internal.jaxp.SAXParserFactoryImpl`. This is the Sun JDK default and works well with Jetty. It avoids issues with alternative parsers in the classpath.

`org.mortbay.xml.XmlParser.NotValidating`: This instructs Jetty to not validate the XML configuration data. You should set it to `false`. Validated XML is good, and these are parsed only at job start time.

These properties may be set by passing them as arguments to the JVM via the `-Dproperty=value` syntax, or by calling `System.setProperty(name,value);` when performing setup for the test.

Hadoop Core servers rely on Jetty to provide web services, which are then available for internal and external use. If the Jetty JAR is missing from the classpath, the servers will not start. Listing 7-1 shows the log lines that indicate this.

Listing 7-1. *Log Lines That Indicate No Jetty JAR Is in the Unit Test Classpath*

```
java.io.IOException: Problem starting http server
Caused by: org.mortbay.util.MultiException[java.lang.reflect. ➥
InvocationTargetException, java.lang.reflect.InvocationTargetException, ➥
java.lang.reflect.InvocationTargetException]
```

The Hadoop Core Jetty configurations also require the JARs that are in the `lib/jetty-ext` directory of the installation. If they are not present in the classpath of the unit test, non-descriptive failure-to-start error messages will be generated.

Listings 7-2 through 7-5 shows the various log entries indicating that specific Jetty JARs are missing from the unit test classpath. In these listings, only the relevant exception lines are shown; the stack traces have been removed to aid clarity. The string *XXXXX* represents some TCP port number.

Listing 7-2. *Log Lines That Indicate jetty-ext/commons-el.jar Is Not in the Unit Test Classpath*

```
java.io.IOException: Call to /0.0.0.0:XXXXX failed on local exception: ➥
Connection refused: no further information
Caused by: java.net.ConnectException: Connection refused: no further information
```

Listing 7-3. *Log Lines That Indicate jetty-ext/jasper-runtime.jar Is Not in the Unit Test Classpath*

```
java.lang.NoClassDefFoundError: org/apache/jasper/JasperException
Caused by: java.lang.ClassNotFoundException: org.apache.jasper.JasperException
```

Listing 7-4. *Log Lines That Indicate jetty-ext/jasper-compiler.jar Is Not in the Unit Test Classpath*

```
java.io.IOException: Problem starting http server
Caused by: org.mortbay.util.MultiException[java.lang.ClassNotFoundException: ➥
org.apache.jasper.servlet.JspServlet, java.lang.ClassNotFoundException: ➥
org.apache.jasper.servlet.JspServlet, java.lang.ClassNotFoundException: ➥
org.apache.jasper.servlet.JspServlet]
```

Listing 7-5. *Log Lines That Indicate jetty-ext/jsp-api.jar Is Not in the Unit Test Classpath*

```
java.lang.NoClassDefFoundError: javax/servlet/jsp/JspFactory
Caused by: java.lang.ClassNotFoundException: javax.servlet.jsp.JspFactory
```

The Hadoop Core JAR Is Missing or Malformed

Particularly for IDE developers, the classpath may have the Hadoop source tree rather than the Hadoop Core JAR. In this case, various required configuration files may be missing, resulting in unexpected failures. Listings 7-6 and 7-7 show the log lines that indicate missing configuration files.

Listing 7-6. *Log Lines That Indicate the hadoop-default.xml File Is Missing or Malformed*

```
java.lang.NullPointerException
    at org.apache.hadoop.hdfs.server.namenode. ➥
    FSNamesystem.close(FSNamesystem.java:523)
    at org.apache.hadoop.hdfs.server.namenode. ➥
    FSNamesystem.<init>(FSNamesystem.java:293)
```

Listing 7-7. *Log Lines That Indicate the Java System Property* `hadoop.log.dir` *Is Unset*

```
ERROR mapred.MiniMRCluster: Job tracker crashed
java.lang.NullPointerException
    at java.io.File.<init>(Unknown Source)
    at org.apache.hadoop.mapred.JobHistory.init(JobHistory.java:143)
    at org.apache.hadoop.mapred.JobTracker.<init>(JobTracker.java:1110)
    at org.apache.hadoop.mapred.JobTracker.startTracker(JobTracker.java:143)
    at org.apache.hadoop.mapred. ➥
    MiniMRCluster$JobTrackerRunner.run(MiniMRCluster.java:96)
    at java.lang.Thread.run(Unknown Source)
```

The `MiniDFSCluster` creates the directories for HDFS storage in the path `build/test/data/dfs/data`, or `build\test\data\dfs\data` under Windows. It will attempt to remove the directory before starting. Listing 7-8 shows the error message that results if the directories cannot be deleted. The typical reason for the failure is that a prior instance of `MiniDFSCluster` is still running.

Listing 7-8. *Log Lines That Indicate a Unit Test Is Already in Progress*

```
java.io.IOException: Cannot remove data directory: build\test\data\dfs\data
    at org.apache.hadoop.hdfs.MiniDFSCluster.<init>(MiniDFSCluster.java:263)
    at org.apache.hadoop.hdfs.MiniDFSCluster.<init>(MiniDFSCluster.java:119)
    at org.apache.hadoop.mapred.ClusterMapReduceTestCase. ➥
    startCluster(ClusterMapReduceTestCase.java:81)
```

The Virtual Cluster Failed to Start

ClusterMapReduceTestCase builds a virtual Hadoop cluster on which to run the test cases. By default, this virtual cluster starts six server processes: one NameNode, one JobTracker, two DataNodes, and two TaskTrackers. If the NameNode or the JobTracker did not start, the tests cannot be run.

The HDFS portion of the cluster is started first. It is composed of one NameNode and two DataNodes. If the HDFS fails to start, the MapReduce portion is not started.

The NameNode is kind enough to actually report that it is up, as in this example log line:

```
namenode.NameNode: Namenode up at: localhost/127.0.0.1:XXXXX
```

The DataNodes' state must be deduced from the log messages. The log messages in Listing 7-9 indicate successful startup. (Your timestamps and port allocations will vary.)

Listing 7-9. *Log Lines That Indicate Both DataNodes Are Running*

```
Starting DataNode 0 with dfs.data.dir:
➥ build\test\data\dfs\data\data1,build\test\data\dfs\data\data2
...
INFO datanode.DataNode: New storage id DS-222715038-192.168.1.12-2232- ➥
1236829361312 is assigned to data-node 127.0.0.1:2232
INFO datanode.DataNode: DatanodeRegistration(127.0.0.1:2232, ➥
storageID=DS-222715038-192.168.1.12-2232-1236829361312, infoPort=2233, ➥
ipcPort=2234)In DataNode.run, data = FSDataset{dirpath= ➥
'C:\Documents and Settings\Jason\My Documents\HadoopBook\ ➥
code\examples\build\test\data\dfs\data\data1\current, ➥
C:\Documents and Settings\Jason\My Documents\HadoopBook\ ➥
code\examples\build\test\data\dfs\data\data2\current'}
INFO datanode.DataNode: using BLOCKREPORT_INTERVAL of 3600000msec ➥
Initial delay:0msec

Starting DataNode 1 with dfs.data.dir: ➥
build\test\data\dfs\data\data3,build\test\data\dfs\data\data4
...
```

```
INFO datanode.DataNode: New storage id DS-2049952137-192.168.1.12-2239- ➥
1236829361718 is assigned to data-node 127.0.0.1:2239
INFO datanode.DataNode: DatanodeRegistration(127.0.0.1:2239, ➥
storageID=DS-2049952137-192.168.1.12-2239-1236829361718, infoPort=2240, ➥
 ipcPort=2241)In DataNode.run, data = FSDataset{dirpath= ➥
'C:\Documents and Settings\Jason\My Documents\HadoopBook\ ➥
code\examples\build\test\data\dfs\data\data3\current, ➥
C:\Documents and Settings\Jason\My Documents\HadoopBook\ ➥
code\examples\build\test\data\dfs\data\data4\current'}
INFO datanode.DataNode: using BLOCKREPORT_INTERVAL of 3600000msec ➥
Initial delay: 0msec

Waiting for the Mini HDFS Cluster to start...
INFO datanode.DataNode: BlockReport of 0 blocks got processed in 0 msecs
INFO datanode.DataNode: Starting Periodic block scanner.
INFO datanode.DataNode: BlockReport of 0 blocks got processed in 0 msecs
INFO datanode.DataNode: Starting Periodic block scanner.
```

The two DataNodes are started after the NameNode is started, and each has a `Starting` line followed by a final line that indicates the `BLOCKREPORT_INTERVAL`. If the lines containing `BLOCKREPORT_INTERVAL` are missing, the DataNode did not start.

As with the HDFS portion, the JobTracker informs you directly that it is running, but the TaskTracker status must be deduced from the logs. Here is the message from a successfully started JobTracker:

```
INFO mapred.JobTracker: Starting RUNNING
```

Listing 7-10 shows the log lines that show the TaskTracker is running. You need two sets of these for full service.

Listing 7-10. *Log Lines That Indicate a TaskTracker Is Running*

```
mapred.TaskTracker: TaskTracker up at: 0.0.0.0/0.0.0.0:2262
mapred.TaskTracker: Starting tracker tracker_host1.foo.com:0.0.0.0/0.0.0.0:2262
```

If the `hadoop.log.dir` system property is unset, a subprocess of the virtual cluster may crash and leave the test case in limbo. Listing 7-11 shows this error.

Listing 7-11. *A Virtual Cluster Server Process Has Crashed*

```
ERROR mapred.MiniMRCluster: Job tracker crashed
java.lang.NullPointerException
    at java.io.File.<init>(Unknown Source)
    at org.apache.hadoop.mapred.JobHistory.init(JobHistory.java:143)
    at org.apache.hadoop.mapred.JobTracker.<init>(JobTracker.java:1110)
    at org.apache.hadoop.mapred.JobTracker.startTracker(JobTracker.java:143)
    at org.apache.hadoop.mapred. ➥
    MiniMRCluster$JobTrackerRunner.run(MiniMRCluster.java:96)
    at java.lang.Thread.run(Unknown Source)
```

There should be no ERROR level log messages.

The Eclipse framework provides a decent way to run individual or class-based Hadoop unit tests, as well as simple debugging. Occasionally, some state can get lost, particularly in the Windows environment, and the virtual cluster will fail to start. The indication of this will be a series of messages stating that a connection attempt has failed. In particular, if you see connection failure messages that have /0.0.0.0:, it is an indication that Eclipse needs a restart, as shown in Listing 7-12.

Listing 7-12. *Log Lines That Indicate Eclipse Has Lost State and Needs to Be Restarted*

```
INFO ipc.Client: Retrying connect to server: /0.0.0.0:9100. Already tried 0 time(s).
INFO ipc.Client: Retrying connect to server: /0.0.0.0:9100. Already tried 1 time(s).
INFO ipc.Client: Retrying connect to server: /0.0.0.0:9100. Already tried 2 time(s).
INFO ipc.Client: Retrying connect to server: /0.0.0.0:9100. Already tried 3 time(s).
INFO ipc.Client: Retrying connect to server: /0.0.0.0:9100. Already tried 4 time(s).
INFO ipc.Client: Retrying connect to server: /0.0.0.0:9100. Already tried 5 time(s).
INFO ipc.Client: Retrying connect to server: /0.0.0.0:9100. Already tried 6 time(s).
INFO ipc.Client: Retrying connect to server: /0.0.0.0:9100. Already tried 7 time(s).
INFO ipc.Client: Retrying connect to server: /0.0.0.0:9100. Already tried 8 time(s).
INFO ipc.Client: Retrying connect to server: /0.0.0.0:9100. Already tried 9 time(s).
```

Simpler Testing and Debugging with ClusterMapReduceDelegate

When running tests, it is very helpful to be able to interact with the test cluster HDFS, to access the Web GUIs of the various servers, and to examine the log files. This book's downloadable code includes a class com.apress.hadoop.mapred.test.ClusterMapReduceDelegate that provides a wrapper around the Hadoop Core test framework class ClusterMapReduceTestCase. This delegate class provides a JUnit 4-friendly way to build test classes, and exposes information useful to understanding what is happening in your test. All the test case classes discussed here extend the class ClusterMapReduceDelegate.

Core Methods of ClusterMapReduceDelegate

Table 7-1 lists the core methods that any unit test interacting with Hadoop will use. In particular, all JobConf objects used by the test cases and classes being tested must be children of the JobConf object returned by the createJobConf() method.

All tests will need to call the methods setupBeforeClass() and tearDownAfterClass() at least once to start and stop the virtual cluster. Any test case that needs to create files in HDFS or access files in HDFS will need to call the method getFileSystem() to get a file system object to use for the interactions.

Table 7-1. *Core Methods of ClusterMapReduceDelegate*

Method	When to Use	What It Does
`setupBeforeClass()`	In the `@BeforeClass` method.	Ensures that the required Hadoop system properties are set to sensible values if they are unset. It will then start the virtual cluster, with 2 TaskTrackers, 2 DataNodes, a NameNode, and a JobTracker. This method will throw an exception or possibly hang if the virtual cluster does not start.
`logConfiguration (JobConf conf, Logger log)`	If you need detailed information on how the virtual cluster is configured.	Dumps the key parameters out of `conf`, and the virtual cluster NameNode and DFS configuration objects to log at level info.
`Configuration getHDFSConfiguration()`	If you or your test need to interact with NameNode. This object has the key parameters such as `fs.default.name` and `dfs.http.address`.	Returns the virtual cluster's HDFS configuration object.
`Configuration getJobTrackerConfiguration()`	If you or your test need to interact with the JobTracker. This object has key parameters such as `mapred.job.tracker`, `mapred.job.tracker.http.address` and `mapred.system.dir`.	Returns the configuration object used by the `MiniMRCluster` private base class, which sets up the JobTracker for the virtual cluster.
`Path getTestRootDir()`	To determine the root path of the test in HDFS, when you need to examine data files.	Returns the path of the test case in HDFS.
`FileSystem getFileSystem()`	When the test case need to create files or otherwise interact with HDFS.	Returns a file system object constructed for the virtual cluster's HDFS file system.
`JobConf createJobConf ()`	*Must be used to create the JobConf object's used by your test cases.*	Creates a `JobConf` object that is correctly configured for the virtual cluster.
`void tearDownAfterClass ()`	In the `@AfterClass` method of your test class.	Stops the virtual cluster.
`void logDefaults (Logger log)`	If you have a problem with the test case or cluster starting.	Writes the locations of the Core Hadoop JAR files and the critical parameters to log at the info level.

Configuration Parameters for Interacting with Virtual Clusters

Several core parameters are needed for the tester and the test cases, when interacting with the virtual cluster. Table 7-2 details the parameter names, how to get their values, and what to do with them. When debugging test cases, it is very useful to know where the log files are being written, and the web addresses for the NameNode and JobTracker. For accessing files in the virtual cluster's HDFS, the HDFS file system URL must be available.

Table 7-2. *Important Configuration Parameters for Interacting with the Virtual Cluster*

Parameter	What It Is	How to Get It	What to Do with It
`hadoop.log.dir`	The path to the directory log files are written to	`System.getProperties ("hadoop.log.dir");`	Look in this directory for cluster log files, such as user logs for the per task log files.
`fs.default.name`	The URL for HDFS	`getFileSystem(). getUri();`	Use this URL to interact with the virtual HDFS from the command line via `bin/hadoop dfs -fs URL` file operations.
`mapred.job.tracker. http.address`	The URL for the virtual cluster JobTracker web interface.	`createJobConf(). get("mapred.job. tracker.http. address");`	Use this URL to view the state of running and finished jobs in the virtual cluster.
`dfs.http.address`	The URL for the virtual cluster NameNode web interface.	`getHDFS Configuration().get ("dfs.http.address");`	Use this URL to view the state of the virtual cluster HDFS.

Writing a Test Case: SimpleUnitTest

The sample `SimpleUnitTest` test case simply starts a cluster, writes a single file to the cluster HDFS, and reads that file back, verifying the contents are correct. This section will walk through building this unit test. The full code for this example is in the file `SimpleUnitTest.java` of package `com.apress.hadoopbook.examples.ch7` in the downloadable code for this book.

The TestCase Class Declaration

In Listing 7-13, the test case class extends `ClusterMapReduceDelegate` and a `Logger` object is declared. Normally, the `Logger` would be from the class being tested to better enable control over the logging levels. In this sample test case, there is no class being tested, so a logger is created.

Listing 7-13. *Class Declaration from SimpleUnitTest.java*

```
/** This simple unit test exists to demonstrate the creation and teardown of a
 * virtual cluster and the writing of a test case that uses the created cluster.
 *
 * @author Jason
 *
 */
import com.apress.hadoop.mapred.test.ClusterMapReduceDelegate;
public class SimpleUnitTest extends ClusterMapReduceDelegate {
    public static Logger LOG = Logger.getLogger(SimpleUnitTest.class);
```

The Cluster Start Method

The `startVirtualCluster()`method in `SimpleUnitTest`, shown in Listing 7-14, is used to start the virtual cluster and verify that the cluster has started successfully. `startVirtualCluster()` uses the JUnit 4 annotation `@BeforeClass` to indicate to the JUnit framework that this method must be run one time only, and before any of the test cases in the class are launched. The test also makes two JUnit `assertNotNull` checks, to verify that the cluster configuration information is available. If there are any failures, an exception should be thrown. The JUnit framework will catch the exception and mark the test set as failed.

The `setupTestClass()` call is a method on the `ClusterMapReduceDelegate` and actually starts the virtual cluster and collects the configuration information. It also will set a small number of required system parameters in the configuration object returned by `getConf()`, if those parameters are currently unset.

Listing 7-14. *Cluster Setup Method with JUnit 4 @BeforeClass Annotation*

```
/** This is the JUnit4 before class, cluster initialization method.
 *
 * This method starts the cluster and performs simple validation of the working ➥
   state of the cluster.
 *
 * Under some failure cases usually related to incorrect CLASSPATH ➥
   configuration, this method may never complete.
 *
 * If all of the test cases in your file can share a cluster, use ➥
   the @BeforeClass annotation.
 * @throws Exception
 */
@BeforeClass
public static void startVirtualCluster() throws Exception
{
    /** Turn down the cluster logging to filter the noise out. Do this if ➥
           the test is basically working. */

    setupTestClass();
    /** Verify that there is a JobConf object for the cluster. */
    assertNotNull("Cluster initialized Correctly", getConf());
    /** Verify that the file system object is available. */
    assertNotNull("Cluster has a file system", getFs());
}
```

The Cluster Stop Method

`SimpleUnitTest.stopVirtualCluster()`,shown in Listing 7-15, uses the JUnit 4 annotation `@AfterClass` to indicate to the JUnit framework that it is to be called after the last test in the class has been run. It is essentially the `finally` clause for the test class.

teardownTestCase() is a method on the ClusterMapReduceDelegate class that will terminate the virtual cluster and clear the cached cluster information.

Listing 7-15. *Cluster Stop Method with JUnit 4 @AfterClass Annotation*

```
/** This is the JUnit4 after class tear down method.
 * This stops the cluster, and would perform any needed cleanup.
 * If all of the test cases in a file can share a cluster use the @AfterClass ➥
   annotation.
 * @throws Exception
 */
@AfterClass
public static void stopVirtualCluster() throws Exception {
    teardownTestClass();
}
```

The Actual Test

The unit test, shown in Listing 7-16, writes a string to a newly created file in the virtual cluster HDFS, and then reads the string back from the file to verify that the same string can be read back. The test has the standard stylized framework you will see in all of the sample code. Any code that allocates objects that hold system-level file descriptors is done in a try block. The try block has a finally clause where the system-level file descriptors are closed. This pattern, if rigorously applied, will greatly reduce job failures when the jobs are running at large scales.

Listing 7-16. *The Actual Test Code with the JUnit 4 @Test Annotation*

```
/** A very simple unit test that uses the virtual cluster.
 *
 * The test case writes a single file to HDFS and reads it back, verifying ➥
   the file contents.
 *
 * @throws Exception
 */
@Test
public void createFileInHdfs() throws Exception
{
    final FileSystem fs = getFs();
    assertEquals( "File System is hdfs", "hdfs", fs.getUri().getScheme());

    Path testFile = new Path("testFile");
    FSDataOutputStream out = null;
    FSDataInputStream in = null;
    final String testData = "HelloWorld";
```

```
    try {
        /** Create our test file and write our test string to it. The writeUTF ➥
         method writes some header information to the file. */
        out = fs.create(testFile,false);
        out.writeUTF(testData);
        /** With HDFS the file really doesn't exist until after it has been
          * closed. */
        out.close();
        out = null;

        /** Verify that the file exists. Open it and read the data back
          * and verify the data. */
        assertTrue( "Test File " + testFile + " exists", fs.exists(testFile));
        in = fs.open(testFile);
        String readBack = in.readUTF();
        assertEquals("Read our test data back: " + testData, testData, readBack);
        in.close();
        in=null;

    } finally {
        /** Our traditional finally when descriptors were opened
          * to ensure they are closed. */
        Utils.closeIf(out);
        Utils.closeIf(in);
    }
}
```

The test grabs the FileSystem object out of the base class using the following call:

```
final FileSystem fs = getFs();
```

This ensures that the file operations will be on the virtual cluster's HDFS.

As a double layer of paranoia, the following line verifies that the file system is in fact an HDFS file system:

```
assertEquals( "File System is hdfs", "hdfs", fs.getUri().getScheme());
```

The following line will create testFile, if there is not an existing file by that name. The file handle is returned and stored in out.

```
out = fs.create(testFile,false);
```

The following line writes the string testData with a small header to testFile.

```
out.writeUTF(testData);
```

At least through Hadoop 0.19.1, files do not really become available until after they are closed; therefore, the file is closed via out.close(). Since the try block will also call the close() method on out, if out is not set to null, out is set to null to avoid a duplicate close().

The following line will trigger an exception if `fs.exists(testFile)` is not `true`. This verifies that the file exists in the file system.

```
assertTrue( "Test File " + testFile + " exists", fs.exists(testFile));
```

At this point in the test case, the file has been created and is known to exist. The test case will now open the file and read the contents, verifying that the contents are exactly what was written.

To open the file, rather than to create it, the following line is used:

```
in = fs.open(testFile);
```

The first line in the next snippet reads back one UTF8 string, and the next line verifies that the expected data was read.

```
String readBack = in.readUTF();
assertEquals("Read our test data back: " + testData, testData, readBack);
```

A Test Case That Launches a MapReduce Job

In this example, we will go over a test case that actually calls a MapReduce job and examines the output. My favorite initial testing tool is the `PiEstimator` example in the `hadoop-<rel>-examples` JAR, which is the class for which this unit test is built. The `PiEstimator` class, as it stands, is not unit test-friendly, and very little information can be extracted. The only thing that can be done to verify the result is to examine the estimated value of pi.

As is common in unit tests, this test case declares that it is in the same package as the class under test:

```
package org.apache.hadoop.examples;
```

The full text file is `PiEstimatorTest.java`.

The `PiEstimator` test class started life as a copy of `SimpleUnitTest,java`. This copy was then modified to highlight the relevant details for the `PiEstimator` test case. The `PiEstimator.startVirtualCluster` method has been modified to reduce the logging verbosity of the virtual cluster server processes, as shown in Listing 7-17.

Listing 7-17. *Reduction in Logging Level for the Virtual Cluster*

```
/** Turn down the cluster logging to filter the noise out.
  * Do this if the test is basically working. */
final String rootLogLevel = System.getProperty("virtual.cluster.logLevel","WARN");
final String testLogLevel = System.getProperty("test.log.level", "INFO");
LOG.info("Setting Log Level to " + rootLogLevel);
LogManager.getRootLogger().setLevel(Level.toLevel(rootLogLevel));

/** Turn up the logging on this class and the delegate. */
LOG.setLevel(Level.toLevel(testLogLevel));
ClusterMapReduceDelegate.LOG.setLevel(Level.toLevel(testLogLevel));
```

No changes have been made to the `stopVirtualCluster()` method.

Note If you wish to interact with the HDFS or the JobTracker web interface, it is necessary to put a breakpoint on `stopVirtualCluster()`, to prevent the cluster from being torn down. When the debugger breaks there, the various servers are still running and available. The NameNode web GUI is not known to work correctly with Hadoop 0.19.0 under the virtual cluster.

The actual test case is very different from `SimpleUnitTest,java`. The method under test is `launch()`, and the test method is `testLaunch()`. The preamble, shown in Listing 7-18, logs a couple of key pieces of information to enable the test runner to interact with the virtual cluster services. The JobTracker URL will let you interact with the running or finished job, examine the task outputs, and look at the job counters. The HDFS URL will let you interact directly with the file system to view the data files.

Note When interacting with the JobTracker web interface, the URLs for the task log files are generated with fictitious names of the form `hostX.foo.com`. Replace the fictitious hostname with `localhost`, and you will be able to fetch the task logs. Alter `http://host0.foo.com:3126/tasklog?taskid=attempt_200903130041_0001_m_000000_0&all=true` to `http://localhost:3126/tasklog?taskid=attempt_200903130041_0001_m_000000_0&all=true`.

Listing 7-18. *Test Member Preamble with Useful Debugging Information*

```
@Test
public void testLaunch() throws Exception {
    final FileSystem fs = getFs();
    final JobConf testBaseConf = getConf();
    LOG.info( "The HDFS url is " + fs.getUri());
    LOG.info( "The Jobtracker URL is " + getJobtrackerURL());
    LOG.info( "The Namenode URL is " + getNamenodeURL());
```

Listing 7-19 constructs a new `JobConf` object out of the test default configuration via `JobConf conf = new JobConf(testBaseConf)`. This is a highly recommended practice, as the object the method `ClusterMapReduceDelegate.getConf()` returns is shared by all test cases of that virtual cluster instance. It is actually the `JobConf` object returned by `ClusterMapReduceTestCase.createJobConf()` method. As a debugging nicety, the next line tells the TaskTrackers to send all task output to the console of the process that submitted the job. This is set using the configuration parameter `jobclient.output.filter` to `ALL`.

Tip It is always wise to ensure that the JAR that contains your MapReduce classes is part of the classpath for tasks, and the `conf.setJarByClass(PiEstimator.class)` call ensures that.

THE MAGIC OF TASK OUTPUT FILTERING

The configuration parameter `jobclient.output.filter` specifies what output, if any, from the tasks are printed on the console of the job submitter. The valid values are as follows:

- `ALL`: Return all task output.
- `NONE`: Return no task output.
- `KILLED`: Return output from tasks that are killed.
- `FAILED`: Return output from tasks that failed.
- `SUCCEEDED`: Return output from tasks that succeeded

The default value is `FAILED`, and only failed tasks have their output printed.

Listing 7-19. *Set Up the JobConf Object for the Class Tested*

```
/** Make a new {@link JobConf} object that is set up to
  * ensure that the jar containing {@link PiEstimator}
 * is available to the TaskTrackers.
 *
 * Note: It is very bad practice to modify the configuration given back by getConf()
 * as the returned object is shared among all tests in the Test file.
 */
JobConf conf = new JobConf(testBaseConf);
/** Make all task output come to the console of the unit test. */
conf.set("jobclient.output.filter","ALL");

/** Ensure that hadoop- -examples.jar is pushed into the DistributedCache
  * and made available to the TaskTrackers.
 *
 */
conf.setJarByClass(PiEstimator.class);
```

The `PiEstimator` instance needs to be created and configured, as shown in Listing 7-20.

Listing 7-20. *Preparing the PiEstimator Instance to Be Run*

```
/** Create the PiEstimator object and initialize it with our conf object. */
PiEstimator toTest = new PiEstimator();
toTest.setConf(conf);
```

The `launch()` method is invoked, and the results are tested, as shown in Listing 7-21. This requires knowledge of the proper arguments to the method.

Listing 7-21. *Actually Calling the PiEstimator.launch() Method and Testing the Result*

```
int maps = 10;
long samples = 1000;
double result = toTest.launch(maps, samples, null, null);
LOG.info( "The computed result for pi is " + result);
assertTrue( "Result Pi >3 ", result > 3);
assertTrue( "Result Pi <4 ", result < 4);
```

This completes the walk-through of a unit test that invokes a MapReduce job.

Running the Debugger on MapReduce Jobs

There are several basic strategies for running a MapReduce job under the debugger. Here, we'll start with simplest and move to the more complex methods.

INCREASING THE MAPREDUCE JOB TIMEOUT LENGTH

When running MapReduce jobs under a debugger, it is important to drastically increase the value of the configuration key `mapred.task.timeout`. The default value is `600000`, or 10 minutes. When you are single-stepping through a map or reduce task, it is common for more than 10 minutes to pass. If the value has not been lengthened, the task you are debugging will be killed.

You can set the task timeout length to a large long value via the `bin/hadoop jar` command line, using `-Dmapred.task.timeout`, after the main class specification, if the job uses the `GenericOptionsParser` that the `ToolRunner` class provides. For example, to set a 2-hour timeout, uses the value `7200000`, as follows:

```
bin/hadoop jar job.jar main.class -Dmapred.task.timeout=7200000 other arguments
```

This value is parsed as a long, so values up to `9223372036854775807` will work on 32-bit JVMs.

Running an Entire MapReduce Job in a Single JVM

The normal process of job submission involves the `JobClient` class storing all of the relevant information about the job in HDFS, and then making an RPC call to the JobTracker to submit the job for execution. The `JobClient` determines the address of the JobTracker through the configuration value of the configuration key `mapred.job.tracker`. This configuration key may have the value of `local`, in which case the entire MapReduce job will be run in the JVM that is submitting the job. This is ideal for debugging small-scale problems.

There are some restrictions to this technique. The cause of most of the restrictions is that the map and reduce tasks do not run in their own JVMs. There is no way to change the JVM working directory, classpath, or other command-line configured options. A significant result of this is that the `DistributedCache` behavior is very different. If your job relies on the `DistributedCache`, this method will not be a good debugging choice. The lifetime of your classes will be longer than you expect, and you may experience unexpected results due to prior variable initialization for static variables.

The number of reduce tasks is limited to zero or one. If your job requires more than one reduce task, this method will not be a good debugging choice.

The example in this section uses our old friend `PiEstimator`, the example that comes with Hadoop Core, as the MapReduce job.

Figures 7-1 through 7-4 are guides to configuring a run/debug profile for a MapReduce application so that it may be run using the `LocalJobRunner`, JobTracker, in a single JVM.

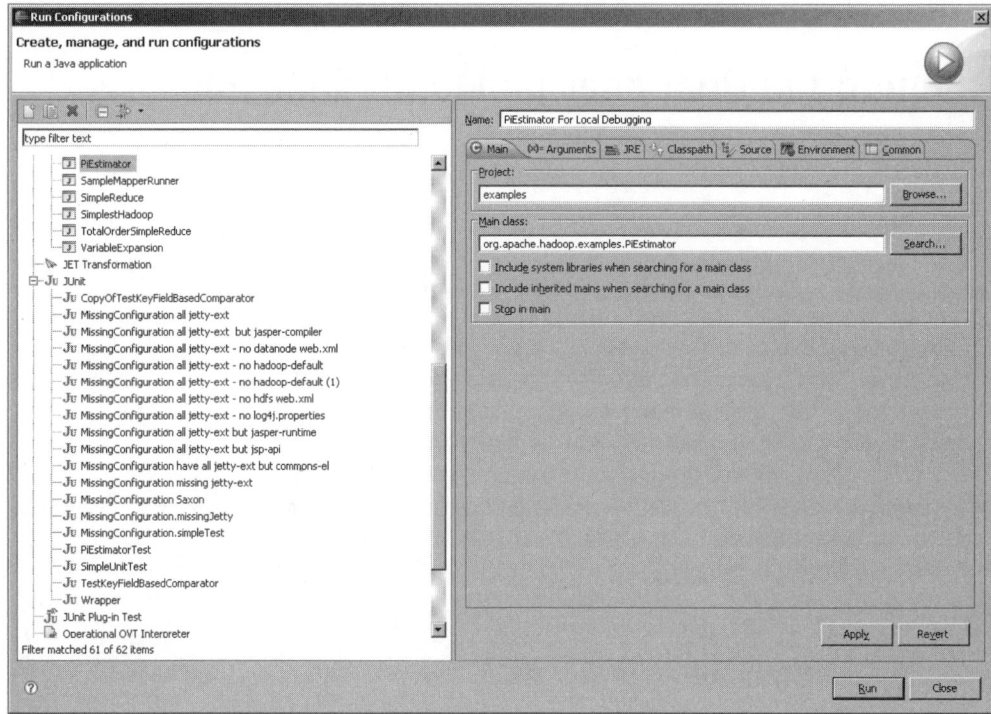

Figure 7-1. *Creating a run configuration for the PiEstimator in Eclipse*

The classpath must include the Core JAR, the example JAR (since the test case comes from this JAR), and all of the JARs from `lib` and `lib/jetty-ext`, as shown in Figure 7-1. You must explicitly specify all of the JARs required for your job here. In this example, Eclipse is not configured to load the native compression codec libraries.

What is the point of using a visual debugger if the source code is not available? You will need to set up any source paths required for your application, as shown in Figure 7-3.

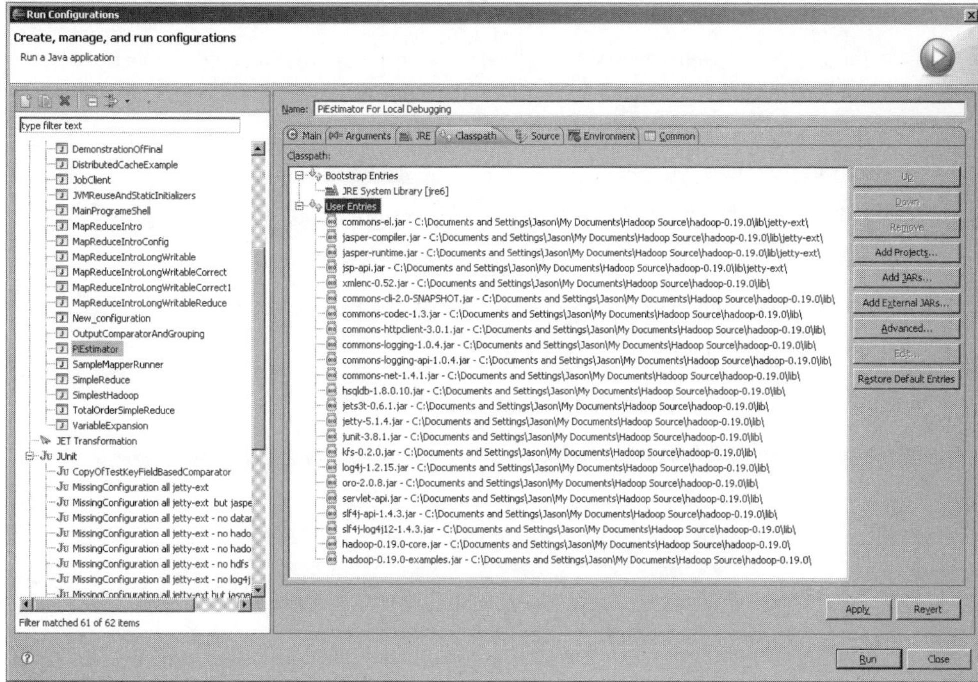

Figure 7-2. *Configuring the classpath*

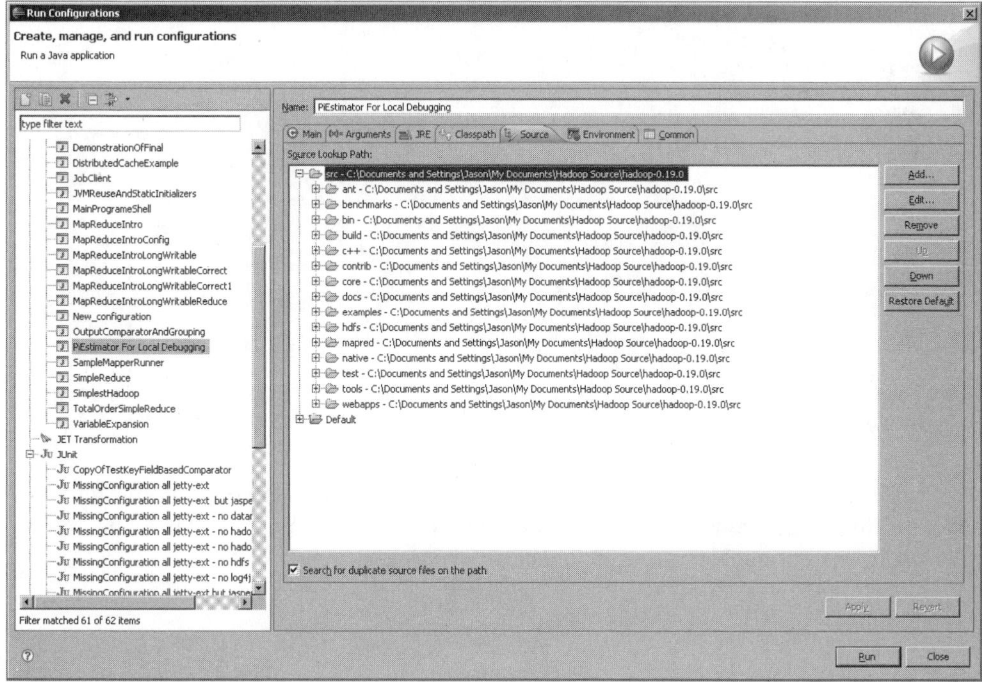

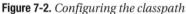

Figure 7-3. *Configuring the source path*

You also need to set up the following arguments, as shown in Figure 7-4:

- `-Dmapred.task.timeout=7200000`: Ensure that your tasks are not killed by the framework for not responding.
- `-Dmapred.job.tracker=local`: Run the MapReduce job using the `LocalJobRunner`, entirely in this JVM.
- `-Dfs.default.name=file:///`: Use the local file system for all storage.
- `-Dhadoop.tmp.dir=/tmp/pidebug`: Store all working files in this directory.
- `-Dio.sort.mb=2`: Allocate only 2MB for the merge-sort working space.
- `2`: Run two map tasks.
- `1000`: Generate 1,000 random samples in each map task.

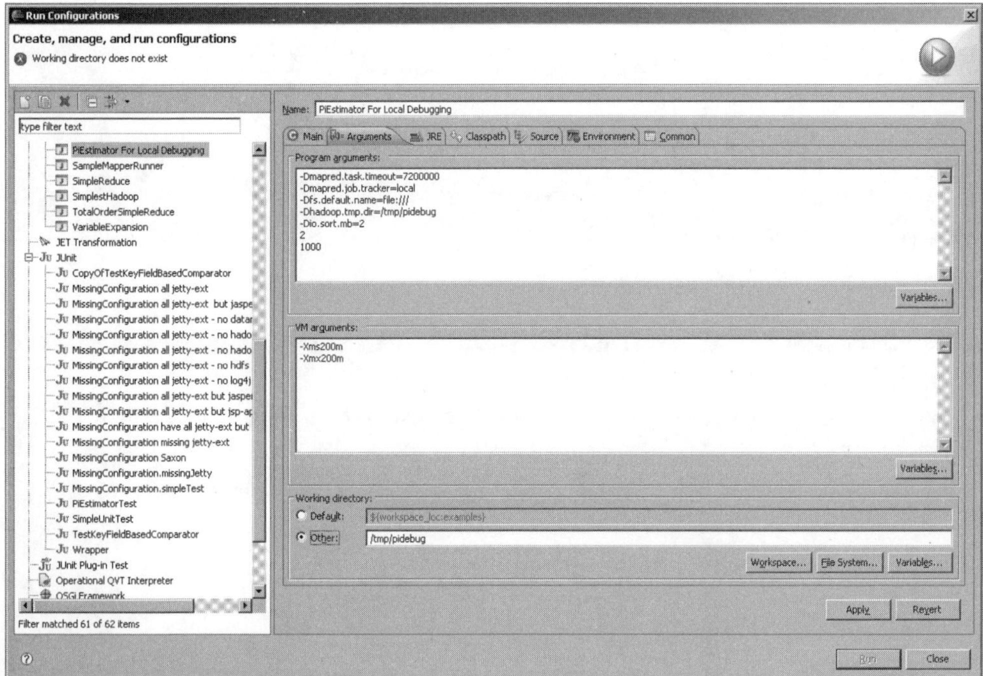

Figure 7-4. *Configuring the command-line arguments for debugging. Note the need to make the working directory.*

Once the directory `/tmp/pidebug` has been made, the Run button in the bottom-right corner of the Run Configurations window will be active. Click this button just to verify that the job will work in this setting. The output should be as follows:

```
Number of Maps = 2 Samples per Map = 1000
Wrote input for Map #0
Wrote input for Map #1
Starting Job
jvm.JvmMetrics: Initializing JVM Metrics with processName=JobTracker, sessionId=
mapred.FileInputFormat: Total input paths to process : 2
mapred.JobClient: Running job: job_local_0001
mapred.FileInputFormat: Total input paths to process : 2
mapred.MapTask: numReduceTasks: 1
mapred.MapTask: io.sort.mb = 2
mapred.MapTask: data buffer = 1593843/1992304
mapred.MapTask: record buffer = 5242/6553
mapred.MapTask: Starting flush of map output
mapred.MapTask: Finished spill 0
mapred.TaskRunner: Task:attempt_local_0001_m_000000_0 is done. ➥
And is in the process of commiting
mapred.LocalJobRunner: Generated 1 samples.
mapred.TaskRunner: Task 'attempt_local_0001_m_000000_0' done.
mapred.MapTask: numReduceTasks: 1
mapred.MapTask: io.sort.mb = 2
mapred.MapTask: data buffer = 1593843/1992304
mapred.MapTask: record buffer = 5242/6553
mapred.MapTask: Starting flush of map output
mapred.MapTask: Finished spill 0
mapred.TaskRunner: Task:attempt_local_0001_m_000001_0 is done. ➥
And is in the process of commiting
mapred.LocalJobRunner: Generated 1 samples.
mapred.TaskRunner: Task 'attempt_local_0001_m_000001_0' done.
mapred.Merger: Merging 2 sorted segments
mapred.Merger: Down to the last merge-pass, with 2 segments left ➥
of total size: 76 bytes
mapred.TaskRunner: Task:attempt_local_0001_r_000000_0 is done. ➥
And is in the process of commiting
mapred.LocalJobRunner:
mapred.TaskRunner: Task attempt_local_0001_r_000000_0 is allowed to commit now
mapred.FileOutputCommitter: Saved output of task 'attempt_local_0001_r_ ➥
000000_0' to file:/C:/tmp/pidebug/test-mini-mr/out
mapred.LocalJobRunner: reduce > reduce
mapred.TaskRunner: Task 'attempt_local_0001_r_000000_0' done.
mapred.JobClient: Job complete: job_local_0001
mapred.JobClient: Counters: 11
mapred.JobClient:   File Systems
mapred.JobClient:     Local bytes read=450833
mapred.JobClient:     Local bytes written=503689
```

```
mapred.JobClient:   Map-Reduce Framework
mapred.JobClient:     Reduce input groups=2
mapred.JobClient:     Combine output records=0
mapred.JobClient:     Map input records=2
mapred.JobClient:     Reduce output records=0
mapred.JobClient:     Map output bytes=64
mapred.JobClient:     Map input bytes=48
mapred.JobClient:     Combine input records=0
mapred.JobClient:     Map output records=4
Job Finished in 1.547 seconds
mapred.JobClient:     Reduce input records=4
Estimated value of PI is 3.132
```

The final line, with the estimated value of pi, indicates that the environment is correctly configured.

For Figure 7-5, a breakpoint was set in the map task, and the job launched via the Debug As Java Application menu item.

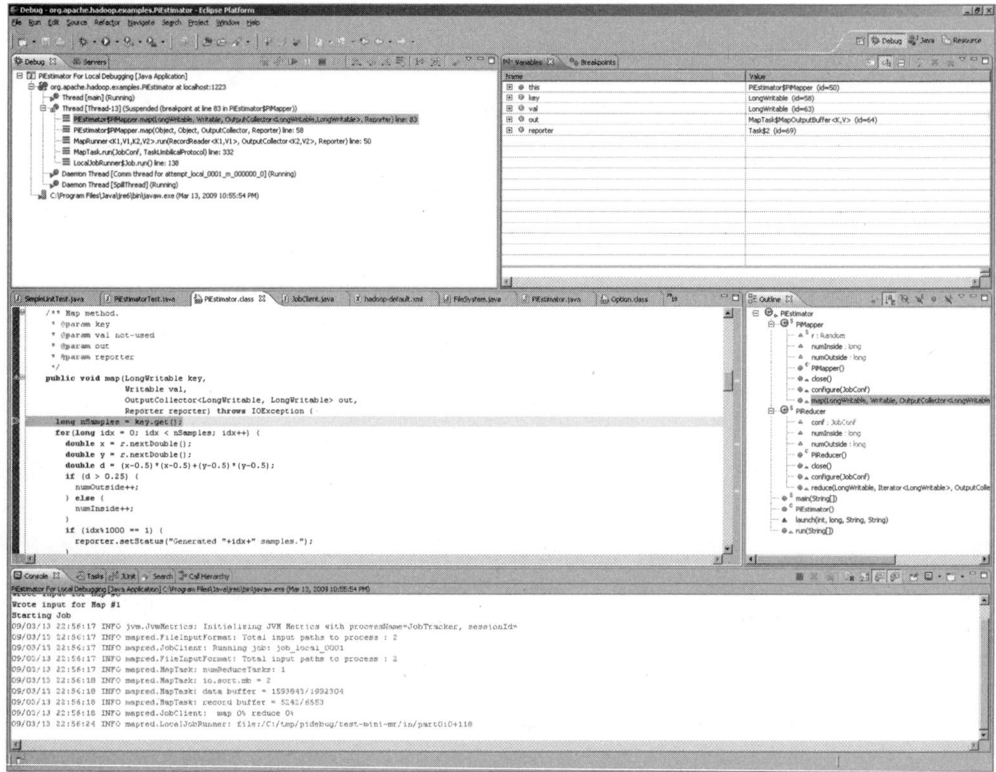

Figure 7-5. *Eclipse with PiEstimator stopped in the map task*

The job is configured to use the local file system for all working storage, and you may examine the files using the normal file system tools.

In this case, with the PiEstimator test stopped in the first line of the first map task, you can see that the working directory in the local file system contains a number of interesting files.

```
mapred/local/localRunner/.job_local_0001.xml.crc
mapred/local/localRunner/.split.dta.crc
mapred/local/localRunner/job_local_0001.xml
mapred/local/localRunner/split.dta
mapred/system/job_local_0001/.job.jar.crc
mapred/system/job_local_0001/.job.split.crc
mapred/system/job_local_0001/.job.xml.crc
mapred/system/job_local_0001/job.jar
mapred/system/job_local_0001/job.split
mapred/system/job_local_0001/job.xml
test-mini-mr/in/.part0.crc
test-mini-mr/in/.part1.crc
test-mini-mr/in/part0
test-mini-mr/in/part1
```

These files contain the following:

- mapred/local/localRunner/job_local_0001.xml contains the textual representation of the localized JobConf object for the task in progress, which is the task that is stopped in the debugger.
- mapred/system/job_local_0001/job.xml contains the textual representation of the JobConf object for the job. The file names that end in .crc are checksum files written out by the framework to provide data integrity checking.
- mapred/local/localRunner/split.dta contains the input split class and the data fields of the split that this task is to use as input.
- mapred/system/job_local_0001/job.split contains the list of input splits. It is prepared by the class JobClient as part of the job submission process.
- test-mini-mr/in/part0 and test-mini-mr/in/part1 are the input files to the job, prepared by the PiEstimator class. The PiEstimator class writes out SequenceFiles for its input, and these binary files may be examined with the command-line Hadoop tool via bin/hadoop dfs -fs file:/// -text /tmp/pidebug/ test-mini-mr/in/part1.

■ **Note** -fs file:/// indicates that the local file system is to be used, and -text causes the input file to be read as a SequenceFile and the key/value pairs printed via their respective toString() methods.

The input files part0 and part1 each has a single record of 1000, 0. Let's examine the contents of part0 of the input to demonstrate this.

```
> bin/hadoop dfs -fs file:/// -text /tmp/pidebug/test-mini-mr/in/part0
```

```
1000    0
```

Debugging a Task Running on a Cluster

Hadoop Core, as of version 0.19.0, does not provide any tools to specify which tasks of a job to enable Java debugging services on, nor does Hadoop Core provide a way to indicate on which host and port such a task might be listening for remote debugging connections. To debug tasks running on a cluster, the JVM parameters for the task have remote debugging enabled via the command-line arguments. The core issue is to arrange for the JVM of the server or task that is to be debugged to have the additional command-line arguments that enable the Java Platform Debugger Architecture (JPDA) servers. Table 7-3 describes the parameters to the JPDA debugging agent, `agentlib:jdwp`, that must be enabled in the task JVM to allow connections from debuggers.

■ **Tip** The Sun VM Invocation Options guide at `http://java.sun.com/javase/6/docs/technotes/guides/jpda/conninv.html#Invocation` provides many details on how the command-line arguments for debugging may be constructed.

Table 7-3. *Configuration Parameters for the Debugger Agent*

Parameter	Description
`suspend`	Set this to `suspend=y` to suspend the task on start; otherwise, the task will run normally until a debugger connects to it.
`address`	Set this to `address=0.0.0.0` to have the debugger agent bind to the machine wildcard address at an allocated port. This address is specific to IP version 4.
`launch`	A program to invoke when the debugger initializes. The program receives two arguments: the transport, `dt_socket`, and the allocated port. Only a program name is accepted as a value. This program may be used to provide notification that the task has engaged the debugger agent and the port it is listening on for a debugger connection.
`onthrow`	Do not initialize the debugger agent until an exception with the parameter value is thrown. In the example `onthrow=java.io.IOException`, the debugger agent would not initialize and bind a port until an `IOException` was thrown. At this point, any program defined by `launch` would be executed. Program execution will be stopped, while the agent waits for a debugger connection.
`onuncaught`	Works like `onthrow`, except that the exception must not be caught.

This section will consider only using the debugger configured for TCP/socket-based transport, à *la* remote debugging, in Eclipse.

The configuration parameter that needs to be set is `mapred.child.java.opts`, which may be set at the job level. The JPDA invocation arguments need to be added to the value for this parameter. The parameters that are most relevant to the user are `suspend` and `address`. The `suspend=y` setting forces the JVM to be stopped just before the `main()` method is invoked. The value `suspend=n` may also be used, in which case the JVM will run normally. The parameter `address=0.0.0.0` instructs the debugger interface in the JVM to allocate a free port and to listen for connections on the wildcard address of the local machine. The port that is allocated will be printed on the standard output.

■ **Note** Unless the cluster is configured for one map task per machine, and the job has no reduce tasks specified, there will be multiple tasks running on any given TaskTracker machine. This precluded specifying a fixed port as part of the address parameter. If only a single task will run at a time, a port may be specified via `address=host:port`, where `host` is optional.

Let's start a Hadoop job with remote debugging enabled.

```
HADOOP_OPTS=-agentlib:jdwp=transport=dt_socket,server=y,suspend=y,address=0.0.0.0 ➥
bin/hadoop  dfs -fs file:/// -text /tmp/pidebug/test-mini-mr/in/part0
```

```
Listening for transport dt_socket at address: 59348
```

It is best to have only the option `suspend` set to `y` for all of the task JVMs, isolate this option to a single machine in the cluster, to avoid requiring extensive interaction with each task, at minimum, connecting to the task in the debugger to resume execution.

Figures 7-6, 7-7, and 7-8 show our friend `PiEstimator` being run on a small cluster. The JPDA arguments will be passed via the command line to the child JVM, and JVM reuse will be explicitly disabled to avoid complications. Figure 7-6 shows the Eclipse setup for a remote debugging session.

In the dialog box in Figure 7-7, New has been clicked, and the title of the debug session changed to `Remote PiEstimator 53075`. The Source tab has been configured identically to the earlier source configuration shown in Figure 7-3, with the `hadoop/src` directory and its subfolders. Then you click the Apply and Debug buttons, to save this session and connect to the task. Figure 7-8 shows the `PiEstimator` task in the debugger, stopped at a breakpoint in the `map()` method.

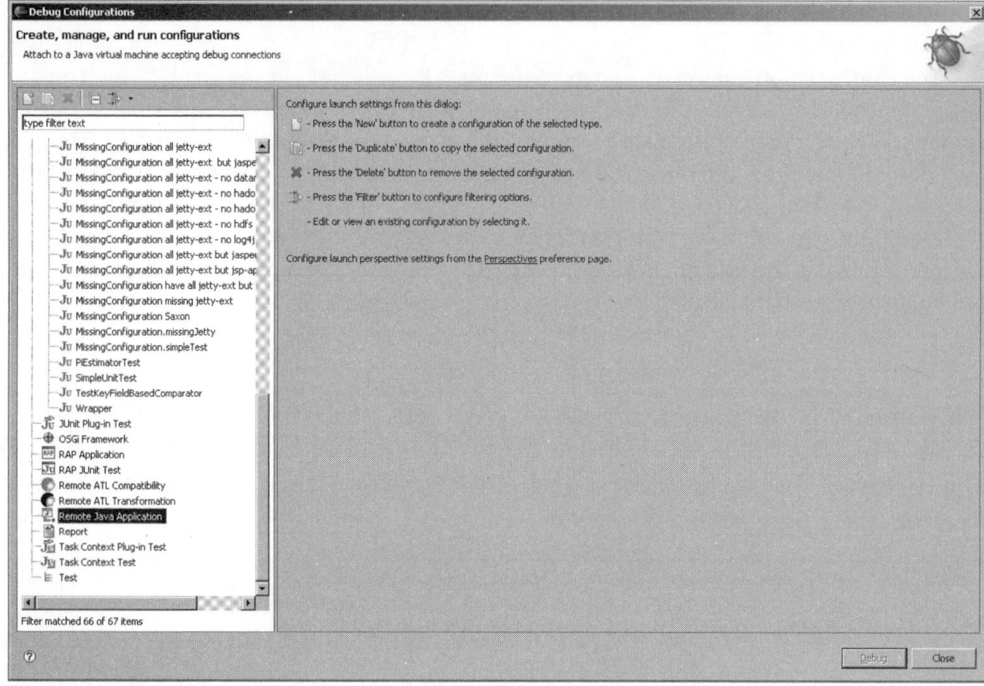

Figure 7-6. *Setting up Eclipse for a remote debugging session*

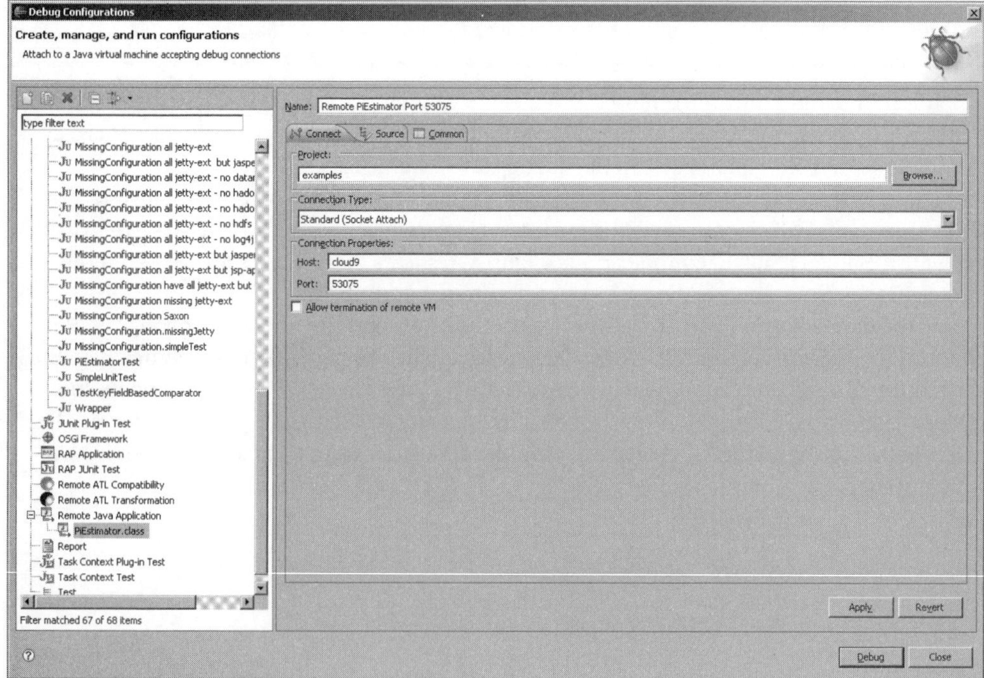

Figure 7-7. *Specifying the port and host to connect to. This is configured per task and needs to be determined from task log files.*

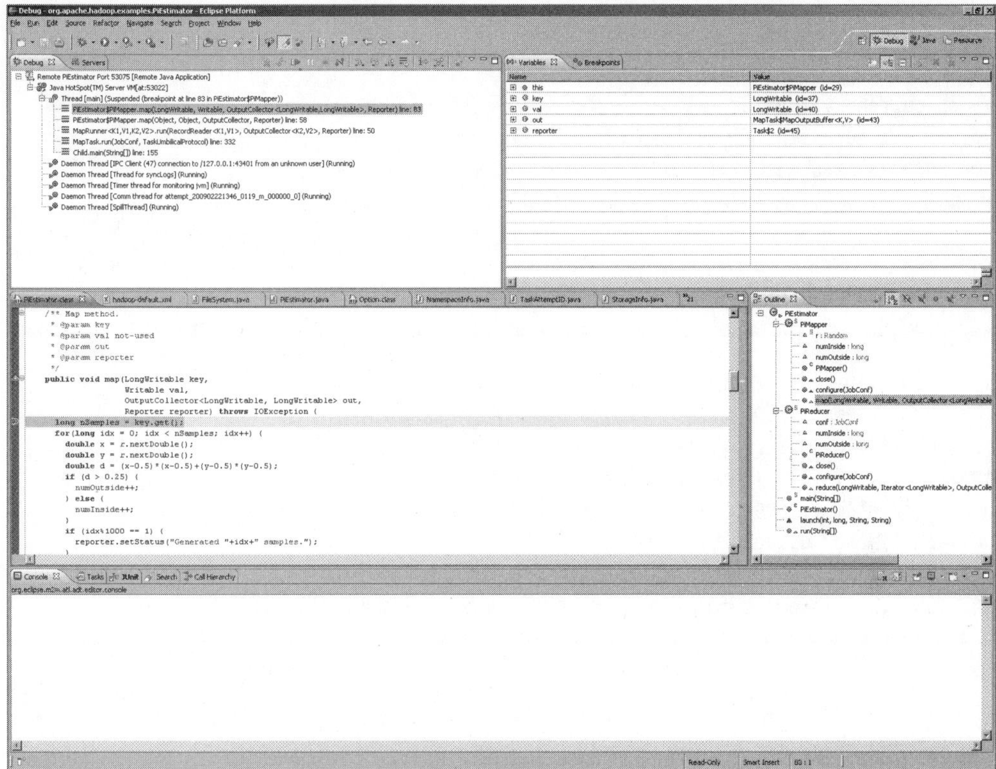

Figure 7-8. *A remotely connected session to a map task*

To determine the port to connect to, if no `launch` program has been provided, requires finding the per-task stdout log file. At present, this step is manual and requires issuing shell commands on one of the cluster machines.

The job ID must be determined. This is available via the JobTracker web interface or via the command-line tool:

```
bin/hadoop job -list
```

In this case, two tasks are suspended and waiting for the debugger, one at 192.168.1.2:43004 and the other at 192.168.1.2:35403. These values are what you would put into the Host and Port fields of the Remote Debugging configuration dialog box. The `slaves.sh` command will execute its command-line arguments as a `shell` command on each machine in the `conf/slaves` file, the output of these commands will have the generating host's IP address prefixed to the output lines.

Let's see an example of determining the ports and hosts of the suspended tasks:

```
> bin/hadoop job -list
```

```
1 job currently running
JobId   State   StartTime   UserName    Priority    SchedulingInfo
job_200902221346_0119   1   1237025200426   jason   NORMAL
```

```
> cd logs
../bin/slaves.sh cat $PWD/userlogs/*200902221346_0119*/stdout
```

```
192.168.1.119: cat: /home/jason/src/hadoop-0.19.0/logs/ ➥
userlogs/attempt_200902221346_0119_r_000000_0/stdout: ➥
No such file or directory ➥
192.168.1.119: cat: /home/jason/src/hadoop-0.19.0/logs/ ➥
userlogs/attempt_200902221346_0119_r_000002_0/stdout: ➥
No such file or directory
192.168.1.2: Listening for transport dt_socket at address: 43004
192.168.1.2: Listening for transport dt_socket at address: 35403
```

■ **Note** The port numbers change for each task, and there may be inconsistency in the ports shown in the screenshots and those mentioned in the text.

Rerunning a Failed Task

The `IsolationRunner` provides a way of rerunning a task out of a failed job. Normally, the framework immediately removes all local task specific data when a task finishes.

Configuring the Job or Cluster to Save the Task Local Working Directory

Two configuration keys provide some control over the ability to rerun a task. The value of the configuration key `keep.task.files.pattern` is a Java regular expression, which is matched against task files. Any task that matches this pattern will not have its task files removed. The effective code used for this is as follows:

```
alwaysKeepTaskFiles =
  Pattern.matches(conf.getKeepTaskFilesPattern(), task.getTaskID().toString()
```

To save the results of all map tasks, a pattern `*_m.*` would work. To match all reduces tasks, `*_r_*`.

The other option is to set the value of the configuration key `keep.failed.tasks.files` to `true`. Any task that fails will not be subject to cleanup.

Caution Nothing will reclaim this space, which may be quite large. Do this with care and clean up afterwards.

Determining the Location of the Task Local Working Directory

The root directory set for the local working areas is stored in the configuration under the key `mapred.local.dir`, and may be a comma-separated list of directories. Running the `find` command on this set of directories looking for files named `job.xml` will present a set of candidate tasks to be run via the `IsolationRunner`.

```
> find /tmp/hadoop-0.19.0-jason/mapred/local/ -wholename '*attempt*/job.xml' -print
```

```
/tmp/hadoop-0.19.0-jason/mapred/local/taskTracker/jobcache/ ➥
job_200902221346_0119/attempt_200902221346_0119_m_000000_0/job.xml
```

Running a Job with a Keep Pattern and Debugging via the IsolationRunner

Again, this example uses our old friend the `PiEstimator` job. It demonstrates how to run it so that the map task local file space is left in intact. Then you find the `job.xml` files that can be run via the `IsolationRunner` and run one of them in a way that will enable the use of the debugger.

Here's how to put it all together for the `IsolationRunner`:

```
> bin/hadoop jar hadoop-0.19.0-examples.jar pi ➥
-D keep.task.files.pattern=".*_m.*" 2 1000
```

```
...cd
Estimated value of PI is 3.102
```

```
> find /tmp/hadoop-0.19.0-jason/mapred/local/ -wholename '*attempt*/job.xml' -print
```

```
/tmp/hadoop-0.19.0-jason/mapred/local/taskTracker/jobcache/ ➥
job_200902221346_0120/attempt_200902221346_0120_m_000002_0/job.xml
/tmp/hadoop-0.19.0-jason/mapred/local/taskTracker/jobcache/ ➥
job_200902221346_0120/attempt_200902221346_0120_m_000003_0/job.xml
> cd /tmp/hadoop-0.19.0-jason/mapred/local/taskTracker/jobcache/ ➥
job_200902221346_0120/attempt_200902221346_0120_m_000003_0/
```

```
> HADOOP_OPTS=-agentlib:jdwp=transport=dt_socket,server=y,address=0.0.0.0 ➥
~/src/hadoop-0.19.0/bin/hadoop org.apache.hadoop.mapred.IsolationRunner ../job.xml
```

```
Listening for transport dt_socket at address: 54990
```

Once the child JVM is configured and waiting, Eclipse must be configured to connect to it. Figure 7-9 shows the Eclipse Debug Configuration window for setting up a remote debugging connection. The Host field must be filled in with the host on which the JVM is running, and the Port field filled in with the value from the `Listening for transport dt_socket at address:` *XXXXX* output line from the JVM. Figure 7-10 shows Eclipse connected to the running JVM of the `PiExample` test case.

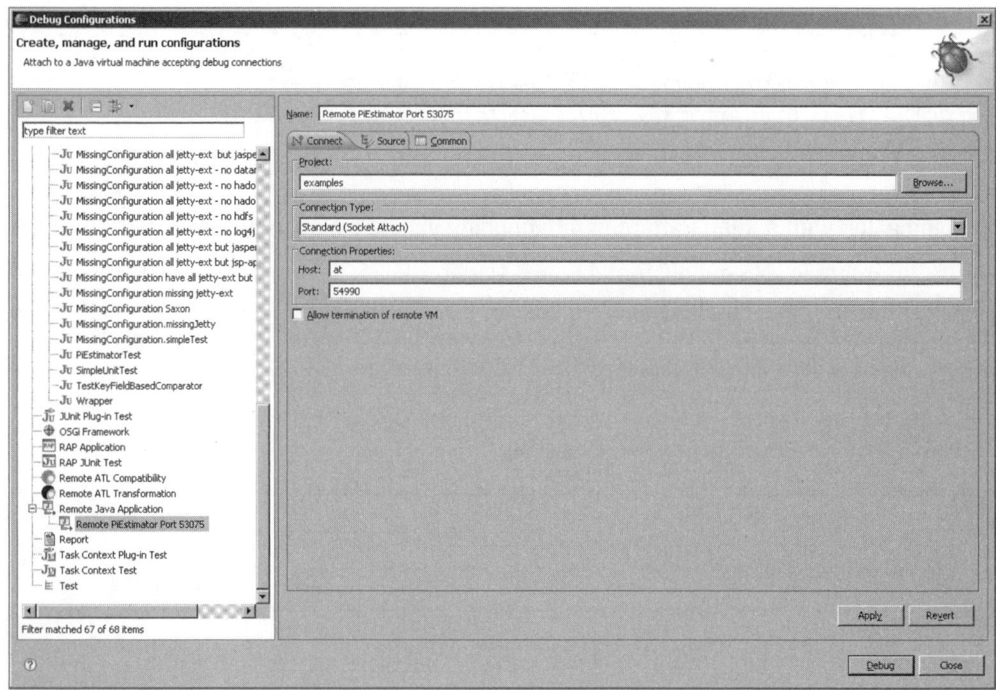

Figure 7-9. *Eclipse setup to connect to port 54990 on the specified host*

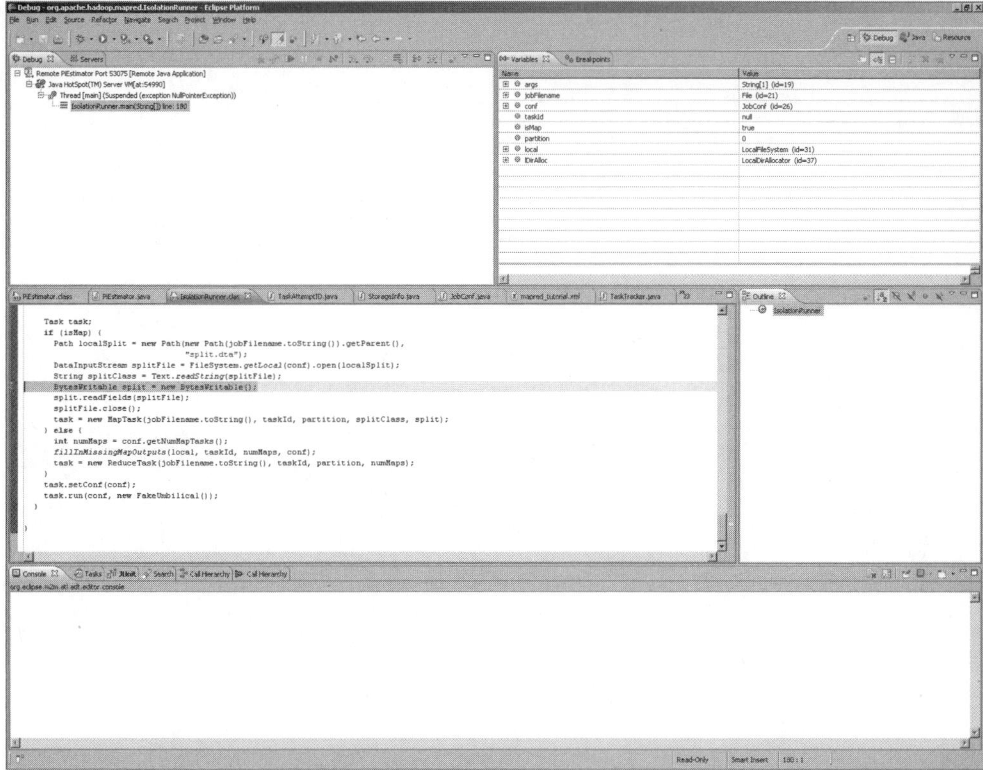

Figure 7-10. *Eclipse connected to the IsolationRunner*

Now you may set your breakpoints, debug, and explore.

Summary

It is very important to know that your code is correct. This chapter has provided you with the techniques needed to build unit tests for your MapReduce jobs. MapReduce applications can also be test-driven.

This chapter also demonstrated three ways to run MapReduce applications under the Eclipse debugger. This will enable you to understand problems that occur only at scale on your large clusters, as well as explore how things work directly on your local development machine.

CHAPTER 8

Advanced and Alternate MapReduce Techniques

This chapter discusses techniques for handling larger jobs with more complex requirements. In particular, the section on map-side joins covers the case in which the input data is already sorted, and the section on chaining discusses ways of adding additional mapper classes to a job without passing all the job data through the network multiple times.

The traditional MapReduce job involves providing a pair of Java classes to handle the map and reduce tasks: reading a set of textual input files using `KeyValueTextInputFormat` or `SequenceFileInputFormat`, and writing the sorted results set out using `TextOutputFormat` or `SequenceFileOutputFormat`. The framework will schedule the map tasks if possible so that each map task's input is local, and provides several ways of reducing the volume of output that must pass over the network to the reduce tasks. This is a good pattern for many (although not all) applications.

There are other options available to Hadoop Core users, either by changing the pattern of the job or by providing the ability to use other languages, such as C++ or Perl and Python, for mapping and reducing.

Streaming: Running Custom MapReduce Jobs from the Command Line

The streaming API allows users to configure and submit complete MapReduce jobs using the command line. As an added bonus, streaming provides the ability to use external programs as any of the job's mapper, combiner, or reducer. The job is a traditional MapReduce job, with the framework handling input splitting, scheduling map tasks, scheduling input split pairs to run, shuffling and sorting map outputs, scheduling reduce tasks to run, and then writing the reduce output to the Hadoop Distributed File System (HDFS).

In the following example, we will demonstrate how to run a simple streaming job to sort all the input records of a dataset using MapReduce. The argument informs the `bin/hadoop` script that the streaming JAR is to be used. Mapper and reduce are defaulted to the identity versions. `-inputformat org.apache.hadoop.mapred.KeyValueTextInputFormat` causes the `KeyValueTextInputFormat` class to be used to provide the key/value pairs from the input. The use of this input format requires the output key format be set to Text via `-D mapred.output.key.class=org.apache.hadoop.io.Text`.

```
bin/hadoop jar ./contrib/streaming/hadoop-0.19.0-streaming.jar -D➥
mapred.output.key.class=org.apache.hadoop.io.Text➥
-inputformat org.apache.hadoop.mapred.KeyValueTextInputFormat➥
-numReduceTasks 2 -input words -output next
```

```
packageJobJar: [/tmp/hadoop-0.19.0-jason/hadoop-unjar11738/]➥
[] /tmp/streamjob11739.jar tmpDir=null
mapred.FileInputFormat: Total input paths to process : 1
streaming.StreamJob: getLocalDirs(): [/tmp/hadoop-0.19.0-jason/mapred/local]
streaming.StreamJob: Running job: job_200902221346_0136
streaming.StreamJob: To kill this job, run:
streaming.StreamJob: /home/jason/src/hadoop-0.19.0/bin/../bin/hadoop job➥
-Dmapred.job.tracker=cloud9:8021 -kill job_200902221346_0136
streaming.StreamJob: Tracking URL: ➥
http://192.168.1.2:50030/jobdetails.jsp?jobid=job_200902221346_0136
streaming.StreamJob:  map 0%  reduce 0%
streaming.StreamJob:  map 50%  reduce 0%
streaming.StreamJob:  map 100%  reduce 0%
streaming.StreamJob:  map 100%  reduce 8%
streaming.StreamJob:  map 100%  reduce 17%
streaming.StreamJob:  map 100%  reduce 58%
streaming.StreamJob:  map 100%  reduce 100%
streaming.StreamJob: Job complete: job_200902221346_0136
streaming.StreamJob: Output: next
```

Hadoop streaming provides the user with the ability to use arbitrary programs for a job's map and reduce methods. The framework handles a streaming job like any other MapReduce job. The job might specify that an executable be used as the map processor and for the reduce processor. Each task will start an instance of the applicable executable and write an applicable representation of the input key/value pairs to the executable. The standard output of the executable is parsed as textual key/value pairs. The executable being run for the reduce task will given an input line for each value in the reduce value iterator, composed of the key and that value.

The following example uses `/bin/cat` as the mapper and a Perl program to produce line counts of distinct lines from the input set. The argument `-file /tmp/wordCount.pl` causes the file `/tmp/wordCount.pl` to be copied into HDFS and then made available to the map and reduce tasks in their current working directory. The argument `-reducer "/usr/bin/perl -w wordCount.pl"` causes the Perl program `wordCount.pl` to be used to perform the reduce.

```
bin/hadoop jar ./contrib/streaming/hadoop-0.19.0-streaming.jar -input words➥
-output next -file /tmp/wordCount.pl -mapper /bin/cat➥
-reducer "/usr/bin/perl -w wordCount.pl"
```

```
packageJobJar: [/tmp/wordCount.pl, /tmp/hadoop-0.19.0-jason/hadoop-unjar57851/]➥
[] /tmp/streamjob57852.jar tmpDir=null
09/03/15 15:20:40 INFO mapred.FileInputFormat: Total input paths to process : 1
09/03/15 15:20:40 INFO streaming.StreamJob: getLocalDirs():➥
[/tmp/hadoop-0.19.0-jason/mapred/local]
09/03/15 15:20:40 INFO streaming.StreamJob: Running job: job_200902221346_0139
09/03/15 15:20:40 INFO streaming.StreamJob: To kill this job, run:
09/03/15 15:20:40 INFO streaming.StreamJob: ➥
/home/jason/src/hadoop-0.19.0/bin/../bin/hadoop job➥
-Dmapred.job.tracker=cloud9:8021 -kill job_200902221346_0139
09/03/15 15:20:40 INFO streaming.StreamJob: Tracking URL:➥
http://192.168.1.2:50030/jobdetails.jsp?jobid=job_200902221346_0139
09/03/15 15:20:41 INFO streaming.StreamJob:  map 0%  reduce 0%
09/03/15 15:20:53 INFO streaming.StreamJob:  map 50%  reduce 0%
09/03/15 15:20:55 INFO streaming.StreamJob:  map 100%  reduce 0%

09/03/15 15:21:13 INFO streaming.StreamJob:  map 100%  reduce 100%
09/03/15 15:21:14 INFO streaming.StreamJob: Job complete: job_200902221346_0139
09/03/15 15:21:14 INFO streaming.StreamJob: Output: next
```

```
cat /tmp/wordCount.pl
```

```
#! /usr/bin/perl -w

use strict;

# The reduce is just passed the same key with
# each value in the value group, so keep track of
# the last key seen to determine when a new value group has started.

my $lastSeenKey;
my $currentCount = 0;

while( <> ) {
    chomp;

    my $currentKey = $_;

    if ($lastSeenKey && $lastSeenKey ne $currentKey) {
        # emit the record for the previous key
        print $currentCount, "\t", $lastSeenKey, "\n";
        $currentCount = 1;
        $lastSeenKey = $currentKey;
```

```
    } else {
        # save currentKey away just in case it hasn't been saved
        $lastSeenKey = $currentKey;
        $currentCount++;

    }
}

# make sure that the count for the last key in the input is emitted
if ($lastSeenKey) {
    print $currentCount, "\t", $lastSeenKey, "\n";
}
```

Hadoop streaming is a wonderful tool. I had a large dataset composed of many input files in one compression format, and the data needed to be compressed in a different format. The author ran a streaming job, with the map executable set to `/bin/cat` and the minimum input split size set to `Long.MAX_VALUE`, and enabled map output compression of the required type. In a few minutes the cluster had uncompressed and recompressed the data files.

The following streaming example takes input from the directory `words`, uses `/bin/cat` as the map executable, has no reduce phase, and compresses the job output using the `GzipCodec`. The `IdentityMapper` could have been used just as easily, but the use of `/bin/cat` is just plain fun.

```
bin/hadoop jar ./contrib/streaming/hadoop-0.19.0-streaming.jar➥
-D mapred.output.compress=true➥
-D mapred.output.compression.codec=org.apache.hadoop.io.compress.GzipCodec➥
-D mapred.min.split.size=111111111111 -input words➥
-mapper /bin/cat -numReduceTasks 0  -output next
```

```
packageJobJar: [/tmp/hadoop-0.19.0-jason/hadoop-unjar13326/]➥
[] /tmp/streamjob13327.jar tmpDir=null
mapred.FileInputFormat: Total input paths to process : 1
streaming.StreamJob: getLocalDirs(): [/tmp/hadoop-0.19.0-jason/mapred/local]
streaming.StreamJob: Running job: job_200902221346_0125
streaming.StreamJob: To kill this job, run:
streaming.StreamJob: /home/jason/src/hadoop-0.19.0/bin/../bin/hadoop job➥
-Dmapred.job.tracker=cloud9:8021 -kill job_200902221346_0125
streaming.StreamJob: Tracking URL: ➥
http://192.168.1.2:50030/jobdetails.jsp?jobid=job_200902221346_0125
streaming.StreamJob:  map 0%  reduce 0%
streaming.StreamJob:  map 100%  reduce 0%
streaming.StreamJob: Job complete: job_200902221346_0125
streaming.StreamJob: Output: next
```

```
bin/hadoop dfs -ls next
```

```
Found 2 items
drwxr-xr-x   - jason supergroup          0 2009-03-15 01:59 /user/jason/next/_logs
-rw-r--r--   3 jason supergroup    1496397 2009-03-15 01:59➥
/user/jason/next/part-00000.gz
```

■ **Note** The streaming API explicitly forces the output key/value classes to be text. If Java classes are used for the mapper, combiner, or reducer, the `InputFormat` used must produce text key/value pairs, or the `mapred.map.output.key.class` and `mapred.map.output.value.class` configuration key/values must be explicitly set to the class names of the key/value classes via `-D mapred.map.output.key.class=java.class.name` or `-D mapred.map.output.value.class=java.class.name`.

JYTHON: A WAY OF INTERACTING WITH JAVA CLASSES IN PYTHON

The Jython Project, `http://www.jython.org/`, provides an implementation of Python written in Java. Not all Python features are available. There are additional language constructs that allow the addition of arbitrary Java classes into the namespace of the Jython applications. There are also additional primitive operators for interacting with native Java types, and a transparent translation between the Java String class and the Python string class.

The Hadoop Core distribution provides a Jython example MapReduce application in `src/examples/python/WordCount.py`. People have good results having Python applications used by Hadoop streaming.

Streaming Command-Line Arguments

The streaming command-line interface provides a rich set of command-line arguments for controlling the execution of your streaming job. The standard Hadoop `GenericOptionsParser` arguments are also supported. Table 8-1 describes the streaming-specific command-line arguments, and Figure 8-1 details how the job records are transformed between records and key/value pairs as the framework passes the records through the steps of a streaming job.

Table 8-1. *Streaming Specific Command-Line Arguments*

Flag	Value	Description	Java Equivalent
`-input`	Required	The file or directory to use as input. This flag sets the input location for the MapReduce job. It may be given multiple times to provide multiple input paths.	`FileInputFormat.addInputPath()`
`-output`	Required	The directory to use for output. This flag sets the directory that output files will be written to. The directory must not exist prior to job start, and will be created by the framework for the job.	`FileOutputFormat.setOutputPath()`
`-mapper`	`org.apache.hadoop.mapred.lib.IdentityMapper`	A Java class name or an executable file. This flag is used as the mapper for the map tasks.	`JobConf.setMapperClass()`
`-combiner`	None	A Java class name or an executable file. This flag is used as the combiner for the map output.	`JobConf.setCombinerClass()`
`-reducer`	`org.apache.hadoop.mapred.lib.IdentityReducer`	A Java class name or an executable file. This flag is used as the reducer for the reduce tasks.	`JobConf.setReducerClass()`
`-file`	None	A file to be made available locally to each task. This flag is often used to pass the executable to be used for the mapper, combiner, or reducer to the tasks. The executable will be stored in the current working directory of the task.	`DistributedCache.addCacheFile()` with symlink
`-inputformat`	`org.apache.hadoop.mapred.TextInputFormat`	The class name of the handler that will split and read the input files and provide key/value pairs for the mapper. There is special handling for the fully qualified class names of `TextInputFormat`, `KeyValueTextInputFormat`, `SequenceFileInputFormat`, and `SequenceFileAsTextInputFormat`. The default `TextInputFormat` is most efficient because the individual input lines are not split into key/value pairs.	`JobConf.setInputForm`
`-outputformat`	`org.apache.hadoop.mapred.TextOutputFormat`	The class that will be used write the output files for the job.	
`-partitioner`	`org.apache.hadoop.mapred.lib.HashPartitioner`	The class that will be used to determine which reduce any given key is sent to.	`JobConf.setPartitionerClass()`
`-numReduceTasks`	1	The number of reduce tasks to run.	`JobConf.setNumReduceTasks()`

Flag	Value	Description	Java Equivalent
-inputreader	None	Custom class to read records from input files. This is currently used only in the framework for the org.apache.hadoop.streaming.StreamXmlRecordReader, which lets splits be defined by a beginning and ending XML tag. The argument StreamXmlRecord,begin=BEGIN_STRING,end=END_STRING will result in input splits composed of the text found in files between BEGIN_STRING and END_STRING. (See Listing 8-1 for an example.)	
-cmdenv	None	Key/value pairs to set in the process environment before starting an executable mapper, combiner, or reducer.	
-mapdebug	None	A script to invoke when a map task fails.	JobConf.setMapDebugScript()
-reducedebug	None	A script to invoke when a reduce task fails.	JobConf.setReduceDebugScript()
-verbose	None	Used to turn on verbose output for the streaming framework.	

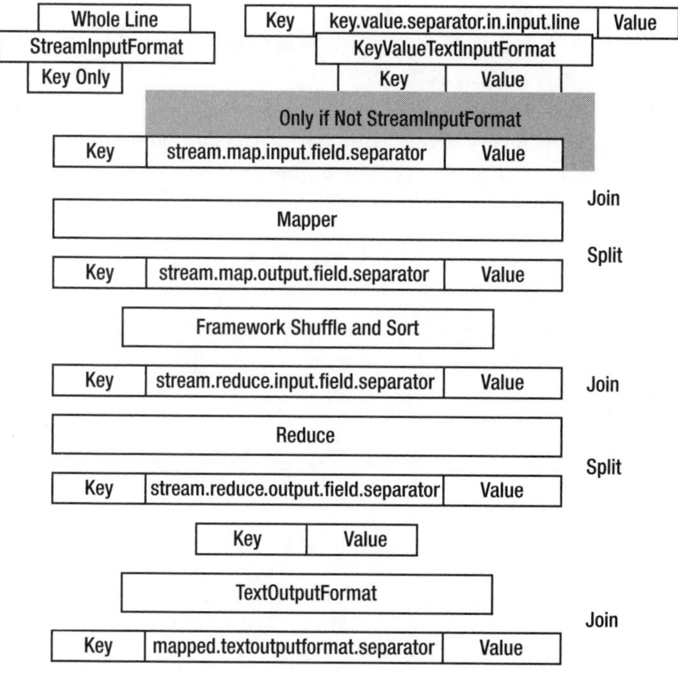

Figure 8-1. *How key/value pairs are split and joined in a streaming job*

Using -inputreader org.apache.hadoop.streaming.StreamXmlRecordReader

The `-inputreader` command-line flag is an unusual input format handler that provides two core features. The keys that are emitted by the `StreamXMLRecordReader` class contain only the text that is between a beginning and an ending marker, inclusive. Any text in the file that is not between a beginning and an ending marker is ignored.

When run on an Ant build file, the following example produces a sorted list of the target blocks within the XML file. The block selection is based on the `begin=<target name` and `end=</target>` values. The keys produced will contain all the text of the block, starting with `<target name` and ending with `</target>`. This text will include any line separator sequences that are present in the original file in the block. The value is the empty string.

It is best to use `StreamXmlRecordReader` with a Java-based mapper because the key/value pairs may contain line separators. Listing 8-1 demonstrates the use of the `StreamXMLRecordReader`, Listing 8-2 provides the input, and Listing 8-3 provides the output.

Listing 8-1. *Using the StreamRecordReader*

```
bin/hadoop jar ./contrib/streaming/hadoop-0.19.0-streaming.jar ➥
-Dmapred.mapoutput.key.class=org.apache.hadoop.io.Text -inputreader➥
"org.apache.hadoop.streaming.StreamXmlRecordReader,begin=➥
<target name,end=➥
</target>" -input xml_test.xml -output next
```

```
mapred.FileInputFormat: Total input paths to process : 1
streaming.StreamJob: getLocalDirs(): [/tmp/hadoop-0.19.0-jason/mapred/local]
streaming.StreamJob: Running job: job_200902221346_0144
streaming.StreamJob: To kill this job, run:
streaming.StreamJob: /home/jason/src/hadoop-0.19.0/bin/../bin/hadoop job➥
-Dmapred.job.tracker=cloud9:8021 -kill job_200902221346_0144➥
streaming.StreamJob: Tracking URL: http://192.168.1.2:50030/➥
jobdetails.jsp?jobid=job_200902221346_0144➥
streaming.StreamJob:  map 0%  reduce 0%
streaming.StreamJob:  map 50%  reduce 0%
streaming.StreamJob:  map 100%  reduce 0%
streaming.StreamJob:  map 100%  reduce 100%
streaming.StreamJob: Job complete: job_200902221346_0144
streaming.StreamJob: Output: next
```

Listing 8-2. *StreamXMLRecordReader Sample Input, xml_text.xml*

```
<xml>
<target name="part 05">part 05</target>
<target name="part 04"><name>part 04</name></target>
<target name="part 03"><name>part 03</name><value>More things</value></target>
<target name="part 02">
    <name>part 02</name>
    <description>Multi line
    text block>
    </description>
</target>
<target name="part 01"><name>part 01</name>
</target>
</xml>
```

Listing 8-3. *StreamRecordReader Output, Next/Part-00000*

```
<target name="part 01"><name>part 01</name>
</target>
<target name="part 02">
    <name>part 02</name>
    <description>Multi line
    text block>
    </description>
</target>
<target name="part 03"><name>part 03</name><value>More things</value></target>
<target name="part 04"><name>part 04</name></target>
<target name="part 05">part 05</target>
```

The `StreamXmlRecordReader` parameters are described in Table 8-2. Two parameters, `begin` and `end`, are required. There is some ability to control how much read ahead is done when looking for a match end.

It is possible to control the maximum size of a key. If the parameter `slowmatch=true` is provided, the framework will attempt to exclude recognizing the beginning and ending text if they are within a CDATA block. The framework will look ahead only in `lookahead` bytes, which by default is equal to twice `maxrec` bytes, or 50,000.

Table 8-2. *Control Parameters for StreamXMLRecordReader*

Parameter	Default	Description
`begin`	None	Required, String, or Java regular expression used to match the beginning of a block of interest. The value is interpreted as a regular expression if `slowmatch` is `true`.
`end`	None	Required, String, or Java regular expression used to match the end of a block of interest. The value is interpreted as a regular expression if `slowmatch` is `true`.
`slowmatch`	`False`	Attempts to exclude beginning or ending matches that are in CDATA blocks.
`maxrec`	50000	Used only when `slowmatch` is `true`. The record reader will look forward only the maximum of `maxrec` or `lookahead` bytes for the end of a CDATA block.
`lookahead`	`2*maxrec`	Used only when `slowmatch` is `true`. The record reader will look forward only the maximum of `maxrec` or `lookahead` bytes for the end of a CDATA block.

For more information, look at the excellent tutorial on using streaming at the Hadoop Core web site: `http://hadoop.apache.org/core/docs/current/streaming.html`.

Using Pipes

Hadoop Core provides a set of APIs for use by other languages that allow a reasonably rich interaction with the Hadoop framework. There are libraries available for C++. The C++ interface lends itself to usage by Simplified Wrapper and Interface Generator (SWIG) to generate other language interfaces.

The usage of the pipes APIs are outside of the scope of this book (refer to the wordcount-simple.cc example in your distribution and the tutorial in the Hadoop wiki: `http://wiki.apache.org/hadoop/C++WordCount`).

SWIG

SWIG (`http://www.swig.org/`) is a tool for building language interfaces to C and C++ code.

The C++ APIs for interacting with MapReduce are located in the directory `src/c++/pipes/api/hadoop` of the Hadoop Core distribution.

Using Counters in Streaming and Pipes Jobs

The framework monitors the standard error stream of the mapper and reducer processes. Any line read from the standard error stream that starts with the string `reporter:` is considered by the framework as an interaction command. As of Hadoop 0.19.0, there are two commands honored: `counter:` and `status:`.

The actual command control string is configurable. The default value is `reporter:`, but the value of `stream.stderr.reporter.prefix` will be used if set. The standard error stream is read via a `LineReader`, and a command must be one whole line as returned by the `LineReader.readLine()` method. We strongly recommend that you follow the same practice with counters in your streaming jobs as you do with regular jobs. A counter record should be emitted for each input record, for each output record, for each record that is invalid, and for each crash or exception when possible. The more detail about the job provided by the counters, the more understandable the job behavior is. A line output to the standard error stream, of the form `reporter:counter:UserCounters,InputLines,1`, will increment a counter `InputLines` in the counter group `UserCounters`. If the job was run with the command-line parameter `-D stream.stderr.reporter.prefix=`*`mylog:`*, the line would be *`mylog:`*`counter:UserCounters,InputLines,1`.

■ **Note** The value specified for the `stream.stderr.reporter.prefix` configuration key is the entire prefix string, the framework will use that exact string as the prefix, and the text that comes afterward must be `counter:group,counter,increment`. The colon character is not added by the framework as a separator.

Using the reporter:counter:group,counter,increment Command

A line of the form `reporter:counter:group,counter,increment` is converted by the framework into a call on the `Reporter` object of the following:

```
reporter.incrCounter("group", "counter", increment);
```

The parameter `increment` must be a whole number between `LONG.MIN_VALUE` and `Long.MAX_VALUE`. The `group` and `counter` parameters must not have the comma character in them.

Using the reporter:status:message Command

A line of the form `reporter:status:message` is converted by the framework into a call on the `Reporter` object of the following:

```
reporter.setStatus(message);
```

Alternative Methods for Accessing HDFS

Hadoop Core provides two tools: `libhdfs`, a native shared library, and `fuse-dfs`, built upon `libhdfs` to allow non-Hadoop–aware Java programs to access the HDFS file system.

libhdfs

`libhdfs` provides native access to HDFS for applications that can use it. The library provides application writers with a set of methods for interacting with HDFS. The methods in turn use JNI to actually interact with an embedded Java Virtual Machine (JVM) which actually interacts with HDFS. Table 8-3 provides a summary of the methods available.

Table 8-3. *Summary of the Methods Provided by libhdfs*

Method Name	Description
`hdfsConnect()`	Connects to a NameNode.
`hdfsDisconnect()`	Disconnects from a NameNode.
`hdfsOpenFile()`	Opens an HDFS file.
`hdfsCloseFile()`	Closes an HDFS file.
`hdfsExists()`	Tests for the existence of an HDFS file.
`hdfsSeek()`	Seeks only a read-only HDFS file.
`hdfsTell()`	Tells the current offset of the open HDFS file.
`hdfsRead()`	Reads a block of data from an HDFS file from the current offset point.
`hdfsPread()`	Reads a block of data from at HDFS file starting at a specified offset.
`hdfsWrite()`	Writes a block of data to an HDFS file.
`hdfsFlush()`	Flushes pending data. This method is subject to underlying support for flush in HDFS (not available through Hadoop 0.19.0).
`hdfsAvailable()`	Notes the number of bytes of data available to read without blocking from the open file.
`hdfsCopy()`	Copies a file from one file system to another.
`hdfsMove()`	Moves a file from one file system to another.
`hdfsDelete()`	Deletes a file from HDFS.
`hdfsRename()`	Renames a file in HDFS.
`hdfsGetWorkingDirectory()`	Returns the working directory of the HDFS file system.
`hdfsSetWorkingDirectory()`	Sets the working directory of the HDFS file system.
`hdfsCreateDirectory()`	Creates a directory in HDFS.
`hdfsSetReplication()`	Sets the replication count for an HDFS file.
`hdfsListDirectory()`	Lists the entries in an HDFS directory.
`hdfsGetPathInfo()`	Gets file status information for a path.
`hdfsFreeFileInfo()`	Frees the returned file status information, the pointer returned by `hdfsListDirectory()` and `hdfsGetPathInfo()`.
`hdfsGetHosts()`	Returns the list of DataNode hostnames where the particular block of the specified path are stored.
`hdfsFreeHosts()`	Frees the pointer returned by `hdfsGetHosts()`.
`hdfsGetDefaultBlockSize()`	Returns the basic HDFS block size.
`hdfsGetCapacity()`	Returns the raw storage capacity of the HDFS file system.
`hdfsGetUsed()`	Returns the raw size of all of the files in the HDFS file system.

`libhdfs` is compiled as part of the normal build process. A Linux i386 version is provided in the distribution in the directory `libhdfs`. If you need to build a custom version of the library or want to experiment with it, you need to force the build system to compile it. The command to cause `libhdfs` to be compiled is the following:

```
ant -Dlibhdfs=1 compile-libhdfs
```

If the `libhdfs` property is not set, Ant will not compile `libhdfs`.

Tip Any application using `libhdfs` must have a set `CLASSPATH` environment variable that includes the `hadoop-<rel>-core.jar` file and the JARs in the `lib` directory of the Hadoop distribution, and a shared library loading path that includes the `libjvm.so` shared library from the Java Development Kit (JDK). If they are not preset, the embedded JVM will either fail to launch or the class loader of the embedded JVM will not be able to load the Hadoop classes required to provide HDFS file service.

---

fuse-dfs

The `fuse-dfs` application uses `libhdfs` and the Filesystem in Userspace (FUSE) APIs to make HDFS file systems appear to be a locally mounted file system on the host machine. It allows arbitrary programs to access data that is stored in HDFS.

USERSPACE FILE SYSTEMS

The SourceForge project FUSE, `http://fuse.sourceforge.net/`, provides a set of APIs that allow programs written to those APIs to be mounted as host-level file systems.

There is no prebuilt version of `fuse-dfs` bundled into the distribution. In the `src/contrib` subtree is a package called `fuse-dfs`. The README file in `src/contrib/fuse-dfs/README` provides details and requirements. The i386 version may be compiled via the following:

```
ant compile-contrib -Dlibhdfs=1 -Dfusedfs=1
```

The preceding compile command will populate the directory `build/contrib/fuse_dfs`.

Note The `fuse-dfs` compilation environment will compile only for the i386 OS architecture. If X86_64 is required for 64-bit JVMs, the `OS_ARCH` variable must be manually modified in `src/c++/libhdfs/Makefile` and set to `amd64`.

The fuse-dfs package requires a modern Linux kernel with the FUSE module, fuse.ko, loaded. To actually mount an HDFS file system, the environment variables listed in Table 8-4 must be set correctly.

Table 8-4. *fuse_dfs Required Environment Variables*

Variable	Required Element	Used By
LD_LIBRARY_PATH	This variable must have the paths to the directories containing the following shared libraries: libjvm.so from the Java Runtime Environment (JRE), libhdfs.so from the Hadoop distribution, and libfuse.so from the system FUSE implementation.	The fuse_dfs program to load the required shared libraries.
JAVA_HOME	The path to the system JRE or JDK installation.	The fuse_dfs_wrapper.sh script to set up the runtime environment for the fuse_dfs program.
OS_ARCH	This variable determines which version of libjvm.so is used. The OS compilation architecture of libjvm.so, libhdfs, and fuse_dfs must be identical. The current choices are i386 and amd64.	The fuse_dfs_wrapper.sh script to set up the runtime environment for the fuse_dfs program.
CLASSPATH	This variable must have the JARs from the lib directory of the distribution and the core JAR.	The libjvm.so shared library will use this for the classpath of the JVM that is embedded the fuse_dfs program.

■ **Note** The author has used fuse_dfs in Hadoop 0.16.0 successfully. In Hadoop 0.19.0, the FUSE mounts produced corrupted directory listings. fuse_dfs appears to work correctly in Hadoop 0.19.1.

Mounting an HDFS File System Using fuse_dfs

After successfully compiling the fuse_dfs package via the following command, the directory build/contrib./fuse_dfs will be populated:

```
ant compile-contrib -Dlibhdfs=1 -Dfusedfs=1
```

The directory should contain at least the files fuse_dfs and fuse_dfs_wrapper.sh. The fuse_dfs_wrapper.sh script makes some assumptions that are not generally applicable and may not work for most installations without modification. The core configuration requires that the LD_LIBRARY_PATH environment variable include the directories that libjvm.so and libhdfs.so are resident in, and that the CLASSPATH has the Hadoop Core JAR and the support JARs present.

Listing 8-4, if run from the Hadoop installation root or with the environment variable HADOOP_HOME set to the installation root, will produce the correct settings for LD_LIBRARY_PATH and CLASSPATH.

Listing 8-4. *Script to Compute the Correct LD_LIBRARY_PATH and CLASSPATH Environment Variables for fuse_dfs, setup_fuse_dfs.sh*

```
#! /bin/sh

if [ -z "${HADOOP_HOME}" -a -r bin/hadoop-config.sh ]; then
    (echo -n "This script must run from the hadoop installation"
     echo "directory, or have HADOOP_HOME set in the environment") 1>&2
    exit 1
fi

if [ ! -z "${HADOOP_HOME}" -a -d "${HADOOP_HOME}" ]; then
    cd "${HADOOP_HOME}"
    if [ $? -ne 0 ]; then
        echo "Unable to cd to HADOOP_HOME [$HADOOP_HOME]" 1>&2
        exit 1
    fi
fi


if [ ! -r bin/hadoop-config.sh ]; then
    echo "Unable to find the hadoop-config.sh script" 1>&2
    exit 1
fi

export HADOOP_HOME=$PWD
HADOOP_CONF_DIR="${HADOOP_CONF_DIR:-$HADOOP_HOME/conf}"

if [ -z "${JAVA_HOME}" ]; then
    echo "JAVA_HOME is not set" 1>&2
    exit 1
fi

## Cut from bin/hadoop, to ensure classpath is the same as running installation

if [ -f "${HADOOP_CONF_DIR}/hadoop-env.sh" ]; then
  . "${HADOOP_CONF_DIR}/hadoop-env.sh"
fi


# CLASSPATH initially contains $HADOOP_CONF_DIR
CLASSPATH="${HADOOP_CONF_DIR}"
CLASSPATH=${CLASSPATH}:$JAVA_HOME/lib/tools.jar
```

```
# for developers, add Hadoop classes to CLASSPATH
if [ -d "$HADOOP_HOME/build/classes" ]; then
  CLASSPATH=${CLASSPATH}:$HADOOP_HOME/build/classes
fi

for f in $HADOOP_HOME/hadoop-*-core.jar; do
  CLASSPATH=${CLASSPATH}:$f;
done

# add libs to CLASSPATH
for f in $HADOOP_HOME/lib/*.jar; do
  CLASSPATH=${CLASSPATH}:$f;
done

for f in $HADOOP_HOME/lib/jetty-ext/*.jar; do
  CLASSPATH=${CLASSPATH}:$f;
done

LIBJVM=`find -L $JAVA_HOME -wholename '*/server/libjvm.so' -print | tail -1`
if [ -z "${LIBJVM}" ]; then
    echo "Unable to find libjvm.so in JAVA_HOME $JAVA_HOME" 1>&2
    exit 1
fi

# prefer the libhdfs in build
LIBHDFS=`find $PWD/libhdfs $PWD/build  -iname libhdfs.so -print | tail -1`
if [ -z "${LIBHDFS}" ]; then
    echo "Unable to find libhdfs.so in libhdfs or build" 1>&2
fi

if [ -z "${LD_LIBRARY_PATH}" ]; then
    LD_LIBRARY_PATH="`dirname "${LIBJVM}"`:`dirname "${LIBHDFS}"`"
else
 LD_LIBRARY_PATH="`dirname "${LIBJVM}"`:`dirname "${LIBHDFS}"`":"${LD_LIBRARY_PATH}"
fi
echo "export CLASSPATH='${CLASSPATH}'"
echo "export LD_LIBRARY_PATH='${LD_LIBRARY_PATH}'"
```

After the runtime environment is correctly configured, the fuse_dfs program can be run by using the command-line arguments shown in Table 8-5.

Table 8-5. *fuse_dfs Command-Line Arguments*

Argument	Default	Suggested Value	Description
server	None required	NameNode hostname	The server to connect to for HDFS servers.
port	None required	NameNode port	The port that the NameNode listens for HDFS requests on.
entry_timeout	60	-	The cache timeout for names.
attribute_timeout	60	-	The cache timeout for attributes.
protected	None	/user:/tmp	The list of exact paths that fuse_dfs will not delete or move.
rdbuffer	10485760	10485760	The size of the buffer used for reading from HDFS.
private	None	None	Allows only the user running fuse_dfs to access the file system.
ro	N/A	N/A	Mounts the file system read-only.
rw	N/A	N/A	Mounts the file system read-write.
debug	N/A	N/A	Enables debugging messages and runs in the foreground.
initchecks	N/A	N/A	Performs environment checks and logs results on startup.
nopermissions	enabled	enabled	Does not do permission checking; permission checking not supported as of Hadoop 0.19.0.
big_writes	None	enabled	Configures fuse_dfs to use large writes.
usetrash	enabled	enabled	Uses the trash directory when deleting files.
notrash	disabled	disabled	Does not use the trash directory when deleting files. Does not work in Hadoop 0.19.0.

The arguments in Table 8-6 are arguments that are passed to the underlying FUSE implementation, not handled directly by fuse_dfs.

Table 8-6. *Selected FUSE Command-Line Arguments*

Argument	Default	Suggested Value	Description
allow_other	None	enabled	Allows access to other users.
allow_root	None	disabled	Allows only root access.
nonempty	None	disabled	Allows mounts over non-empty file or directory.
fsname	None	None	Sets the file system name for /etc/mtab.
subtype	None	None	Sets file system type for /etc/mtab.
direct_io	None	None	Uses direct I/O instead of buffered I/O.
kernel_cache	None	None	Caches files in kernel.
[no]auto_cache	None	None	Enables caching based on modification times (off).

The following command will mount a read-only HDFS file system with debugging on. The fs.default.name for the file system being mounted is hdfs://cloud9:8020. The mount point for the file system is /mnt/hdfs, and the arguments after the /mnt/hdfs are passed to the FUSE subsystem. These are reasonable arguments for mounting an HDFS file system:

```
./fuse_dfs -oserver=cloud9 -oport=8020 -oro  -oinitchecks -oallow_other /mnt/hdfs➥
-o fsname="HDFS" -o debug
```

It is possible to set up a Linux system so that an HDFS is mounted at system start time by updating the system /etc/fstab file with a mount request for an HDFS file system. To set up system-managed mounts via /etc/fstab, a script /bin/fuse_dfs must be created that sets up the environment and then passes the command-line arguments to the actual fuse_dfs program. This script just sets up the CLASSPATH environment variable and the LD_LIBRARY_PATH variable as the script in Listing 8-4 does.

A candidate line for use in /etc/fstab is to mount an HDFS file system at system initialization time. The mount script, for the /etc/fstab entry in Listing 8-5, would be passed four arguments. To actually auto mount, the following could be added: ${HADOOP_HOME}/build/contrib/fuse-dfs/fuse_dfs "$3" "$4" "$1" "$2". And the script could be placed in /bin (see the script bin_fuse_dfs in the examples).

Listing 8-5. *A Candidate Mount Line for /etc/fstab to Mount an HDFS File System*

```
fuse_dfs#dfs://at:9020 /mnt/hdfs fuse rw,usetrash,allow_other,initchecks 0 0
```

Alternate MapReduce Techniques

The traditional MapReduce job reads a set of input data, performs some transformations in the map phase, sorts the results, performs another transformation in the reduce phase, and writes a set of output data. The sorting stage requires data to be transferred across the network and also requires the computational expense of sorting. In addition, the input data is read from and the output data is written to HDFS. The overhead involved in passing data between HDFS and the map phase, and the overhead involved in moving the data during the sort stage, and the writing of data to HDFS at the end of the job result in application design patterns that have large complex map methods and potentially complex reduce methods, to minimize the number of times the data is passed through the cluster.

Many processes require multiple steps, some of which require a reduce phase, leaving at least one input to the next job step already sorted. Having to re-sort this data may use significant cluster resources.

The following section goes into detail about a variety of techniques that are helpful for special situations.

Chaining: Efficiently Connecting Multiple Map and/or Reduce Steps

New in Hadoop 0.19.0 is the ability to connect several map tasks together in a chain. Prior to the chaining feature, the user was forced to either construct large map methods or run multiple jobs as a pipeline, with all the additional I/O overhead. Figure 8-2 provides a graphical depiction of the flow of key/value pairs through a job that uses chaining.

The chaining feature constructs a pipeline, internal to the task, which feeds each key/value pair from each `output.collect` to the map method of the next mapper in the chain. The map task may be a chain, and the reduce task may have a chain as a post processor.

This allows for the construction of simple mapper classes that do one thing well, as well as the ability to rapidly modify a chain to support additional or different features.

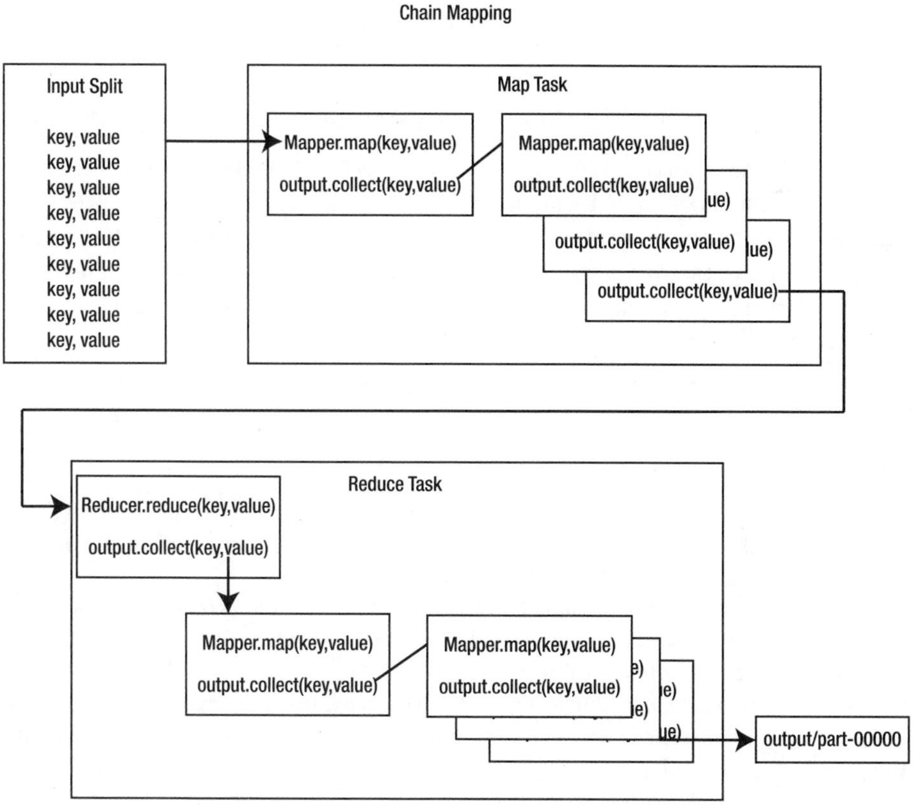

Chain Mapping

Input Split
key, value
key, value
key, value
key, value
key, value
key, value
key, value
key, value
key, value

Map Task
Mapper.map(key,value)
output.collect(key,value)
Mapper.map(key,value)
output.collect(key,value)
output.collect(key,value)
output.collect(key,value)

Reduce Task
Reducer.reduce(key,value)
output.collect(key,value)
Mapper.map(key,value)
output.collect(key,value)
Mapper.map(key,value)
output.collect(key,value)
output/part-00000

Figure 8-2. *Chain mapping*

Note At least through Hadoop 0.19.0, it is not possible to run the chain mapper through the streaming APIs.

Configuring for Chains

There are two possible chains that can be established for a job: the map task can be a chain or the reduce task can have a chain.

Passing Key/Value Pairs by Value or by Reference

Part of the contract for key/value pair management with a mapper or reducer is that the contents of the key/value are not modified during a call to `output.collect( key, value )`. The framework serializes the key/value into the output format for the particular task, and the `output.collect()` method returns with the contents of the key/value object unchanged.

With chaining, each key/value pair passed to the `output.collect()` method is the input to the next mapper in the chain. During job configuration, when a mapper is being added to a chain, the style of key/value passage is specified, either by value or by reference.

Passing by reference eliminates a serialization and deserialization for the key/value, a potential speed increase. If the `Mapper.map()` method uses the key or value method after the `output.collect()` call, subtle errors may occur if the key or value has been modified by a subsequent mapper.

Note If pass by reference is enabled, some level of verification needs to be in place to ensure that no use of the key/value object is made after a call to `output.collect` or that no mapper in the chain that receives the key or value reference modifies the contents. Any compliance failures in this implicit contract will cause difficulties in diagnosing problems. This is especially difficult because the configuration for pass by reference is remote from the `Mapper.map()` method that has to determine whether the pass by reference is safe and by the fact that the mapper class might be unaware about being part of a chain.

Type Checking for Chained Keys and Values

The standard Hadoop framework verifies that the type of key/value pairs being passed to the map or reduce method are the classes configured for the map or reduce. If they are not, an exception will be thrown.

The `RecordReaders` for the input split will throw an `IOException` for "wrong key class" or "wrong value class", and the `OutputCollector` will throw an `IOException` for "Type mismatch in key from map" or "Type mismatch in value from map".

At least as of Hadoop 0.19.0, the chaining code does not explicitly check the runtime types of the key/value pairs being passed between elements in the chain. The types are checked only during the job configuration phase.

Per Chain Item Job Configuration Objects

The chaining interface provides a way for each item in the chain to receive custom configuration parameters. It is recommended that these custom configurations be light configurations, which have only the special parameters for that item. For a map task, the chain will have only mapper items. For a reduce task, the chain will have a leading reducer item and then some number of mapper items.

At task runtime, a `JobConf` object is made for each item. This `JobConf` object is constructed by making a copy of the localized task `JobConf` object and then copying each key/value pair out of the per map configuration into the copy. This modified copy is then passed to the configure method of the item.

How the close() Method Is Called for Items in a Chain

The mapper `close()` methods are called in order. The reducer close is called after all the mapper `close()` methods have completed. If any `close()` method throws an exception, no further `close()` methods are run.

■ **Caution** This is the `close()` behavior, as of Hadoop 0.19.0.

Configuring Mapper Tasks to be a Chain

A mapper task is either a normal map task or a chain. The configuration of one excludes the configuration for the other. A call to the `JobConf.setMapperClass()` method after a chain has been configured will disable the chain.

The framework provides a class, `org.apache.hadoop.mapred.lib.ChainMapper`, which provides the `addMapper()` method. Table 8-7 details the parameters, and Listing 8-6 provides the declaration. The `addMapper()` method configures the mapper tasks to be run as chains and appends the specified mapper class to the end of the current chain mapper task chain.

Table 8-7. *The ChainMapper.addMapper Parameters*

Type	Parameter	Modified	Description
JobConf	job	true	The per job JobConf object.
Class<? extends Mapper <K1 V1 K2 V2>>	klass	false	The mapper class to be run. A call, job.setMapperClass (ChainMapper.class), will be made by this method.
Class<? extends K1>	inputKeyClass	false	The input key class; must be the type of output key of the previous chain item or the type of key for this task if it is the first item in the chain.
Class<? extends V1>	inputValueClass	false	The input value class; must be the type of output value of the previous chain item or the type of value for this task if it is the first item in the chain.
Class<? extends K2>	outputKeyClass	false	The output key class. A call, job.setMapOutputKeyClass(outputKeyClass), will be made in this method.
Class<? extends V2>	outputValueClass	false	The output value class. A call, job.setMapOutputValueClass(outputValueClass), will be made in this method.
boolean	byValue	false	If false, *klass*, the mapper class does not use the key or value objects after the call to output.collect, or no map later in the chain will modify the values, and the key/value will be passed by reference instead of copied via serialization.
JobConf	mapperConf	true	The configuration object that provides custom configuration data for this mapper instance at mapper runtime. The input and output classes will be stored in this object. Any keys present will override the corresponding values in the task's localized JobConf object.

■ **Note** For each addMapper() call, the mapperConf object should be constructed via JobConf mapperConf = new JobConf(false). This will minimize the possibility that a configuration value will step on a task's localized value. Any custom parameters may then be set on the mapperConf object before the addMapper() call. The clear() method may be used to reset a mapperConf for use in the next call to addMapper(). As of Hadoop 0.19.0, this parameter should not be passed as a null because a full JobConf object will be initialized.

Listing 8-6. *The ChainMapper.addMapper() method Declaration with JavaDoc*

```
/**
   * Adds a mapper class to the chain job's JobConf.
   * <p/>
   * It has to be specified how key and values are passed from one element of
   * the chain to the next, by value or by reference. If a mapper leverages the
   * assumed semantics that the key and values are not modified by the collector
   * 'by value' must be used. If the mapper does not expect this semantics, as
   * an optimization to avoid serialization and deserialization 'by reference'
   * can be used.
   * <p/>
   * For the added mapper the configuration given for it,
   * <code>mapperConf</code>, have precedence over the job's JobConf. This
   * precedence is in effect when the task is running.
   * <p/>
   * IMPORTANT: There is no need to specify the output key/value classes for the
   * ChainMapper, this is done by the addMapper for the last mapper in the chain
   * <p/>
   *
   * @param job              job's JobConf to add the mapper class.
   * @param klass            the mapper class to add.
   * @param inputKeyClass    mapper input key class.
   * @param inputValueClass  mapper input value class.
   * @param outputKeyClass   mapper output key class.
   * @param outputValueClass mapper output value class.
   * @param byValue          indicates if key/values should be passed by value
   * to the next mapper in the chain, if any.
   * @param mapperConf       a JobConf with the configuration for the mapper
   * class. It is recommended to use a JobConf without default values using the
   * <code>JobConf(boolean loadDefaults)</code> constructor with FALSE.
   */
  public static <K1, V1, K2, V2> void addMapper(JobConf job,
                         Class<? extends Mapper<K1, V1, K2, V2>> klass,
                         Class<? extends K1> inputKeyClass,
                         Class<? extends V1> inputValueClass,
                         Class<? extends K2> outputKeyClass,
                         Class<? extends V2> outputValueClass,
                         boolean byValue, JobConf mapperConf)
```

Configuring the Reducer Tasks to Be Chains

Configuring the reducer phase is very similar to the configuration of the mapper phase with one additional requirement: the job configuration step must make a call to `ChainReducer.setReducer()` before adding any mappers to the reducer chain. Table 8-8 describes the parameters for `ChainReducer.setReducer()`, and Table 8-9 describes the parameters for `ChainReducer.addMapper()`. The other minor difference is that the `ChainReducer.addMapper()` method must be used in place of the `ChainMapper.addmapper()`method.

Table 8-8. *ChainReducer.setReducer Parameters*

Type	Parameter	Modified	Description
`JobConf`	job	`true`	The per job `JobConf` object.
`Class<? extends Reducer<K1 V1 K2 V2>>`	*klass*	`false`	The reducer class to be run. A call, `job.setReducerClass (ChainReducer.class)`, will be made by this method.
`Class<? extends K1>`	`inputKeyClass`	`false`	The input key class; must be the type of output key of the previous chain item or the type of key for this task if it is the first item in the chain.
`Class<? extends V1>`	`inputValueClass`	`false`	The input value class; must be the type of output value of the previous chain item or the type of value for this task if it is the first item in the chain.
`Class<? extends K2>`	`outputKeyClass`	`false`	The output key class. A call, `job.setOutputKeyClass(outputKeyClass)`, will be made in this method.
`Class<? extends V2>`	`outputValueClass`	`false`	The output value class. A call, `job.setOutputValueClass(outputValueClass)`, will be made in this method.
`boolean`	`byValue`	`false`	If `false`, *klass*, the reducer class does not use the key or value objects after the call to `output.collect()`, or no map later in the chain will modify the values, and the key/value will be passed by reference instead of copied via serialization.
`JobConf`	`reducerConf`	`true`	The configuration object that provides custom configuration data for this reducer instance at reducer runtime. A `null` may be passed. The input and output classes will be stored in this object. Any keys present will override the corresponding values in the task's localized `JobConf` object.

Table 8-9. *ChainReducer.addMapper Parameters*

Type	Parameter	Modified	Description
JobConf	job	true	The per job JobConf object.
Class<? extends Mapper<K1 V1 K2 V2>>	*klass*	false	The mapper class to be run.
Class<? extends K1>	inputKeyClass	false	The input key class; must be the type of output key of the previous chain item or the type of key for this task if it is the first item in the chain.
Class<? extends V1>	inputValueClass	false	The input value class; must be the type of output value of the previous chain item or the type of value for this task if it is the first item in the chain.
Class<? extends K2>	outputKeyClass	false	The output key class. A call, job.setOutputKeyClass(outputKeyClass), will be made in this method.
Class<? extends V2>	outputValueClass	false	The output value class. A call, job.setOutputValueClass(outputValueClass), will be made in this method.
boolean	byValue	false	If false, *klass*, the mapper class does not use the key or value objects after the call to output.collect(), or no map later in the chain will modify the values, and the key/value will be passed by reference instead of copied via serialization.
JobConf	mapperConf	true	The configuration object that provides custom configuration data for this mapper instance at mapper runtime. The input and output classes will be stored in this object. Any keys present will override the corresponding values in the task's localized JobConf object.

Note It is important to use ChainMapper.addMapper(), ChainReducer.setReducer(), and ChainReducer.addMapper() instead of the public methods Chain.addMapper() and Chain.setReducer(). The chain methods do not configure the job level chaining configuration parameters.

The provided code in ChainMappingExample and ChainMappingExampleMapperReducer provides a simple example of chain mapping that is structured to help you understand the order of events in your chain. The sample code sets a chain mapping job. The maps and the reduce have a particular id to help distinguish them. The actual ordering information has to be extracted from the job log.

Table 8-10 demonstrates running the ChainMappingExample and details the exact sequence of the method invocation on the mapper and reducer classes. The assumptions are that the hadoopprobook and commons-lang JARs are in the current working directory. The construct 2>&1 forces the standard error output to go to the same descriptor as the standard output.

```
HADOOP_CLASSPATH=./commons-lang-2.4.jar hadoop jar hadoopprobook.jar➥
com.apress.hadoopbook.examples.ch8.ChainMappingExample -jt local➥
-verbose -logLevel INFO -tsr ALL --deleteOutput 2>&1 | grep ': Event'
```

Table 8-10. *Event Ordering in the ChainMappingExample*

Sequence Number	Task	Method
0	Master map	constructor()
1	Master map	configure()
2	Map 1	constructor()
3	Map 1	configure()
4	Map 2	constructor()
5	Map 2	configure()
6	Master map	map() Key0
7	Map 1	map() Key0
8	Map 2	map() Key0
9	Master map	map() Key1
10	Map 1	map() Key1
11	Map 2	map() Key1
12	Master map	map() Key2
13	Map 1	map() Key2
14	Map 2	map() Key2
15	Master map	map() Key3
16	Map 1	map() Key3
17	Map 2	map() Key3
18	Master map	map() Key4
19	Map 1	map() Key4
20	Map 2	map() Key4
21	Master map	close()
22	Map 1	close()
23	Map 2	close()
24	Reduce 1	constructor()

25	Reduce 1	`configure()`
26	Reduce 2	`constructor()`
27	Reduce 2	`configure()`
28	Master Reduce	`constructor()`
29	Master Reduce	`configure()`
30	Master Reduce	`reduce()` Key0
31	Reduce 1	`map()` Key0
32	Reduce 2	`map()` Key0
33	Master Reduce	`reduce()` Key1
34	Reduce 1	`map()` Key1
35	Reduce 2	`map()` Key1
36	Master Reduce	`reduce()` Key2
37	Reduce 1	`map()` Key2
38	Reduce 2	`map()` Key2
39	Master Reduce	`reduce()` Key3
40	Reduce 1	`map()` Key3
41	Reduce 2	`map()` Key3
42	Master Reduce	`reduce()` Key4
43	Reduce 1	`map()` Key4
44	Reduce 2	`map()` Key4
45	Reduce 1	`close()`
46	Reduce 2	`close()`
47	Master Reduce	`close()`

Map-side Join: Sequentially Reading Data from Multiple Sorted Inputs

In a traditional MapReduce job, the framework sorts all data for a reduce task before presenting the keys sequentially to the reduce task. If the input data is already sorted, traditional MapReduce requires that the full map shuffle and sort process take place before the reduce task receives the sorted keys.

Map-side joins provide a way for a map task to receive keys in sequential order and to receive all the values associated with each key (very similar to a reduce task). The map task reads the data directly from HDFS and no reduce is needed, which greatly reduces cluster loading. The author processed a dataset that was reduced from 5 hours to 12 minutes by converting the job to use map-side joins.

The map-side join provides a framework for performing operations on multiple sorted datasets. Although the individual map tasks in a join lose much of the advantage of data locality, the overall job gains due to the potential for the elimination of the reduce phase and/or the great reduction in the amount of data required for the reduce.

Caution There are several constraints on when map-side joins may be used, and the cluster loses capability to manage data locality for the map tasks (see Table 8-11). There are also bugs in the join code that cause unpredictable behavior if there are more than 31 tables in a join; see `https://issues.apache.org/jira/browse/HADOOP-5589` and `https://issues.apache.org/jira/browse/HADOOP-5571`.

The author has used map-side joins extensively in large-scale web crawls to eliminate recently crawled URLs from the set of freshly harvested URLs being prepared for fetching.

As of Hadoop 0.19.0, the join package supports full inner and outer joins. All joins are full table scans at present; one optimization currently missing from the join package is the capability to use the indexes supplied with `org.apache.hadoop.io.MapFile` to skip over unneeded records in datasets.

In the following section, the term *dataset* is used to refer to one join item in the set of elements being joined. The dataset can be an actual dataset or the result of a join.

Note As of at least Hadoop 0.19.0, the joins handle only keys that implement `WritableComparable` and values that implement `Writable`. The join framework has not been updated to handle arbitrary key/value classes.

Examining Join Datasets

A join dataset is specified by providing a dataset name and an `InputFormat` class. A dataset is the set of input splits that an `InputFormat` will produce when given the name as an argument. The `mapred.min.split.size` is set to `Long.MAX_VALUE` before the `InputFormat.getSplits()` method is called. The goal is to force the `InputFormat` not to split individual data files, thereby ensuring that each returned split contains the entirety of a single reduce task output, or partition. The directory and the partition files it contains is a dataset. Using the join package imposes the following limitations on your application.

Table 8-11. *Limitations on Datasets Used in Joins*

Limitation	Why
All datasets must be sorted using the same comparator.	The sort ordering of the data in each dataset must be identical for datasets to be joined.
All datasets must be partitioned using the same partitioner.	A given key has to be in the same partition in each dataset so that all partitions that can hold a key are joined together.
The number of partitions in the datasets must be identical.	A given key has to be in the same partition in each dataset so that all partitions that can hold a key are joined together.

Limitation	Why
The `InputFormat` must return the input splits in `Partitioner` order.	The `OutputPartitioner` class returns a partition number for each key, which determines the reduce task each key is assigned to. This partition number is commonly used to construct the file name of the reduce output partition, `part-%05d`. The file name is the string `part-`, followed by a 0 padded five-digit number, which is the reduce output partition. At split time, no information is readily available to determine what partition number the split was originally a part of, so the ordinal number in the `InputSplit` array, returned by the `InputSplit[] InputFormat.getSplits()` method, is used as a surrogate for the partition number. For any given key in the `N`th input split returned by an `InputFormat.getSplits` call, if that key could be present in another dataset, it would be present only in the `N`th split returned by that dataset's `InputFormat.getSplits` call.

■ **Note** The map-side join has no simple way to discover what reduce partition a split was created as. The `InputFormat`'s split routine is called with the minimum split size set to `Long.MAX_VALUE`, under the assumption that this will cause each split returned to be one complete input partition. The map-side join assumes that the `InputSplit` arrays returned by each dataset's `InputFormat.getSplits()` returns the splits, or partitions in the same partition order (i.e., any given single index slice through arrays of splits will return a set of splits in which all the keys in each set belong to the same partition). If this assumption of equivalent ordering is incorrect, the behavior of the map-side join will be incorrect, and this failure will be detectable only by examining the output data.

Under the Covers: How a Join Works

The customary output of a MapReduce job that has a reduce phase is a single directory with `N` files of the form `part-00000` through `part-0*N-1`. When a `FileInputFormat`-based `InputFormat` is given that output directory as input, and the `mapred.min.split.size` is set to `Long.MAX_VALUE`, `N` input splits will be generated one for each part file or partition.

For `FileInputFormat`-based datasets, the input splits are returned as an array, in partition file name lexical order (e.g., `part-00000` is first in the array, followed by `part-00001`, and so on).

For each dataset specified in the join, the input splits of the dataset are collected. If the number of input splits returned by each dataset's `InputFormat` is not identical, the framework throws an exception of the form `IOException("Inconsistent split cardinality from child N, Y/Z")` where `N` is the ordinal number of the dataset, per the input specification; `Y` is the expected number of splits or partition; and `Z` is the number of splits provided by the Nth dataset's `InputFormat`.

For each single index slice of the `InputSplit` arrays, a `WrappedRecordReader` is constructed. The `WrappedRecordReaderClass` implements the interface `org.apache.hadoop.mapred.join.ComposableRecordReader` and provides the standard `RecordReader` function of `next(K key, V value)`. The set of `ComposableRecordReaders` that are to be used for a particular join are bundled into a `JoinRecordReader`, which also implements the interface `ComposableRecordReader`.

The basic `JoinRecordReader.next(key, value)` method returns the keys of the entire set of keys present in the `WrappedRecordReaders` in `OutputComparator` order. The `value` is a `TupleWritable` object, which contains each value associated with the key across the set of `WrappedRecordReaders`, and information about which `WrappedRecordReader` the value originated in. A `JoinRecordReader` can have any `ComposableRecordReader` implementer as one of its inputs; by default, they are `WrappedRecordReaders` and `JoinRecordReaders`.

Each map task is given a `JoinRecordReader` from the outermost join as the task input record reader and receives the key/value sets of the join one by one in the map method. In a simple case, this `JoinRecordReader` will have `N` `WrappedRecordReaders` from slice `N` of the original `InputSplit` arrays. The default outer join behavior will receive each key in the input split set, in the sort order with all the values for that key. The map method behaves very much like a traditional reduce.

Types of Joins Supported

The join framework comes with support for three types of joins: `outer`, `inner`, and `override`. Joins can be made on direct input datasets or on the results of joining input datasets; arbitrary deep nesting of this joining structure is supported.

Inner Join

The *inner join* is a traditional database-style inner join. The map method will be called with a key/value set only if every dataset in the join contains the key. The `TupleWritable` value will contain a value for every dataset in the join.

Outer Join

The *outer join* is a traditional database-style outer join. The map method will be called for every key in the set of datasets being joined. The `TupleWritable` value will contain values for only those datasets that have a value for this key.

Override Join

The *override join* is unusual in that the there will only ever be one value passed to the map method. In the inner and other joins there will be a set of values passed to the map method. The override join maps a call to the map method with each key in the input split set and with that single value from the rightmost input split or join that has a value for the key.

The use of this join style requires that you order your input datasets (from least to most important). For any given key, your map method will be given the value from the most important dataset that contains the key.

Composing Your Own Join Operators

The join framework provides a mechanism for defining additional operators. The configuration key `mapred.join.define.`*`YOUR_OPERATOR`* must be set to the class name of a class that implements the `ComposableRecordReader` interface. The string *`YOUR_OPERATOR`* in the key definition must be replaced with the name of the custom join operation. *`YOUR_OPERATOR`* can then be passed as the `op` parameter to the compose methods that accept `op`, and used anywhere that the predefined operators, `inner`, `outer`, and `override`, are used.

Details of a Join Specification

A join specification is an operator and a set of data sources. The predefined operators are `inner`, `outer`, and `override` to correspond with the join types. A data source is either a table statement or a join specification. A table statement is a string `tbl(input.format.class.name,"path")`. The comma character is used to separate data sources; parentheses, `()`, are used to group the data sources for an operator. Table 8-12 provides examples of several join data source specifications.

Table 8-12. *Data Source Examples*

Data Source	Description
`tbl(org.apache.hadoop.mapred.KeyValueTextInputFormat,"textSource")`	A data source located in or at the path `textSource` that contain records that are to be parsed by the `KeyValueTextInputFormat` class.
`inner(tbl(org.apache.hadoop.mapred.SequenceFileAsTextInputFormat,"sequence"),tbl(org.apache.hadoop.mapred.KeyValueTextInputFormat,"textSource"))`	A data source composed of the inner join of the data in sequence file format at or in `sequence`, and the textual data at or in `textSource`. The key/value classes read from `sequence` are converted into `Text`.
`override(inner(tbl(org.apache.hadoop.mapred.SequenceFileAsTextInputFormat,"sequence"),tbl(org.apache.hadoop.mapred.KeyValueTextInputFormat,"textSource")), tbl(org.apache.hadoop.mapred.KeyValueTextInputFormat,"priority")`	A composite data source composed of a nested inner join of `sequence` and `textSource`, joined with `priority`, and with a preference for values from `priority` if multiple sources in the join have values for a given key.

Handling Duplicate Keys in a Dataset

For a join in which a table in the join has duplicate key/value pairs, the map method will be called one time for each possible permutation of the key/value pairs. For example, suppose that a join of two tables is made. Table 1 has two records (`1,a` and `1,b`), and Table 2 has one record (`1, c`). The map method will be called twice with key `1`; once with a tuple `a,c`; and once with a tuple `b,c`.

Composing a Join Specification

The framework provides three helper methods, all named `compose()`, which build either a full join specification or build the input specification for a particular dataset in the join. Two of the methods construct a full join specification and are used when all the datasets within the join have the same `InputFormat`. These two methods differ only in accepting `String` or `Path` objects for the dataset locations. The third is used to construct a table statement for a dataset that includes a specified `InputFormat` and requires the application developer to aggregate the results into a full join specification. The methods are provided via the `CompositeInputFormat` class.

String CompositeInputFormat.compose(Class<? extends InputFormat> inf, String path)

This method produces a table statement from an input format class object and a path to a dataset. The fully qualified class name of `inf` will be used in the returned table statement. This method does not produce a full join statement. It is commonly used when building a join statement from input datasets that have different input formats. (Refer to Table 8-12, items 2 and 3, for examples of complete join statements.)

Here's a sample use of this method:

```
CompositeInputFormat.compose(KeyValueTextInputFormat.class,"mydata");
```

```
tbl(org.apache.hadoop.KeyValueTextInputFormat,"mydata")
```

String CompositeInputFormat.compose(String op, Class<? extends InputFormat> inf, String... path)

This method produces a full join statement. The resulting string can be stored in the configuration under the key `mapred.join.expr` or used as a nested join within another join statement.

Here's a sample use of this method:

```
CompositeInputFormat.compose( "inner", KeyValueTextInputFormat, ➥
"maptest_a.txt","maptest_b.txt","maptest_c.txt" );
```

```
inner(tbl(org.apache.hadoop.mapred.KeyValueTextInputFormat,"maptest_a.txt"),➥
tbl(org.apache.hadoop.mapred.KeyValueTextInputFormat,"maptest_b.txt"),➥
tbl(org.apache.hadoop.mapred.KeyValueTextInputFormat,"maptest_c.txt"))
```

String CompositeInputFormat.compose(String op, Class<? extends InputFormat> inf, Path... path)

This method is identical to the `String` variant except that `Path` objects instead of `String` objects provide the table paths.

Building and Running a Join

There are two critical pieces of engaging the join behavior: the input format must be set to `CompositeInputFormat.class`, and the key `mapred.join.expr` must have a value that is a valid join specification. Optionally, the mapper, reducer, reduce count, and output key/value classes may be set.

The mapper key class will be the key class of the leftmost data source, and the key classes of all data sources should be identical. The mapper value class will be `TupleWritable` for inner, outer, and user-defined join operators. For the `override` join operator, the mapper value class will be the value class of the data sources.

In Listing 8-7, note that the quote characters surrounding the path names are escaped.

Listing 8-7. *Synthetic Example of Configuring a Join Map Job*

```
/** All of the outputs are Text. */
conf.setOutputFormat(TextOutputFormat.class);
conf.setOutputKeyClass(Text.class);
conf.setOutputValueClass(Text.class);
conf.setMapperClass(MyMap.class);
/** setting the input format to {@link CompositeInputFormat}
  * is the trigger for the map-side join behavior. */
conf.setInputFormat(CompositeInputFormat.class);
conf.set("mapred.join.expr",
"override(tbl(org.apache.hadoop.mapred.KeyValueTextInputFormat, ➥
\"maptest_a.txt\"),tbl(org.apache.hadoop.mapred.KeyValueTextInputFormat, ➥
\"maptest_b.txt\"),tbl(org.apache.hadoop.mapred.KeyValueTextInputFormat, ➥
\"maptest_c.txt\"))");
```

Synthetic Example of Configuring a Join Map Job Using the Compose Helper

```
/** All of the outputs are Text. */
conf.setOutputFormat(TextOutputFormat.class);
conf.setOutputKeyClass(Text.class);
conf.setOutputValueClass(Text.class);
conf.setMapperClass(MyMap.class);
/** setting the input format to {@link CompositeInputFormat}
  * is the trigger for the map-side join behavior. */
conf.setInputFormat(CompositeInputFormat.class);
conf.set("mapred.join.expr",CompositeInputFormat.compose("override",➥
KeyValueTextInputFormat.class, "maptest_a.txt",➥
"maptest_b.txt", "maptest_c.txt"));
```

The Magic of the TupleWritable in the Mapper.map() Method

The map method for the inner and outer join has a value class of `TupleWritable`, and each call to the map method presents one join result row. The `TupleWritable` class provides a number of ways to understand the shape of the join result row. Listing 8-8 provides a sample mapper that demonstrates the use of `TupleWritable.size()`, `TupleWriter.iterator()`, `TupleWritable.has()`, and `TupleWritable.get()` methods. Table 8-13 provides a description of these methods.

Table 8-13. *TupleWritable Methods for Interacting with the Join Result Row*

Method	Argument	Description
`boolean has(int i)`	The ordinal number of a dataset.	Returns `true` if that dataset provides a value to this result row.
`Writable get(int i)`	The ordinal number of a dataset.	Returns the value object that the dataset has provided to this result row. The object returned by `get` will be reinitialized on the next call to get. The application will need to make a copy of the contents before calling `get()` again if the contents need to exist past the next call to `get()`.
`int size()`		Returns the number of datasets in the join. Only the top-level datasets are counted, even if the dataset is the result of many nested joins. This method is used to provide an index limit for loops through the values using `has` and `get`. `for( int i = 0; i < tuple.size(); i++ ) if ( tuple.has(i) )...`
`Iterator<Writable> iterator()`		Returns an `iterator` through the values that are present. For any dataset that did not contribute a value to this result record, the `iterator` will skip over that dataset.

■ **Note** A dataset may provide a `null` value to a join result record if the dataset is composed only of keys. Using the construct `get(i)==null` will not correctly indicate that dataset `i` did not have the join result record key present; only the call `having(i)` is sufficient.

Listing 8-8. *A Sample Mapper*

```
package com.apress.hadoopbook.examples.ch8;

import java.io.IOException;

import org.apache.hadoop.io.Text;
import org.apache.hadoop.io.Writable;
import org.apache.hadoop.mapred.MapReduceBase;
import org.apache.hadoop.mapred.Mapper;
import org.apache.hadoop.mapred.OutputCollector;
import org.apache.hadoop.mapred.Reporter;
import org.apache.hadoop.mapred.join.TupleWritable;
```

```
/** A cut down join mapper that does very little but demonstrates
  * using the TupleWritable
 *
 * @author Jason
 *
 */
class CutDownJoinMapper extends MapReduceBase implements
         Mapper<Text,TupleWritable,Text,Text> {

    Text outputValue = new Text();

    @Override
    public void map(Text key, TupleWritable value,
            OutputCollector<Text, Text> output, Reporter reporter)
            throws IOException {
        try {

            /** The user has two choices here, there is an iterator
             * and a get(i) size option.
             * The down side of the iterator is you don't know what table
             * the value item comes from.
             */

            /** Gratuitous demonstration of using the TupleWritable iterator. */
            int valueCountTotal = 0;
            for( @SuppressWarnings("unused") Writable item : value) {
                valueCountTotal++;
            }
            reporter.incrCounter("Map Value Count Histogram", key.toString() +
                                             " " + valueCountTotal, 1);


            /** Act like the Identity Mapper. */
            final int max = value.size();
            int valuesOutputCount = 0;
            for( int i = 0; i < max; i++) {
                if (value.has(i)) {
                    // Note, get returns the same object initialized
                    // to the data for the current get
                    output.collect( key, new Text( value.get(i).toString() ) );
                    valuesOutputCount++;
                }
            }
```

```
            assert valueCountTotal == valuesOutputCount :
                    "The iterator must always return the same number of➥
values as a loop monitoring has(i)";
        } catch (Throwable e) {
            reporter.incrCounter("Exceptions", "MapExceptionsTotal", 1);
            MapSideJoinExample.LOG.error( "Failed to handle record for " + key, e);
        }
    }
}
```

Aggregation: A Framework for MapReduce Jobs that Count or Aggregate Data

The Hadoop Core framework provides a package for performing data aggregation jobs. This package may conceptually be thought of as Hadoop streaming for statistics. The analogy is incomplete because some code must be written to use the aggregation services. The aggregation services are provided by classes that implement the interface `org.apache.hadoop.mapred.lib.aggregate.ValueAggregator`. The framework provides a set of aggregator services (see Table 8-14 for descriptions of the predefined aggregator services). The user can define the custom aggregator (see Listing 8-15). Aggregation can be run via Hadoop streaming. The aggregation framework manages the mapper, combiner, and reducer; and the aggregation service produces the correct key/value pairs to pass forward. The user is responsible for parsing the input record and invoking the aggregate service with the record key and count; the record and count are the traditional map task output key/value pairs. Quite often, the key has no meaning for the job and is simply a label for the end user. The count must the textual representation of an object that the aggregator service expects: a number for `DoubleValueSum`, a whole number for the `LongValue` series, an arbitrary string for the `StringValue` series, and a whole number for `UniqueValueCount` and `ValueHistogram`.

Table 8-14. *Predefined Aggregation Services*

Class	Description	Id	Key Value	Count Value	Example Code	Sample
`DoubleValueSum`	Computes the sum of input values. The input values are expected to be doubles and are summed. A single output record per reduce.	`DoubleValueSum`	Label	The number to accumulate in the sum. The behavior is identical to `LongValueSum.pl`, so the `LongValueSum.pl` example is used	-	`DoubleValueSum:LabelTAB37`
`LongValueMax`	Computes the maximum input value. The input values are expected to be longs, and the max value is output. A single output record per reduce.	`LongValueMax`	Label	The number to challenge the current max value with.	`LongMax.pl`	`LongValueMax:LabelTAB37`
`LongValueMin`	Computes the minimum input value. The input values are expected to be longs, and the min value is output. A single output record per reduce.	`LongValueMin`	Label	The number to challenge the current min value with. The behavior is essentially identical to `LongValueMax`, so the `LongMax.pl` example is used.	-	`LongValueMin:LabelTAB3`

Class	Description	Id	Key Value	Count Value	Example Code	Sample
LongValueSum	Computes the long sum of input values. Input values are expected to be longs, and the sum is output. A single output record per reduce.	LongValueSum	Label	The number to add to the sum.	LongSum.pl	LongValueSum: LabelTAB37
StringValueMax	Computes the lexically greatest input value. The values object's toString() method is invoked, and the resulting String is compared. The lexically largest is output. A single output record per reduce.	StringValueMax	Label	String to challenge the current lexically largest string.	StringMax.pl	StringValueMax: LabelTABMyLong StringCandidate
StringValueMin	Computes the lexically least input value. The values object's toString() method is invoked and the resulting String is compared. The lexically smallest is output. A single output record per reduce.	StringValueMin	Label	String to challenge the current lexically smallest string.	-	StringValueMin: LabelTABMyShort StringCandidate.
UniqValueCount	Computes the set of unique input values. The value object's equals() method is used to determine equality. The set of unique object is output. The configuration parameter aggregate.max.num.unique.values, which defaults to Long.MAX_VALUE, limits the number of unique items accumulated. Any new objects encountered in a map or reduce task past this value are discarded.	UniqValueCount	Object as a string.	Ignored; 1 is acceptable.	UniqValue.pl	UniqValueCount: object.toString() TABignored.
ValueHistogram	Computes a histogram of the occurrence counts of the unique input values. The input value object's equals() method is used to determine equality.	ValueHistogram	The object as a string.	The count of times the object occurred in this record; 1 is usually correct.	LongHistogram.pl	ValueHistogram: object.to String()TAB1.

The code that the user must supply can be supplied as a streaming mapper or via a Java class.

Aggregation Using Streaming

The user-supplied code must take an input record and return an aggregator record. The aggregator record is textually the `id: key\tcount`, where `id` is the aggregator service id, `key` is an applicable key for the job, and `count` is the appropriate value for `key` commonly `1`. Listing 8-9 provides a sample Perl mapper that computes the sums of input files that are sets of long values.

Listing 8-9. *Perl Streaming Mapper for LongValueSum of Input Files Composed of Long Values, LongSum.pl*

```
#! /usr/bin/perl -w

use strict;

eval {
    while(<>) {
        print STDERR "reporter:counter:Map,Input Records,1\n";
        chomp;
        my @parts = split(/\s/, $_); # split on white space
        foreach my $part ( @parts ) {
            print STDERR "reporter:counter:Map,Output Records,1\n";
            print "LongValueSum:SUM\t$part\n";
        }
    }
};
if ($@) {
    print STDERR "reporter:counter:Map,Exceptions,1\n";
}
```

Each reduce task will have a single output value, the key will be the string SUM, and the value will be the sum of all of the long values routed to that reduce task. The streaming command that was used is in Listing 8-10. An input file with white space separated whole numbers must be in `/tmp/numbers`, and the sums will be placed in `/tmp/numbers_sum_output`.

■ **Note** The reducer is defined as aggregate for the streaming job in Listing 8-10.

Listing 8-10. *The Streaming Command to invoke LongSum.pl*

```
bin/hadoop jar contrib/streaming/hadoop-0.19.0-streaming.jar -jt local➥
-fs file:/// -input /tmp/numbers  -output /tmp/numbers_sum_output -verbose➥
-reducer aggregate -mapper LongSum.pl -file /tmp/LongSum.pl
```

If an error shown in Listing 8-11 happens, it generally means that an unrecognized aggregator id has been output by the mapper.

Listing 8-11. *Exception Resulting from an Unrecognized Aggregator Service Id*

```
java.lang.NullPointerException
    at org.apache.hadoop.mapred.lib.aggregate.ValueAggregatorCombiner.➥
reduce(ValueAggregatorCombiner.java:59)
    at org.apache.hadoop.mapred.lib.aggregate.ValueAggregatorCombiner.➥
reduce(ValueAggregatorCombiner.java:34)
```

```
    at org.apache.hadoop.mapred.MapTask$MapOutputBuffer.➥
combineAndSpill(MapTask.java:1106)
    at org.apache.hadoop.mapred.MapTask$MapOutputBuffer.sortAndSpill➥
(MapTask.java:979)
    at org.apache.hadoop.mapred.MapTask$MapOutputBuffer.flush(MapTask.java:832)
    at org.apache.hadoop.mapred.MapTask.run(MapTask.java:333)
    at org.apache.hadoop.mapred.LocalJobRunner$Job.run(LocalJobRunner.java:138)
```

Aggregation Using Java Classes

A Java application that wants to use the Aggregation services must provide a class that implements the class `ValueAggregatorDescriptor`. The framework provides a base class `ValueAggregatorBaseDescriptor` that can be extended. The job must provide a specific implementation of the method `ArrayList<Entry<Text, Text>> generateKeyValPairs(Object key, Object val);`.This method must provide the same service that the Perl examples in the streaming section did. Listing 8-12 provides the Hadoop example of the `AggregateWordCount` implementation. The `generateEntry()` method is provided by the `ValueAggregatorBaseDescriptor` and builds a key of the form `ID:KEY`, where *ID* is `countType` and *KEY* is a word found in the tokenized variable `line`.

Listing 8-12. *Hadoop Example AggregateWordCount's generateKeyValuePairs() Method*

```
public ArrayList<Entry<Text, Text>>
     generateKeyValPairs(Object key, Object val) {
      String countType = LONG_VALUE_SUM;
      ArrayList<Entry<Text, Text>> retv = new ArrayList<Entry<Text, Text>>();
      String line = val.toString();
      StringTokenizer itr = new StringTokenizer(line);
      while (itr.hasMoreTokens()) {
        Entry<Text, Text> e = generateEntry(countType, itr.nextToken(), ONE);
        if (e != null) {
          retv.add(e);
        }
      }
      return retv;
    }
```

The job is launched by calling the `ValueAggregatorJob.createValueAggregatorJob()` method, as shown in Listing 8-13. The command-line arguments accepted in `args` are listed in Table 8-15.

Listing 8-13. *Launching the AggregatorWordCount Example from AggregatorWordCount.java*

```
  public static void main(String[] args) throws IOException {
    JobConf conf = ValueAggregatorJob.createValueAggregatorJob(args
        , new Class[] {WordCountPlugInClass.class});
    JobClient.runJob(conf);
```

Table 8-15. *Command-line Options Handled by the ValueAggregatorJob.createValueAggregatorJob() method.*

Ordinal Position	Optional	Default Value	Description
0	Required	None	The input directory or file to load input records from.
1	Required	None	The output directory to store results in. As with any MapReduce job, this directory must not exist prior to job start and will be created by the framework for the job.
2	Optional	1	The number of reduce tasks.
3	Optional	textinputformat	May be textinputformat or seq, indicating that the records in argument 0, input, are to be handled using `TextInputFormat` or `SequenceFileInputFormat`.
4	Optional	None	An XML file to load as configuration data.
5	Optional	Empty String	The suffix to append to the job name, which is initialized to `ValueAggregatorJob:`.

Specifying the ValueAggregatorDescriptor Class via Configuration Parameters

The Hadoop test class `TestAggregates` provides an example of specifying the `ValueAggregatorDescriptor` class via the configuration instead of using `ValueAggregatorJob.createValueAggregatorJob()`. Listing 8-14 covers the special configuration to use a Java class that implements `ValueAggregatorDescriptor`. The configuration data causes the class `AggregatorTests` to be used. The text `UserDefined` tells the framework that this is a user-defined class. The parameter `aggregator.descriptor.num` tells the framework how many definitions there are. For each descriptor class to be used by the job, a configuration parameter key of the form `aggregator.descriptor.`*#* is defined, where *#* is the ordinal number of the descriptor class, less than the value of `aggregator.descriptor.num`. The value is the two-part text string, `UserDefined`, and the fully qualified class name, with a comma separating the parts. Input records are passed this order to each of the defined classes. In Listing 8-14, there is one class because `aggregator.descriptor.num` is set to `1`, and the class is `org.apache.hadoop.mapred.lib.aggregate.AggregatorTests`, the value of `aggregator.descriptor.`*0*.

Listing 8-14. *How TestAggregates Defines a Custom Aggregator Service.*

```
    job.setInt("aggregator.descriptor.num", 1);
    job.set("aggregator.descriptor.0",➥
"UserDefined,org.apache.hadoop.mapred.lib.aggregate.AggregatorTests");
```

The framework does not have an example for defining a custom value aggregation service. Such a service would need to implement the `ValueAggregator` interface, and jobs using the custom service would have to provide an implementation of `ValueAggregatorDescriptor.generateValueAggregator()` that understands the id of the implemented service type.

Side Effect Files: Map and Reduce Tasks Can Write Additional Output Files

The Hadoop Core framework assumes that individual map and reduce tasks can be killed with impunity, which allows the use of speculative execution and retrying of failed tasks. The framework achieves this by placing the task output in a per-task temporary directory that is deleted if the task fails or is killed, or committed to the job output if the task succeeds. Prior to Hadoop release 0.19.0, this per-task directory was available under the task configuration key `mapred.output.dir`. As of Hadoop 0.19.0, this directory is a function of the `OutputCommitter` the job is using. The default `OutputCommitter` is the `FileOutputCommitter`, which stores the task local output directory in the configuration key `mapred.work.output.dir`, and a getter is defined as `FileOutputFormat.getWorkOutputDir(JobConf conf)`. The `FileOutputCommitter` class will move all files and directories from a successful tasks work output directory to the job output directory.

■ **Tip** Side effect files should have job unique names; the method `FileOutputFormat.getUniqueName(conf,name)` produces unique names. If `fs` is a `FileSystem` object for the job output directory, and `conf` is the `JobConf` object for the task, `FSDataOutputStream sideEffect = fs.create( new Path( FileOutputFormat.getWorkOutputDir(conf), FileOutputFormat.getUniqueName(conf, "side_effect_file")) );`, will create a uniquely named side effect file in the task temporary directory with a base name of `side_effect_file` and return an `FsDataOutputStream` object to the opened file. As of Hadoop 0.19.0, the actual file name is `side_effect_file_`*`{m|r}_partition`*, where *`{m|r}`* stands for a map or reduce task, and *`partition`* is the ordinal number of the map or reduce task.

Tasks that want to create additional output files directly can create them in the temporary output directory.

Tasks can create files in this directory, and the files will be part of the final job output when the tasks succeed. The `OutputCommitter` actually commits the files to the actual job output directory.

Handling Acceptable Failure Rates

Hadoop jobs typically process large volumes of data that originates from some other source. This data, commonly called *dirty data*, is often not perfectly compliant with the data specification. It might also be the case that some input records, while compliant, are unanticipated. These data records can cause a map or a reduce task to hang, crash, or otherwise complete abnormally. By default, the framework will retry the failed task, and the entire job will be terminated if the task does not complete after a number of attempts.

Operationally it is not desirable to have a long running job terminated if only a small number of records are causing problems. New in Hadoop 0.19.0 is the ability to specify that a job can succeed even if a specified number of records cannot be processed.

The framework also allows the job to specify what percentage of the map tasks and what percentage of the reduce tasks must succeed for the job to be considered a success. The default is 100% of the map tasks and 100% of the reduce tasks.

In some applications, there is a threshold for good enough that is less than 100%. In an application the author worked with, there was a piece of legacy code that would catastrophically crash every few thousand records. Due to a variety of business reasons, it was decided not to attempt to fix the legacy application, but instead to just accept those failures.

In the real world of large-scale data processing, often the individual data records are not valuable, and the time value of the transformation result of the dataset is high. In these situations it is acceptable to accept some failing records and or some failing tasks and then let the job complete.

Dealing with Task Failure

The Hadoop framework provides four different mechanisms for dealing with task failure:

- At the highest level, the JobTracker keeps track of the number of tasks that have failed on a particular TaskTracker node on a per-job basis. If this number crosses a threshold, `mapred.max.tracker.failures`, that TaskTracker is blacklisted from executing further tasks for the job.
- The next level is the standard method that most users are familiar with: to retry a failed tasks a number of times, `mapred.map.max.attempts`, for map tasks and `mapred.reduce.max.attempts` for reduce tasks. If any task fails more than the respective number of times, the job is terminated. A task isn't actually considered failed by the JobTracker until it has used up all of its retry attempts.
- The framework also allows the job to specify what percentage of the tasks can fail before the job is terminated. This is normally 0%, but the parameters `mapred.max.map.failures.percent` and `mapred.max.reduce.failures.percent` control the allowed failure percentage.
- The job may also specify that bad record skipping is enabled, as described in the next section.

Skipping Bad Records

You can enable bad record skipping by setting `mapred.skip.map.max.skip.records` and/or `mapred.skip.reduce.max.skip.groups` to a positive nonzero value. The actual value specified is the size of the record block that is acceptable to lose. The smaller the number, the more work the framework might need to do to minimize the dropped records.

The configuration parameter `mapred.skip.attempts.to.start.skipping` determines how many times a task can fail before skip processing is enabled. Skip processing requires that the framework keep track of what record is being processed by the task. For streaming jobs and for jobs that consume multiple records to work on groups of records, the framework cannot track the records; the application developer has to assist the framework in this tracking. For maps and reduces, respectively, there are two configuration parameters and two counters that the application must manage:

- The parameters are `mapred.skip.map.auto.incr.proc.count` and `mapred.skip.reduce.auto.incr.proc.count`. The respective parameter must be set to `false` in the job configuration.
- The application must then increment the respective counter, `SkipBadRecords.COUNTER_MAP_PROCESSED_RECORDS` or `SkipBadRecords.COUNTER_REDUCE_PROCESSED_GROUPS`, for each record processed.

A binary search is used to locate the failing record group within the task. It appears that this search is exhaustive and will continue until the number of task failures is exceeded. For small values of these configuration parameters, increasing the number of task retries is required.

■ **Tip** The number of retries is controlled by the configuration parameters `mapred.map.max.attempts` and `mapred.reduce.max.attempts` with setters `JobConf.setMaxMapAttempts()` and `JobConf.setMaxReduceAttempts()`.

Capacity Scheduler: Execution Queues and Priorities

New in Hadoop 0.19.0 is the Capacity Scheduler. This feature provides somewhat dedicated resource pools, queuing priority, and pool-level access control. In the public documentation, a resource pool is referred to as a *queue*, so that term will be used in this document as well.

A queue has priority access to a specified percentage of the overall cluster task execution slots. When a cluster has unused task execution slots, a job in a queue can use the idle slots, even though these slots are over the queue's priority capacity. If a job with priority access to these resources is started, the over-priority allocation task slots will be reclaimed as needed within a specified time interval by killing the tasks executing on them.

A queue may have an explicit list of users allowed to submit jobs to it. The Capacity Scheduler may also have a list of users allowed to manage the queues.

Enabling the Capacity Scheduler

To enable the Capacity Scheduler, the following parameter must be placed in the `hadoop-site.xml` file for the cluster. As of Hadoop 0.19.0, the Capacity Scheduler JAR is not part of the default runtime classpath. The JAR file is located in `contrib/capacity-scheduler/hadoop-0.19.0-capacity-scheduler.jar` and must be put on the framework classpath. Adding this JAR to the `HADOOP_CLASSPATH` by amending the `conf/hadoop-env.sh` script is sufficient.

Listing 8-15 defines two queues, `default` and `one-small-queue`.

Listing 8-15. *Enabling Capacity Scheduling, XML Block, in hadoop-site.xml*

```
<property>
  <name>mapred.jobtracker.taskScheduler</name>
  <value>org.apache.hadoop.mapred.CapacityTaskScheduler</value>
</property>
<property>
    <name>mapred.capacity-scheduler.reclaimCapacity.interval</name>
    <value>5</value>
    <final>true</final>
    <description>The polling interval to find needed task slots
           that have a freeloader task executiong.</description>
</property>
<property>
    <name>mapred.queue.names</name>
    <value>default,one-small-queue</value>
    <description>The comma separated list of queue names.</description>
    <final>true</final>
</property>
<property>
    <name>mapred.acls.enabled</name>
    <value>false</value>
    <final>true</final>
<description>Are the access control lists enabled,
           for job submission and queue management.</description>
</property>
```

Each queue that the cluster administrator defines must have a configuration block in the `hadoop-site.xml` file. Listing 8-16 defines one queue, `one-small-queue`, with user `jason` and group `wheel` given submission and control permissions. Replace `one-small-queue` with the queue name being configured. These values could be in `hadoop-site.xml`, but the suggested location is in `capacity-scheduler.xml`. Figure 8-3 shows the JobTracker web interface for this queue set.

Listing 8-16. *For Each Queue to be Defined, XML Block in capacity-scheduler.xml*

```
<!-- for each queue, the following set of properties must exist -->
<!--, where one-small-queue is the name of the queue -->
<property>
    <name>mapred.capacity-scheduler.queue.one-small-queue.guaranteed-capacity</name>
    <value>34</value>
    <final>true</final>
    <description>A value between 0 and 100, the percentage
      of the task execution slot that one-small-queue has
            priority for.</description>
</property>
```

```
<property>
    <name>mapred.capacity-scheduler.queue.one-small-queue.reclaim-time-limit</name>
    <value>300</value>
    <description>The time in seconds before a task running on a
          loaned out slot is killed when the slot is needed.</description>
    <final>true</final>
</property>
<property>
    <name>mapred.capacity-scheduler.queue.one-small-queue.supports-priority</name>
    <value>true</value>
    <description>If true, the queue supports priorities for queued
                 jobs.</description>
</property>
<property>
    <name>
       mapred.capacity-scheduler.queue.one-small-queue.minimum-user-limit-percent
    </name>
    <value>100</value>
    <description>The percentage of the resources of this queue any user may
                 use at one time.</description>
</property>
<property>
    <name>mapred.queue.one-small-queue.acl-submit-job</name>
    <value>jason wheel</value>
    <final>true</final>
    <description>Two comma separated lists, separated by a space. The list of
              users and the list of groups. This is the set that may submit
              jobs to one-small-queue</description>
</property>
<property>
    <name>mapred.queue.one-small-queue.acl-administer-job</name>
    <value>jason wheel</value>
    <final>true</final>
    <description>Two comma separated lists, separated by a space. The list of
                 users and the list of groups. This is the set that may kill
                 or change the priority of other users jobs.</description>
</property>
```

The sum of the percentage cluster capacity for all queues must not exceed 100, or the JobTracker will not start, and there will be an exception in the log file:

```
org.apache.hadoop.mapred.JobTracker: java.lang.IllegalArgumentException: ➥
Sum of queue capacities over 100% at SOMEVALUE
```

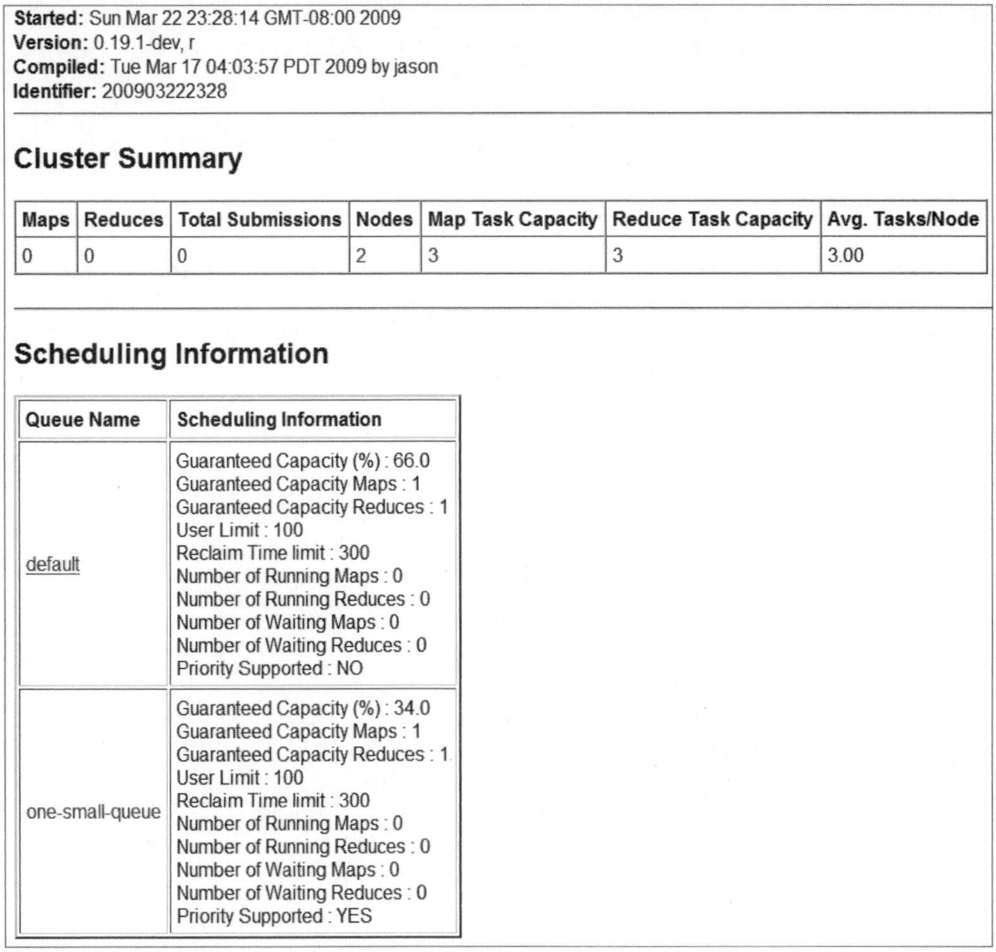

Started: Sun Mar 22 23:28:14 GMT-08:00 2009
Version: 0.19.1-dev, r
Compiled: Tue Mar 17 04:03:57 PDT 2009 by jason
Identifier: 200903222328

Cluster Summary

Maps	Reduces	Total Submissions	Nodes	Map Task Capacity	Reduce Task Capacity	Avg. Tasks/Node
0	0	0	2	3	3	3.00

Scheduling Information

Queue Name	Scheduling Information
default	Guaranteed Capacity (%) : 66.0 Guaranteed Capacity Maps : 1 Guaranteed Capacity Reduces : 1 User Limit : 100 Reclaim Time limit : 300 Number of Running Maps : 0 Number of Running Reduces : 0 Number of Waiting Maps : 0 Number of Waiting Reduces : 0 Priority Supported : NO
one-small-queue	Guaranteed Capacity (%) : 34.0 Guaranteed Capacity Maps : 1 Guaranteed Capacity Reduces : 1 User Limit : 100 Reclaim Time limit : 300 Number of Running Maps : 0 Number of Running Reduces : 0 Number of Waiting Maps : 0 Number of Waiting Reduces : 0 Priority Supported : YES

Figure 8-3. *Screenshot of a JobTracker screen with two queues enabled*

Summary

The Hadoop framework provides a powerful set of tools to enable users to run more than standard MapReduce jobs. This chapter covers a number (but by no means all) of the features. Hadoop is under active development, and new features are being introduced on a regular basis. The Hadoop streaming and aggregator features are powerful and provide the user command-line tools for performing data analysis on large datasets. Chain mapping provides a way to maintain code simplicity and reduce overall data flow through the system by allowing multiple mapper classes to be applied to the data for a job. Map-side joins provide database-style joins that can drastically speed up jobs that process bulk data that is already sorted. There are also a number of features that have become their own Apache projects (see Chapter 10).

CHAPTER 9

Solving Problems with Hadoop

On the Hadoop Core mailing list, a user was wondering about the way to handle a specific style of range query with MapReduce. The application had a search space and incoming search requests. In this chapter, we'll look at a similar setup, as follows:

- The search space dataset has the key *range begin, range end* and the value *search space data*. For simplicity's sake, let's assume that ranges in the search space do not overlap.
- The search request dataset has the key *value* and the value *search request data.*
- The result set for a value that is between *range begin and range end* has the key *value* and the value *search request data, search space data.*

How do you solve this problem with a traditional MapReduce application? That's the focus of this chapter.

There are a couple of overall design goals, and the weights of the different factors will vary by installation and by job. In today's environment, there is an intense pressure to get processes up quickly and evolve them. Given agile business practices and tight budgets, rapid evolution becomes the norm. This practice means that there will be little design time, and the application will be modified, possibly by multiple teams, over a medium to long period of time.

Design Goals

Our overall goal is to have a job that runs reliably and fast. To achieve reliability, we aim for simple code, and implement monitoring to be informed when the algorithms being used are no longer suitable for the scale or patterns of data.

Given that this application is going to evolve rapidly, and eventually be modified, perhaps by different people, each piece of code needs to be simple and clear. This is in direct opposition to the requirement that the map and reduce methods be treated as the deeply nested inner loops that they are and carefully optimized.

The data is expected to be real-world, dirty, and to change over time. Wherever possible, the application must handle malformed records in a graceful manner and report on the malformed rate.

To achieve good performance, the job must minimize underuse of the hardware, by managing how the data is split, partitioned, and compressed and by tuning the number of tasks run per node. To avoid having the network speed become the limiting factor, the transform design must attempt to minimize the number of times the data is written to HDFS and the volume of data passed to reduce tasks.

This example will have as input two datasets. One dataset—the search requests—is composed of Apache log file data in common log format. The other dataset—the search space—is composed of IP address ranges and a network name. The output of the job, shown in Table 9-1, will be a modified common log format with the IP address, the network range, and the network name, in place of the original IP address, for those search requests for which a network was found.

■Note The last two octets of all IP addresses in the log files have been randomized. The command used was `perl -ne 'chomp; if ( $_ !~ /^(\d+\.\d+)\.(\d+\.\d+)\s(.*)$/) { print STDERR "Failure on $_\n"; next; } print $1, ".", int(rand(256)), ".", int(rand(256)), " ", $3, "\n";' < access_log.txt > access_log_randomized.txt---; mv -f access_log_randomized.txt access_log.txt`.

Table 9-1. *Sample Job Output*

Log IP	Net Range Begin	Net Range End	Net Name	Log Record
12.229.91.253	12.0.0.0	12.255.255.255	ATT	- - [18/Nov/2008:14:08:59 -0800] "GET / HTTP/1.1" 404 293 "-" "Mozilla/4.0 (compatible; MSIE 7.0; Windows NT 5.1; .NET CLR 1.0.3705; .NET CLR 1.1.4322; Media Center PC 4.0; .NET CLR 2.0.50727; .NET CLR 3.0.4506.2152; .NET CLR 3.5.30729)"
58.68.24.75	58.68.0.0	58.68.127.255	DWL NET	- - [20/Nov/2008:00:47:53 -0800] "GET / HTTP/1.1" 302 315 "-" "Mozilla/4.0 (compatible; MSIE 8.0; Windows NT 5.1; Trident/4.0; .NET CLR 2.0.50727; .NET CLR 3.0.04506.30; .NET CLR 3.0.04506.648; .NET CLR 3.5.21022)"
59.92.27.113	59.88.0.0	59.99.255.255	BSNLNET	- - [19/Nov/2008:12:38:42 -0800] "GET / HTTP/1.1" 302 309 "-" "Mozilla/5.0 (Windows; U; Windows NT 5.1; en-US; rv:1.9.0.4) Gecko/2008102920 Firefox/3.0.4"
60.50.230.247	60.48.0.0	60.54.255.255	XDSLSTREAMYX	- - [16/Nov/2008:21:06:18 -0800] "GET / HTTP/1.1" 404 293 "-" "Mozilla/4.0 (compatible; MSIE 6.0; Windows 98)"
61.135.0.92	61.135.0.0	61.135.255.255	CNCGROUP BJ	- - [22/Nov/2008:07:07:16 -0800] "GET / HTTP/1.1" 302 309 "-" "Baiduspider+(+http://www.baidu.com/search/spider.htm)"

Note Thanks to Apress for the log file samples.

Design 1: Brute-Force MapReduce

The brute-force MapReduce pattern is generally the quickest to get going and the simplest to manage. The downside is that these jobs quickly become bound by the network speed and the sorting speed for the cluster.

In a brute-force MapReduce, the only time you have ordered data is in the reduce step. This forces all of the data to flow through to the reduce task. There is also the additional complexity that you have multiple record types, which need to be distinguished at reduce time.

The overriding constraint here is ensuring that any given search request record finds all records that it is in range of in the search space.

A Single Reduce Task

If a single reduce task is used, all search request records are guaranteed to be in the same partition as their respective search space records. Table 9-2 defines the comparator behavior for the three cases the comparator will encounter.

Table 9-2. *Comparator Cases*

Type of Item 1	Comparison Region of Item 1	Type of Item 2	Comparison Region of Item 2	Equality Condition
Search request	Entire key	Search request	Entire key	Key_1 equal to key_2
Search request	Entire key	Search space	Begin range	Search request key equal to begin range
Search space	Begin range, end range	Search space	Begin range, end range	Begin $range_1$ equal to begin $range_2$ and end $range_1$ equal to end $range_2$

The input plan for the reduce method is to receive individual records and to manage the join behavior by maintaining memory about previous records. This adds complexity to the reduce method and increases the risk of out-of-memory conditions. To enable the framework to do the aggregation would require having redundant data in the records; the end range would need to be in the value of the search space records. This requirement is driven by the fact that the `OutputCompartor` object receives only the key. A simplification that results from this decision is that, in the first pass, using `Text` is acceptable for the key and value, as the records may be distinguished lexically. In a future step, as a performance optimization, we will implement a key class that provides a `WritableComparator` that handles our keys at the byte level rather than at the object level. Using the byte-level comparator for a complex key opens the door to the key format and the comparator getting out of sync, introducing the possibility of errors.

Note Having `Text` objects for the key and value greatly simplifies the initial debugging of the jobs, as the data can be readily examined by eye.

Key Contents and Comparators

For simplicity in this pass, we are going to use the same object, `Text`, for the keys for both datasets, and `Text` for the values. To do this, a simple encoding must be defined that allows the origination dataset to be determined easily from the text of the key. If there is a way to do this without needing to write a custom comparator, the job can be up and running very quickly. For the stock comparator to work, the keys must lexically compare an order that the reduce method understands and can process with minimal complexity.

In this application, a key is an IPv4 address for a search request record, and a pair of IPv4 addresses for the search space records. If all IP addresses are encoded as a zero-padded, fixed-length hexadecimal string, the primary lexical ordering issue is addressed. This leaves a single issue: lexically, keys for the search requests will sort before a search space key that has a begin range value equal to the key of the search request. In the best of all possible worlds, search request keys would appear in the sorted output, after the search space key that opens the range for the request.

The search space key may simply be the begin range and end range values, with a separator character. There are many simple tools for splitting strings based on a separator character. This has the advantage that if a lexically larger character is used as a suffix for the search request keys, the search request keys will sort after the search space key that defines the relevant range. An example is shown in Table 9-3.

Table 9-3. *Expected Sorting Order for Search Space and Search Request Keys Using a Separator Character for the Space Range and a Suffix Character for the Request Keys*

Address	Key Type	Encoded Key
`220.255.7.213:220.255.7.217`	Space	`dcff07d5:dcff07d9`
`220.255.7.217`	Search	`dcff07d9;`

This can be quickly tested by running a small sample dataset through a streaming job to verify that the data compares the way we expect. A test dataset will be prepared from an Apache log file, with the Perl command in Listing 9-1. The code in this section takes the first field of the access log, commonly an IP address, and converts it to an unsigned integer, which is then printed as an eight-character-wide hexadecimal number, with a semicolon (`;`), as a suffix. A fake range is generated by printing that original value, without the semicolon, with a number ten higher, with a colon (`:`) separating them. A few lines of the output are included. Notice that the output ordering is exactly the reverse of what our application needs.

Note Listing 9-1 is structured to run from within the Cygwin environment, in the `examples` directory, on a Windows installation. Adjust the paths and file names as needed for your local installation.

Listing 9-1. *Generating a Sample Set of IP Addresses and Ranges from an Apache Log File*

```
perl -MSocket -ne 'chomp; my @parts = split(/\s/, $_); my $ip = $parts[0]; ➥
my $val = inet_aton($ip); my $num = unpack( "N", $val); printf "%08x;\n", ➥
 $num; printf ➥"%08x:%08x\n", $num, $num+10;' < access_log.txt | ➥
 sort -u > 'C:\tmp\dataset'
head /cygdrive/c/tmp/dataset
```

```
0c065a60:0c065a6a
0c065a60;
0c067f4d:0c067f57
0c067f4d;
0c06e9c7:0c06e9d1
0c06e9c7;
0c06ef2d:0c06ef37
0c06ef2d;
0c1e1694:0c1e169e
0c1e1694;
```

In the command shown in Listing 9-1, a dataset was prepared with converted IP addresses from an Apache log file. Listing 9-2 runs a streaming job to see how the records will actually be sorted by the default comparator. As you can see from the Listing 9-2 output, the search space records (`0c065a60:0c065a6a`) sort before a search request record that starts with the same address (`0c065a60;`). Success—this is the pattern we were hoping to achieve.

Note Cygwin users are likely to always have an error message that starts with `cygpath: cannot create short name of c:\Documents and Settings\Jason\My Documents\Hadoop Source\hadoop-0.19\logs`. This error may be ignored. Listing 9-2 is structured to run from the Hadoop installation directory.

Listing 9-2. *Running a Streaming Job to Verify Comparator Ordering*

```
bin/hadoop jar contrib/streaming/hadoop-0.19-streaming.jar -D ➥
mapred.job.tracker=local -D fs.default.name=file:/// -input 'C:\tmp\dataset' ➥
-output 'C:\tmp\sorted' -mapper 'C:\cygwin\bin\cat' -reducer 'C:\cygwin\bin\cat' ➥
 -numReduceTasks 1;
```

```
jvm.JvmMetrics: Initializing JVM Metrics with processName=JobTracker, sessionId=
mapred.JobClient: No job jar file set.  User classes may not be found. ➥
See JobConf(Class) or JobConf#setJar(String).
mapred.FileInputFormat: Total input paths to process : 1
streaming.StreamJob: getLocalDirs(): [/tmp/hadoop-Jason/mapred/local]
streaming.StreamJob: Running job: job_local_0001
streaming.StreamJob: Job running in-process (local Hadoop)
mapred.FileInputFormat: Total input paths to process : 1
mapred.MapTask: numReduceTasks: 1
mapred.MapTask: io.sort.mb = 1
mapred.MapTask: data buffer = 796928/996160
mapred.MapTask: record buffer = 2620/3276
streaming.PipeMapRed: PipeMapRed exec [C:\cygwin\bin\cat]
streaming.PipeMapRed: R/W/S=1/0/0 in:NA [rec/s] out:NA [rec/s]
streaming.PipeMapRed: R/W/S=10/0/0 in:NA [rec/s] out:NA [rec/s]
streaming.PipeMapRed: R/W/S=100/0/0 in:NA [rec/s] out:NA [rec/s]
streaming.PipeMapRed: mapRedFinished
streaming.PipeMapRed: Records R/W=616/1
streaming.PipeMapRed: MRErrorThread done
streaming.PipeMapRed: MROutputThread done
mapred.MapTask: Starting flush of map output
mapred.MapTask: Finished spill 0
mapred.TaskRunner: Task:attempt_local_0001_m_000000_0 is done. ➥
 And is in the process of commiting
mapred.LocalJobRunner: Records R/W=616/1
mapred.TaskRunner: Task 'attempt_local_0001_m_000000_0' done.
streaming.PipeMapRed: PipeMapRed exec [C:\cygwin\bin\cat]
mapred.Merger: Merging 1 sorted segments
mapred.Merger: Down to the last merge-pass, with 1 segments ➥
 left of total size: 10474 bytes
streaming.PipeMapRed: R/W/S=1/0/0 in:NA [rec/s] out:NA [rec/s]
streaming.PipeMapRed: R/W/S=10/0/0 in:NA [rec/s] out:NA [rec/s]
streaming.PipeMapRed: R/W/S=100/0/0 in:NA [rec/s] out:NA [rec/s]
streaming.PipeMapRed: mapRedFinished
streaming.PipeMapRed: MRErrorThread done
streaming.PipeMapRed: Records R/W=616/1
streaming.PipeMapRed: MROutputThread done
mapred.TaskRunner: Task:attempt_local_0001_r_000000_0 is done. ➥
 And is in the process of commiting
mapred.LocalJobRunner:
mapred.TaskRunner: Task attempt_local_0001_r_000000_0 is allowed to commit now
mapred.FileOutputCommitter: Saved output of task ➥
 'attempt_local_0001_r_000000_0' to file:/C:/tmp/sorted
mapred.LocalJobRunner: Records R/W=616/1 > reduce
mapred.TaskRunner: Task 'attempt_local_0001_r_000000_0' done.
```

```
streaming.StreamJob:  map 100%  reduce 100%
streaming.StreamJob: Job complete: job_local_0001
streaming.StreamJob: Output: C:\tmp\sorted
```

```
head /cygdrive/c/tmp/sorted/part-00000
```

```
0c065a60:0c065a6a
0c065a60;
0c067f4d:0c067f57
0c067f4d;
0c06e9c7:0c06e9d1
0c06e9c7;
0c06ef2d:0c06ef37
0c06ef2d;
0c1e1694:0c1e169e
0c1e1694;
```

A Helper Class for Keys

Key management is critical for this job, and to help avoid introducing errors later in the application life cycle, a helper class for keys will be provided. The initial version needs to be able to validate, pack, and unpack keys to and from the Text objects.

TASK-SPECIFIC CONFIGURATION PARAMETERS

The Hadoop framework creates a runtime environment for the tasks of the job. In the TaskTracker's local working area, the path set defined by the configuration key, mapred.local.dir, a directory tree is built for the job, which contains the unpacked DistributedCache items, a file job.xml that contains the job configuration, a shared directory for all tasks of the job, and a working directory for the task to be run. An instance of the configuration date is created, and the per-task information modified by adding per-task parameters and adjusting the paths of configuration parameters that have been unpacked into the job or task working areas. The bulk of this localization process is handled by TaskTracker.localizeJob. The following parameters are added or modified for a task as of Hadoop 0.19.0:

- job.local.dir: The directory that will be used as root of the local file system space allocated for this job. JobConf.getJobLocalDir() returns this directory. All tasks of this job running on a TaskTracker node will share this directory. A Java system property of the same name is also set.
- mapred.local.dir: The root of the local file system space for this TaskTracker node.
- map.input.file: For the map task, the input file name, if the input split has a file name.
- map.input.start: For the map task, the starting offset in the map.input.file.

- map.input.length: The amount of data to read from map.input.file, starting from map.input.start.
- mapred.tip.id: The task ID for this task. All task attempts for this task will have the same value for this key.
- mapred.task.id: The task ID for this attempt of this task. The framework will make multiple attempts to complete a task. This value for this key holds the ID of the current attempt instance. In Hadoop 0.19, the value stored under this key is very similar to mapred.tip.id, except that it will have a prefix of attempt_. This is unique per task run.
- mapred.task.is.map: Set to true if this is a map task.
- mapred.task.partition: The partition number for this task, if known. For a map task, this is the ordinal number of the task. For a reduce task, it is both the ordinal number of the reduce task and the result of Partitioner.getPartition(K,V, numPartitions), which will be identical for all key/value pairs passed to this reduce task.
- mapred.job.id: The ID of the job that this task is being run on behalf of.
- mapred.work.output.dir: The task-specific directory that output files will be created in by default. FileOutput.getWorkOutputPath(JobConf conf) provides this value.
- mapred.map.tasks: In the reduce task, the actual number of map tasks that succeeded.
- hadoop.net.static.resolutions: Any hostname/IP address mappings that will override the normal lookup results.
- task.memory.mgmt.enabled: Set to true if the TaskTracker is enforcing memory utilization limits.

In our example, four classes are associated with key handling:

- An interface, KeyHelper<K>
- An abstract class, AbstractKeyHelper<K>
- An implementation, TextKeyHelperWithSeparators, for Text-based keys
- A unit test, TestTextKeyHelperWithSeparators, to verify the expected behavior

These classes provide a way to extract the IP address from a key, shown in Listing 9-3, and to pack IP addresses into a key, shown in Listing 9-4. Two configuration parameters are available: examples.ch9.search.suffix.char, which defines the character to be used as a suffix when encoding a search request IP address, and examples.ch9.range.separator.char, which defines the character to be used to separate a pair of IP addresses in a search space key. These parameters have default values of semicolon (;) and colon (:), respectively. They may be any pair of characters, as long as the range separator character sorts first.

Listing 9-3. *boolean TextKeyHelperWithSeparators.getFromRaw(Text raw)*

```
public boolean getFromRaw(Text raw) {
isValid = false;
hasEndRange = false;
String rawText = raw.toString();
if (rawText.length()==(addressLen+1)
    && rawText.charAt(addressLen)==searchRequestSuffix) {
    String searchRequest = rawText.substring(0, addressLen);
    beginRangeOrKey = Long.valueOf(searchRequest,16);
} else if (rawText.length()==(addressLen*2+1)
            && rawText.charAt(addressLen)==rangeSeparator) {

    String beginRange = rawText.substring( 0, addressLen);
    beginRangeOrKey = Long.valueOf(beginRange,16);
    endRange = Long.valueOf(rawText.substring(addressLen+1,addressLen*2+1),16);

    /** Verify that the begin range is less or equal to the end */
    if (beginRangeOrKey>endRange) {
        if (LOG.isDebugEnabled()) {
 LOG.debug("key [" + rawText + "] length " + rawText.length() + " begin > end "
  + beginRangeOrKey + " " + endRange);
       }
        return false;
    }
    hasEndRange = true;
} else {
    if (LOG.isDebugEnabled()) {
 LOG.debug("key [" + rawText + "] length " + rawText.length() + " invalid");
    }
    /** length is wrong, or the separator or suffix is wrong. */
    return false;
}
isValid = true;
return true;
}
```

In Listing 9-3, the key is converted to a `String` and examined to see if it is one of the two patterns that are accepted. All IP addresses will be encoded as eight hexadecimal digits. If the key is a search request, there will be one IP address and a trailing `searchRequestSuffix` character only, forcing the string to be only nine characters in length. If the key is a search space item, there will be two IP addresses, with a `rangeSeparator` character between them only, forcing the string to be seventeen characters in length. The IP addresses are converted into long values via `Long.valueOf(address,16)`. The `String.substring` method is used for extracting the actual IP address data from the raw string.

If a valid search request or search space definition is found, the helper object is marked valid, `isValid = true`, and `beginRangeOrKey` is set to the first IP address found. If the key contained a search space request, `hasEndRange` is set to `true` and `endRange` is set to the second IP address.

The `setToRaw` method, in Listing 9-4, is used to create and store a value in a key object that correctly encodes either a search request or a search space. If the helper object is not valid, nothing is done, and no indication of this is made. This will open the door to missing errors. Changing this behavior requires rearchitecting the application to provide a visible trace of this error; logging it is not likely to be sufficient. A `StringBuilder` and `Formatter` are `ThreadLocal` instance variables, making this class thread-safe. This is done as a small efficiency and a protection against the day when the helper is used in a multithreaded map task.

Listing 9-4. *void TextKeyHelperWithSeparators.setToRaw(Text raw)*

```
public void setToRaw(Text raw) {
if (!isValid) {
    return;
}
Formatter fmt = keyFormatter.get();
fmt.flush();
StringBuilder sb = keyBuilder.get();
sb.setLength(0);

if(hasEndRange) {
    fmt.format( "%08x%c%08x", beginRangeOrKey, rangeSeparator, endRange );
} else {
    fmt.format( "%08x%c", beginRangeOrKey, searchRequestSuffix);
}
fmt.flush();
raw.set(sb.toString());
}
```

Note It is reasonable to assume that anything written to the log by a task will never have been seen by a human being unless something is visibly wrong with the job. The volume of data is just too large.

The Mapper

With the plan for the comparator handled, it is time to design the mapper. This mapper must handle two tasks:

- For the search requests, the mapper must accept Apache log files and extract a key from the line in the key format, passing the rest of the line as the value.
- The search space items will be stored as straight text, with a tab (`\t`) separating the range from the data. The mapper may distinguish between the two records either from the input file name or by the length of the key.

As a demonstration of using chain mapping, our mapper is going to run a chain to process the incoming values. The first element in the chain will take action only if the incoming record does not look like a search request or search space key, but instead looks like an Apache log file record. This mapper will transform the record into a search request. The next map in the chain will perform validity checking on the keys.

■ **Note** In the next version, the example will use `org.apache.hadoop.mapred.lib.MultipleInputs`, and have the search space dataset be in a `SequenceFile`. For simplicity of debugging, this version uses text records only.

This example has two mapper classes: `ApacheLogTransformMapper` and `KeyValidatingMapper`. Listing 9-5 shows the mapper preamble in `ApacheLogTransformMapper`. This demonstrates our standard practice of having a counter, named `TOTAL INPUT`. This provides a clear indication of how the job is going. The `helper` object parses a string that is either a search request or a search space, returning `true` if the key was recognized. In this preamble, if the `helper` can parse the key, it is just passed forward. As a general rule, we log per-key data only at level debug, as the logging volume will be very large.

Listing 9-5. *The Mapper Preamble, ApacheLogTransformMapper.java*

```
reporter.incrCounter("ApacheLogTransformMapper", "TOTAL INPUT", 1);

if (helper.getFromRaw(key)) {
    reporter.incrCounter("ApacheLogTransformMapper", "ALREADY PREPARED KEYS", 1);
    if (LOG.isDebugEnabled())
        LOG.debug("complete key passed forward untouched [" + key + "]");
    }
    output.collect( key, value );
    return;
}
```

In Listing 9-6, the key was not recognized as a prepared key and is assumed to be an Apache log line. If the input separator for the `TextInputFormat` happens to be a single space:

```
conf.get("key.value.separator.in.input.line", "\t");
```

then the key is assumed to be the IP address. The test `keyValueSeparator.length()==1 && keyValueSeparator.charAt(0)==' '` verifies this.

■ **Note** If the input format happens to not be `KeyValueTextInputFormat`, the configuration key changes in `KeyValueTextInputFormat`, or the default value changes, this code will fail silently.

The method parseAddressIntoKey() will take the IP address and convert it into our established format and pass the new key and the value to the output.

Lisitng 9-6. *The Mapper Log Line Processing Part 1, ApacheLogTransformMapper.java*

```
if (LOG.isDebugEnabled()) { LOG.debug("Working on [" + key + "]"); }
reporter.incrCounter("ApacheLogTransformMapper", "LOG LINES", 1);
String logLine = key.toString();
String keyValueSeparator = conf.get("key.value.separator.in.input.line", "\t");
String ipAddress;
/** The IP address in the standard log file entry is the first field,
  * with a trailing space to separate it from the next field.
 */
if (keyValueSeparator.length()==1 && keyValueSeparator.charAt(0)==' ') {
    /** The key and value are already parsed out. */
    ipAddress = logLine;
    if (parseAddressIntoKey(ipAddress, outputKey, reporter)) {
        reporter.incrCounter("ApacheLogTransformMapper", "VALID LOG LINES", 1);
        if (LOG.isDebugEnabled()) {
          LOG.debug( "Key transforms from [" + key + "] to [" + outputKey + "]");
        }
        output.collect(outputKey, value);
        return;
    }
```

In Listing 9-7, the default case of a raw log line is handled. This code does make the assumption that the keyValueSeparator computed in Listing 9-6, is correct. A complete line is assembled in sb, and then parsed. The IP address is assumed to be the first text in the line and to be terminated by an ASCII space character. This code accepts only IPv4 addresses in the format of four dot-separated octets. Once the correct key and new value are produced, they are output. The use of chain mapping actually reduces the efficiency of the task, but it is nice to have a demonstration.

Listing 9-7. *The Mapper Log Line Processing Part 2, ApacheLogTransformMapper.java*

```
/** For paranoia sake, reassemble the log line and split it ourselves
  * on the first space. */
sb.setLength(0);
sb.append(logLine);
sb.append( keyValueSeparator);
sb.append( value.toString());
logLine = sb.toString();

int indexOfSpace = logLine.indexOf(' ');
if (indexOfSpace< 7 || indexOfSpace > 15) {
                         /** xxx.xxx.xxx.xxx = 15 chars, 1.1.1.1 = 7 chars */
```

```
    if (LOG.isDebugEnabled()) {
 LOG.debug("Log line does not start with an ip address [" + logLine + "]" );
    }
    reporter.incrCounter("ApacheLogTransformMapper", "BAD LOG LINES", 1);
    return;
}

ipAddress = logLine.substring(0,indexOfSpace);
logLine = logLine.substring(indexOfSpace+1);

if (parseAddressIntoKey(ipAddress, outputKey, reporter)) {
    outputValue.set( logLine );
    reporter.incrCounter("ApacheLogTransformMapper", "VALID LOG LINES", 1);
    if (LOG.isDebugEnabled()) {
       LOG.debug( "Key transforms from [" + key + "] to [" + outputKey + "]");
    }
    output.collect( outputKey, outputValue );
    return;
}
```

The KeyValidatingMapper, shown in Listing 9-8, just checks the keys for the proper shape—that they are valid IPv4 addresses—and swaps the search space begin and end range values if begin is greater than end. At this point, all keys are assumed to be valid, and this map verifies that. Several counters are kept to help with sort and long-term monitoring of the job.

Listing 9-8. *KeyValidatingMapper.java*

```
if (!helper.getFromRaw(key)) {
    reporter.incrCounter("KeyvalidatingMapper", "INVALID KEYS", 1);
    return;
}
if (helper.isSearchRequest()) {
    reporter.incrCounter("KeyValidatingMapper", "TOTAL SEARCH", 1);

    if (helper.getSearchRequest()<0 || helper.getSearchRequest()>4294967296L) {
        if (LOG.isDebugEnabled()) {
           LOG.debug("Search Key out of range [" + key + "]");
        }
        reporter.incrCounter("KeyValidatingMapper", "SEARCH OUT OF RANGE", 1);
        return;
    }
    output.collect( key, value);
    return;

} else {
    reporter.incrCounter("KeyValidatingMapper", "TOTAL SPACE", 1);
```

```
        if (helper.getBeginRange()<0||helper.getBeginRange()>4294967296L) {
            reporter.incrCounter("KeyValidatingMapper", "SPACE BEGIN OUT OF RANGE", 1);
            return;
        }
        if (helper.getEndRange()<0||helper.getEndRange()>4294967296L) {
            reporter.incrCounter("KeyValidatingMapper", "SPACE END OUT OF RANGE", 1);
            return;
        }

        /** Verify the ordering of the search space item. */
        if (helper.getBeginRange()<=helper.getEndRange()) {
            output.collect(key, value);
            return;

        } else {
            reporter.incrCounter("KeyValidatingMapper", "SPACE OUT OF ORDER", 1);
            long tmp = helper.getBeginRange();
            helper.setBeginRange(helper.getEndRange());
            helper.setEndRange(tmp);
            helper.setToRaw(outputKey);
            output.collect( outputKey, value);
            return;
        }
    }
```

The Combiner

The combiner is often one of the more complex pieces of a MapReduce job, and it's usually given the least thought. What is the correct behavior for encountering duplicate keys in the map output? For simple aggregation jobs, this is straightforward. In our case, we have two different types of keys, and what to do for a duplicate in either case is unclear.

The first proposal would be to use a `TextArrayWritable`, and just keep all of values. This doesn't provide much of a space saving, compared to just not running a combiner. The second proposal would be to discard duplicates. Neither choice is appealing. A combiner should provide either a significant reduction in I/O volume or a significant reduction in resource use for the reduce phase. Neither of the preceding proposals can provide those. If a custom comparator were written, a combiner might make sense.

In the type of MapReduce application we are working on here, a combiner that suppresses duplicate key/value pairs could be helpful. In our constructed example, we know there are no exact duplicates.

The Reducer

Each reducer task will need to receive a stream of key values, where the range statements will be first in the sorting order. This forces the reducer class to maintain state information about which ranges have been seen, and the value of those ranges. This prior range information is bounded, and ranges may be flushed when the end range value is less than the current input key. As an added bonus, the reduce task is also run as a chain, with a postprocessing map that converts the encoded key formats back into dot-separated octet format. The actual reduce task is performed by `ReducerForStandardComparator.java`, shown in Listing 9-9.

Lisitng 9-9. *ReducerForStandardComparator.java*

```
reporter.incrCounter("ReducerForStandardComparator", "TOTAL KEYS", 1);
if (!helper.getFromRaw(key)) {
    reporter.incrCounter("ReducerForStandardComparator", "BAD KEYS", 1);
    return;
}

if (helper.isSearchSpace()) {
    reporter.incrCounter("ReducerForStandardComparator", "SPACE KEYS", 1);

    /** For simplicity, put all of the values in. */
    while (values.hasNext()) {
        final Text value = values.next();
        reporter.incrCounter("ReducerForStandardComparator", "SPACE VALUES", 1);
        activeRanges.activate( reporter, "ReducerForStandardComparator", ➥
           helper.getBeginRange(), helper.getEndRange(), value.toString());
    }
    return;
}

if (helper.isSearchRequest()) {
    /** First, lets prune the activeRanges. */
    final long searchRequest = helper.getSearchRequest();
    activeRanges.deactivate(searchRequest);

    /** Because the ranges are removed when their end is less than end,
     * and because keys are always sorted after the beginning of a range
     * all active ranges are now 'hits' for this search request.
     */
    int max = activeRanges.size();
    while (values.hasNext()) {
        final Text value = values.next();
        for (int i = 0; i < max; i++) {
            ActiveRanges.Range<String> hit = activeRanges.get(i);
            handleHit( key, output, reporter, value, hit);
        }
    }
}
```

In Listing 9-9, our standard counters are in use. At this point, any invalid key is an indication that something has gone very wrong—data corruption at some level, given the level of verification performed on the keys in earlier steps.

Our algorithm is very simple. We keep a queue of networks, ordered by the network end-of-address range. If the current key is a search request and the current key is larger than the end of a network's address range, the network is removed from the active queue. The call `activeRanges.deactivate(searchRequest)` clears any networks from the `activeRanges` queue that can no longer be matched. If the current key is a search space key, it is added to the set of active ranges, via the following:

```
activeRanges.activate( reporter, "ReducerForStandardComparator",
                       helper.getBeginRange(), helper.getEndRange(),
                       value.toString());
```

At this point, each network in `activeRanges` is a match. A network's end range is guaranteed to be larger than the search request key, and due to our comparator's ordering of the keys, the network begin range must be less than or equal to our search request.

For each log line, `while (values.hasNext())`, an output record is generated for each network, `for (int i = 0; i < max; i++) {`, via the call to `handleHit( key, output, reporter, value, hit)`, which is shown in Listing 9-10.

Listing 9-10. *ReducerForStandardComparator.handleHit*

```
StringBuilder sb = new StringBuilder();
Formatter fmt = new Formatter(sb);

protected void handleHit(Text key,
        OutputCollector<Text, Text> output, Reporter reporter,
              Text value, Range<String> hit) throws IOException {

    /** For this version we leave the end alone. */
    sb.setLength(0);
    fmt.format( "%s\t%s", hit.getValue(), value.toString());
    fmt.flush();
    outputValue.set(sb.toString());
    sb.setLength(0);
    /** Lose the suffix */
    fmt.format("%8.8s\t%08x\t%08x", key.toString(),
       hit.getBegin(), hit.getEnd()); fmt.flush();
    outputKey.set(sb.toString());
    output.collect( outputKey, outputValue );

}
```

In Listing 9-10, a `StringBuilder` and `Formatter` are built. These are used to construct the actual output key and output value. The key will be the original log record IP address, followed by the network begin and end addresses. For ease of parsing, these will be separated by an ASCII tab character. The value is simply the network name, ASCII tab, and the rest of the original log line.

The Driver

The driver, shown in Listing 9-11, builds on our base class, `utils/MainProgramShell.java`, and defines only a small number of methods. This example relies on there being only a single reduce task, as the default partitioner will cause this job to fail. In our next design iteration, we will write a custom partitioner.

All of the examples in this chapter are structured to run on small machines, so the reduce sort space has been reduced from 100MB to 10MB, using the following line:

```
conf.setInt("io.sort.mb", 10);
```

The values for `input` and `output` are set by the use of the command-line flags `--input` and `--output`, respectively. The setup follows the general rule for using the chain, and allocates `dummyConf` to use as the private configuration object for the chained map and reduce tasks. The framework serializes the contents in each call to the `ChainMapper` methods, making it safe to clear `dummyConf` and reuse it.

Listing 9-11. *The Job Setup, BruteForceMapReduceDriver.java*

```
super.customSetup(conf);
conf.setJobName("BruteForceRangeMapReduce");
conf.setNumReduceTasks(1);
conf.setInt("io.sort.mb", 10);
conf.setInputFormat( KeyValueTextInputFormat.class);
for( String input : inputs) {
    if (verbose) {
        LOG.info("Adding input path " + input);
    }
    FileInputFormat.addInputPaths(conf, input);
}
if (verbose) {
    LOG.info( "Setting output path " + output);
}
FileOutputFormat.setOutputPath(conf, new Path(output));
conf.setOutputFormat(TextOutputFormat.class);

JobConf dummyConf = new JobConf(false);
ChainMapper.addMapper(conf, ApacheLogTransformMapper.class,
        Text.class, Text.class, Text.class, Text.class, false, dummyConf);
dummyConf.clear();
ChainMapper.addMapper(conf, KeyValidatingMapper.class, Text.class, Text.class,
        Text.class, Text.class, false, dummyConf);

dummyConf.clear();
ChainReducer.setReducer(conf, ReducerForStandardComparator.class,
        Text.class, Text.class, Text.class, Text.class, false, dummyConf);
dummyConf.clear();
ChainReducer.addMapper(conf, TranslateBackToIPMapper.class,
        Text.class, Text.class, Text.class, Text.class, false, dummyConf);
```

The map and reduce methods used do not modify the passed-in key or value objects; therefore, the chaining framework is being formed to pass keys and values by reference. The second-to-last argument, `false`, in the `ChainMapper.addMapper` and `ChainMapper.setReducer` methods forces this behavior.

All of the mappers and reducers expect `Text` objects for the input key and value, and output `Text`. In an updated version of chaining, in which the key and value objects implement `WritableComparable` and `Writable`, passing `TextKeyHelperWithSeparators` objects for the key would probably be significantly more efficient.

The Pluses and Minuses of the Brute-Force Design

The biggest plus of this design is that it is simple and took about a day to put together. The biggest disadvantages are that all of the data must pass through the mapper and be sorted, and that only a single reduce task may be used. Given that the total number of networks is relatively bounded, if the incoming log records are batched in smaller sizes, this job will run reasonably well and reasonably fast. Without a custom partitioner, this job cannot be made to run with multiple reduce tasks.

Design 2: Custom Partitioner for Segmenting the Address Space

The biggest boost for the brute-force method would be to find a simple way to allow multiple reduce tasks. The standard partitioner uses the hash value of the key, modulus the number of partitions as the partition number. A simple strategy for this application might be to simply segment the IP address range. There is no guarantee that the network ranges will fall cleanly on these segments. There will need to be a mechanism to split search space keys into segment-appropriate boundaries during the job, while putting the full range in the output record. Perhaps simply modifying the format for the search space records to allow for an original range to be part of the record will address this.

■ **Note** This partitioning method is still subject to uneven distributions of the key space resulting in a subset of reduce tasks running much longer. To ameliorate this, the key space may be sampled and the partitioning table built using the sample data, in a manner similar to that done by the Hadoop terasort example.

The Simple IP Range Partitioner

The partitioner class for this example is `SimpleIPRangePartitioner`. The `getPartition()` method, shown in Listing 9-12, simply takes the IP address of a search request key or the begin range address of a search space key and returns the partition for that record.

A SCOPE REDUCTION IN THE PARTITIONER

The original design supported a configurable table to ensure that the records were partitioned approximately evenly. This required a tool to scan the records to generate a distribution map and code to load that map into the partitioner. During the process of actually writing the code, the decision was made that if that feature is needed, it may be implemented later. Instead, each partition gets an approximately even number or span of addresses out of the IPv4 space.

For a job with one reduce task, the span for partition 0 is from `0.0.0.0` to `255.255.255.255`. For a job with two reduce tasks, partition 0 would span from `0.0.0.0` to `127.255.255.255`, and partition 1 would span from `128.0.0.0` to `255.255.255`.

This left a few artifacts in the `SimpleIPRangePartitioner`. A `TreeSet` is used instead of simply maintaining an array of long values. The array of long values would be faster and would greatly reduce object churn.

Listing 9-12. *SimpleIPRangePartitioner.getPartition*

```
@Override
public int getPartition(final Text key, final Text value, final int numPartitions) {
    if (!(helper.getFromRaw(key) && helper.isValid())) {
        throw new IllegalArgumentException("key " + key +
                " cannot be parsed as a network range set");
    }
    /** The IP address that effectively defines this range. */
    final long begin;

    if (helper.isSearchRequest()) {
        begin = helper.getSearchRequest();
    } else {
        begin = helper.getBeginRange();
    }

    /** Find the bucket in ranges that is the lowest bucket
      * that is valued higher than begin.
     * That bucket's partition is the partition for this value.
     */e
    final Entry<Long, Integer> partition = ranges.higherEntry(Long.valueOf(begin));
    /** Stored as a variable for debugging ease */
    final int realPartition = partition.getValue();

assert (helper.isSearchSpace() ? partition.getKey() >= helper.getEndRange() : true)
 : String.format( "search space range end %08x exceeds partition limit %0x8",
    helper.getEndRange(), partition.getKey());

    return realPartition;
}
```

The first step is to initialize the key helper and to determine if the key is actually a valid search space or search request key:

```
if (!(helper.getFromRaw(key) && helper.isValid())) {
```

If the key is valid, the IP address of the search request record or the range begin address of the search space record is stored in begin. Once begin is known, it may be looked up in the table, ranges, that maps addresses to reduce partitions. The table is actually a TreeMap, and entry keys are the ending IP address of the partition. The partition number is the entry value. This data structure allows the following line to provide the entry of the partition that the key/value pair must go to:

```
partition = ranges.ceilingEntry(Long.valueOf(begin));
```

The TreeMap method higherEntry returns the element in ranges where the entry key is closest to begin, while not being less than begin. range end is larger than begin. The value of that entry is the partition number for this key/value pair.

For debugging purposes, the entry is assigned to a local variable, partition. The entry value could simply be returned at this point, but a little checking is done to verify that this key/value pair is a search space record, where the end of the search space is also an address that will be in this partition. No checking is made for the case where ranges.higherValue returns null, as it is assumed that the ranges table spans the full IPv4 address space range.

The ranges table is constructed in the configure() method, shown in Listing 9-13, as this is the first time the number of reduce tasks is known.

Listing 9-13. *SimpleIPRangePartitioner.configure*

```
public void configure(JobConf job) {

conf = job;
/** Now that we have a conf object we can initialize the
  * helper and build ranges, using the number of reduces. */
helper = new PartitionedTextKeyHelperWithSeparators(conf);

final int numPartitions = conf.getNumReduceTasks();

ranges = new TreeMap<Long,Integer>();

long rangeSpan = 4294967296L / numPartitions;

/** The partition that ends at <code>spanned</code> */
int partition = 0;
/** The end of the address space already in ranges. */
long spanned;
/** The value stored is the end of the range, the range
  * starts at the previous value + 0, or for the first value
 * at 0.
```

```
 * Note that the test is less than and not less than or equals,
 * the ranges have to end at 2^32-1 as we only have 32 bits.
 */
for (spanned= rangeSpan; spanned < 4294967296L; spanned += rangeSpan, partition++) {
    ranges.put( spanned, partition );
}
/** First address is 0, last address is 2^32 - 1, make sure we cover
  * all the way to the end of the range if
 * the 2^32/numPartitions is not an integer. The last partition may be a small */
if (spanned>4294967296L-1){
    ranges.put( 4294967296L -1, partition); /** The end range */
}

}
```

The first step is to save a copy of the `JobConf` object into `conf`, our standard practice. The key helper for this example is `PartitionedTextKeyHelperWithSeparators`. This class delegates to the `TextKeyHelperWithSeparators` class for any unrecognized input keys, and handles an extended form for search space keys that provides a way of splitting a search space key across multiple partitions and then assembling the resulting records later.

IPv4 addresses are simply unsigned 32-bit integer values, and the entire space runs from `0` through `4294967295` inclusive. Each partition will span approximately `rangeSpan` addresses, defined as `4294967295L / numPartitions`. The application uses long values to avoid issues with sign extension, as Java does not provide an unsigned integer type.

The variable `spanned` contains the ending IPv4 address of the previous partition. Each pass through the `for` loop adds `rangeSpan` to `spanned` defining the ending address of the next partition and increments the partition number:

```
for (spanned= rangeSpan; spanned < 4294967296L; spanned += rangeSpan, partition++) {
    ranges.put( spanned, partition );
}
```

`ranges.put( 4294967296L -1, partition)` stores the partition end address and partition number in `ranges`. These are currently added in order, which is not optimal for a `TreeMap`, as `TreeMaps` are stored as red-black trees and ordered insertion will result in an unbalanced tree. Casting our gaze into the future, it seems unlikely that there may be more than small hundreds of reduce tasks and a rewrite might be planned to eliminate the use of `TreeMap` and simply use an array.

Search Space Keys for Each Reduce Task That May Contain Matching Keys

The `SimpleIPPartitioner` also provides a method `spanSpaceKeys`, shown in Listing 9-14, which is not part of the partitioner interface. Here, I took a design expedience step that perhaps was not optimal given my later experience. I decided to use the `BruteForceMapReduceDriver` (Listing 9-11), and allow more than one reduce task. To achieve this, each search space record must be replicated so that any partition that could have matching requests each gets a copy of the search space record. The concept is that an addition map will, for each incoming search space record, output a set of search space records such that each reduce partition that could receive a matching search request will receive one of the output search space records. This addition map, RangePartitionTransformingMapper (shown later in Listing 9-16), will be added to the mapper chain.

Listing 9-14. *SimpleIPRangePartitioner.spanSpaceKeys Preamble*

```
public int spanSpaceKeys( PartitionedTextKeyHelperWithSeparators outsideHelper,
        Text forConstructedKeys, final Text value,
        final OutputCollector<Text, Text> output, Reporter reporter)
              throws IOException {

    /** If the key isn't valid bail. */
    if (!outsideHelper.isValid()) {
        reporter.incrCounter("KeySpanning", "Invalid Keys", 1);
        throw new IllegalArgumentException("Cannot span invalid keys");
    }

    /** This could just pass the key forward quietly. */
    if (!outsideHelper.isSearchSpace()) {
        reporter.incrCounter("KeySpanning", "Not Search Space", 1);
        throw new IllegalArgumentException("Cannot span search request keys");
    }

    /** If the passed in key is a regular search space key,
      * set the extended attributes for a spanning search space key. */
    if (!outsideHelper.isHasRealRange()) {
        outsideHelper.setRealRangeBegin(outsideHelper.getBeginRange());
        outsideHelper.setRealRangeEnd(outsideHelper.getEndRange());
    }
```

The first portion of Listing 9-14 handles the setup and validation. The calling convention requires that the caller pass in an initialized key helper (outsideHelper) and the value to output (OutputCollector). The Reporter object (reporter) is used to log metrics and failures. The key helper is checked for validity (outsideHelper.isValid()) and that it contains a search space request (outsideHelper.isSearchSpace()). If either constraint check fails, an exception is thrown. The key helper class for these spanned keys has two additional fields: the actual begin and end of the search space request. The begin and end fields will now be fields for the address span of the partition for which the record is output. outsideHelper.setRealRangeBegin(outsideHelper.getBeginRange()) and outsideHelper.setRealRangeEnd(outsideHelper.getEndRange()) initialize the helper correctly if it is not already set up.

As a quick recap, the search space key contains an IPv4 address range, represented as a beginning and ending address. To enable multiple reduce tasks, the search space records must be available in each reduce task that could receive search requests that would match the search space record. This allows the search space requests to be mixed into the job input with the search requests. Each search space key is split into a set of search space keys, such that each individual key contains that portion of the original range that fits within the range of addresses that will be routed to a specific reduce task. Implicit is that each partition starts with the address after the prior partition and there is no overlap in address space between partitions. Partition 0 is assumed to start at address 0, (0.0.0.0), and the last partition is assumed to end at 4294967295 (255.255.255.255).

The block of code in Listing 9-15 is the part of the spanSpaceKeys method that produces the per-partition keys.

Listing 9-15. *Producing Search Space Keys for the Required Reduce Partitions*

```
NavigableMap<Long, Integer> spannedRanges =
   ranges.tailMap(outsideHelper.getRealRangeBegin(), true);

/** The loop below uses the the begin range of
  * <code>outsideHelper</code> as the start point for the next
 * output record.
 * The end range value is used as a convenience and should not be used in test.
 * The real end and real beginning are always the actual
 * begin and end of the search space request.
 */
helper.setBeginRange( outsideHelper.getRealRangeBegin());
helper.setRealRangeBegin( outsideHelper.getRealRangeBegin());
helper.setEndRange( outsideHelper.getRealRangeEnd());
helper.setRealRangeEnd( outsideHelper.getRealRangeEnd());

int count = 0;
/** The real ranges are untouched, and the begin range is moved up
  * and the end range is just set in the loop.
 * When end range <= the spanEnd no more ranges are spanned.
 * the value of getEndRange() is never valid for use in tests.
 */
if (LOG.isDebugEnabled()) {
    LOG.debug(String.format("Spanning key %x:%x %s", helper.getRealRangeBegin(),
      helper.getRealRangeEnd(),value));
}
for( Map.Entry<Long, Integer> span : spannedRanges.entrySet()) {
    final Long spanEnd = span.getKey();

    /** If the newly adjusted begin range is past the end of our key's range,
      * there will be no more keys output. so finish up */
    if (helper.getBeginRange()>helper.getRealRangeEnd()) {
        helper.isValid = false;
        break; /** Done, no more ranges spanned. We could just
                 * return count from here, but this way there is only one
                 * valid exit point */
    }
```

```
/** This should never happen. */
if (spanEnd.longValue() < helper.getBeginRange()) {
    /** at least a partial span. */
    throw new IOException( String.format(
 "Constraint failure, the partition end %d %x is less than the key begin %d %x",
     spanEnd, spanEnd, helper.getBeginRange(), helper.getBeginRange()) );
}

/** The begin value for the current portion of <code>outsideHelper</code>
  * is inside the span of this partition. We have to assume at this point
  * that it is not before the start of the partition.
 *
 * If the spanEnd >= the getRealRangeEnd, this output key is contained entirely
 * within this partition
 *
 */



/** This case indicates that the end of this partition span is past the end of
  * the real search space request.
 * This is the last key that will be output, the output key end to be the real
 * end and finish
 */
if (spanEnd.longValue()>=helper.getRealRangeEnd()) {
    /** The range of the key only extends to this partition. */
    helper.setEndRange(helper.getRealRangeEnd());
    if (LOG.isDebugEnabled()) {
     LOG.debug(String.format(">= spanEnd %x %x of %x:%x %s",
         spanEnd, helper.getRealRangeEnd(), helper.getRealRangeBegin(),
         helper.getRealRangeEnd(), value ) );
    }

} else {    // There will be at least one more output key after this one

    /** In this case, the search space real end is past the end
      * of this partition, output a record from the
     * begin that was setup on the previous run through here or the initial
     * condition and an end == to the span end
     * and continue our loop
     */
```

```
            // Has to be less than the real end range
            helper.setEndRange(spanEnd.longValue());
            if (LOG.isDebugEnabled()) {
               LOG.debug(String.format(" < spanEnd  %x %x:%x  %x:%x %s", spanEnd,
                  helper.getBeginRange(), helper.getEndRange(),
                  helper.getRealRangeBegin(), helper.getRealRangeEnd(), value ) );
           }


        }

        count++;
        helper.setToRaw(forConstructedKeys);
        output.collect(forConstructedKeys,value);
        helper.setBeginRange(helper.getEndRange()+1); // One past the last record output
        reporter.incrCounter("KeySpanning", "Partition " + span.getValue(), 1);

    }
    reporter.incrCounter("KeySpanning", "OUTPUT KEYS", count);
    return count;
```

The passed-in, parsed-input key is in `outsideHelper`, the working object is `helper`, and the actual begin and end addresses for the network are stored in the real begin (`helper.getRealRangeBegin()`) and real end (`helper.getRealRangeEnd()`) fields of `helper`.

The `helper`, a `PartitionedTextKeyHelperWithSeparators` object, holds both the actual original search space key, using the `realRangeBegin` and `realRangeEnd` fields, and the begin and end address of the range within a partition, in the `begin` and `end` fields. For each partition, the begin `helper.setBeginRange()` and end `helper.setEndRange()` will be set to the address range within that partition that this search space record will match, and the `realRangeBegin` and `realRangeEnd` fields will be untouched.

The variable `spannedRanges` is a subset of ranges that contains only partitions that have an end address larger or equal to the real begin range of the key, and equal to or less than the real end range of the key. Put simply, `spannedRanges` contains the partitions that may contain addresses that would match the passed-in search space record.

The following loop examines each of the candidate partitions in ascending order of the partition end address:

```
for( Map.Entry<Long, Integer> span : spannedRanges.entrySet()) {
```

The variable `spanEnd` contains the ending address for the current partition. It is implicit in the data structures used that `spanEnd` will be greater than or equal to `helper.getBeginRange()` (the beginning address of the portion of the key that has not yet been output to a partition is always available as `helper.getBeginRange()`).

When a per-partition key is to be output, the `helper` is set up with the correct end address for that partition. The end address will either be the last address of the partition, `spanEnd`, or the last address of the actual range, `getRealRangeEnd()`, whichever address is least. If the end address of the output key is less than or equal to the end address of the current partition, no more keys need to be output. The begin field of helper is set to the address after the end of the previous output key, `helper.setBeginRange(spanEnd.longValue()+1`.

The core loop is run once for each potential partition that this key may need to have a record placed. The variable `count` keeps track of the number of records output, and `span` contains the information about the current partition, in particular the end address and the partition number. There are a couple checks: one to see if the partition end addresses are not in ascending order (`spanEnd.longValue() < helper.getBeginRange()`) and another to see if the key has been fully spanned across the partitions (`helper.getBeginRange()>helper.getRealRangeEnd()`).

There are two possible cases:

- The remaining portion of the key fits entirely in the current partition, `span`, `spanEnd.longValue()>=helper.getRealRangeEnd()`. The range end of the helper is set to the applicable end value in this case, `helper.setEndRange(helper.getRealRangeEnd())`.
- The key has address space that extends past the end of `span`. In this case, `helper.setEndRange(spanEnd.longValue())` is called.

The end of the loop actually builds the `Text` object with the appropriate data, `helper.setToRaw(forConstructedKeys)`, and resets begin to the address after the just output key, `helper.setBeginRange(helper.getEndRange()+1)`. Each input search space request now has a record that will be placed by the partitioner into each partition that could have search requests that match.

In `RangePartitionTransformingMapper`, shown in Listing 9-16, is a very simple `map()` method. It initializes the key helper from the passed-in key, `helper.getFromRaw(key)`, and for a valid search space key, calls the `spanSpaceKeys` method of `SimpleIPPartitioner` (search requests are just passed through as output).

Listing 9-16. *RangePartitionTransformingMapper*

```
public void map(Text key, Text value,
        OutputCollector<Text, Text> output, Reporter reporter)
        throws IOException {
    try {
        reporter.incrCounter("RangePartitionTransformingMapper", "INPUT KEYS", 1);
        if (!helper.getFromRaw(key)) {
            reporter.incrCounter("RangePartitionTransformingMapper", "", 1);
            return;
        }
        if (helper.isSearchRequest()) {
            output.collect(key, value);
            reporter.incrCounter(
                "RangePartitionTransformingMapper", "Request Keys", 1);
            return;
```

```
        }
        partitioner.spanSpaceKeys(helper, outputKey, value, output, reporter);
    } catch( Throwable e) {
        throwsIOExcepction( reporter, "RangePartitionTransformingMapper", e);
    }


}
```

The original concept was to take the search requests, feed them through the RangePartitioningTransformingMapper using RangePartitionTransformingMapper as a driver class, convert the search space records into a sorted and partitioned dataset, run another MapReduce job over the incoming search requests, and then perform a map-side join on the resulting datasets. After working with the data for a short time, I realized that the search space was so small that it wasn't worth the extra complexity or time to have an additional step for presorting the search space records. I decided to simply add this mapper as part of the mapper chain, and read the search space records as input with the search request records. The configuration changes to BruteForceMapReduceDriver are shown in the next section.

Helper Class for Keys Modifications

The class PartitionedTextKeyHelperWithSeparators will be the new KeyHelper and will support carrying the original key data, so that the output records can be provided with the actual network range instead of that portion of the network range that fits in this partition. A new record format needs to be designed that can carry the additional data. The key format for the search space keys has been *begin*:*end*, where *begin* and *end* are the first and last addresses of the network, each an eight-digit hexadecimal number. For example, 0.0.0.0 would be 00000000, 255.255.255.255 would be ffffffff, and the search space key representing the entire IPv4 address space would be 00000000:ffffffff. To allow partitioning, the search case keys must match keys in a particular partition. My first idea on how to address this was to just have four values instead of two, with the same separator between each. The full code for that version is in com.apress.hadoopbook.examples.ch9.PartitionedTextKeyHelperWithSeparators.java, available with the rest of the downloadable code for this book.

The code for the first design must be modified to examine a configuration parameter, range.key.helper, and instantiate the value as a class, defaulting to the TextKeyHelperWithSeparators class. Listing 9-17 provides an example of this from ApacheLogTransformMapper.

Listing 9-17. *Modifications to Load a Key Helper Based on the Value of range.key.helper*

```
public void configure(JobConf conf) {
    super.configure(conf);
    helper = ReflectionUtils.newInstance(conf.getClass("range.key.helper",
      TextKeyHelperWithSeparators.class,
            TextKeyHelperWithSeparators.class),conf);


}
```

The existing mapper and reducer classes are modified to instantiate their KeyHelper class based on a configuration property, range.key.helper, defaulting to TextKeyHelperWithSeparators. BruteForceMapReduceDriver is modified to set the range.key.helper configuration parameter value to PartitionedTextKeyHelperWithSeparators when the number of reduce tasks is more than one. This leaves the old behavior intact, while allowing multiple reduce tasks.

In Listing 9-18, the configuration key range.key.helper is set to be our partitioning class by conf.setClass("range.key.helper", PartitionedTextKeyHelperWithSeparators.class, KeyHelper.class), and an additional map is placed in the chain, to span the search space keys:

```
ChainMapper.addMapper(conf, RangePartitionTransformingMapper.class, Text.class,
                      Text.class, Text.class, Text.class, false, dummyConf)
```

Listing 9-18. *Modifications to the Setup Method in BruteForceMapReduceDriver.java*

```
if (conf.getNumReduceTasks()!=1) {
    /** If more that one reduce is to be run, the spanning partitioner must be used.
      */
    conf.setClass("range.key.helper", PartitionedTextKeyHelperWithSeparators.class,
                  KeyHelper.class);

    /** Add in the map that takes incoming search space records and spans them
      * across the partitions */
    ChainMapper.addMapper(conf, RangePartitionTransformingMapper.class,
            Text.class, Text.class, Text.class, Text.class, false, dummyConf);
    dummyConf.clear();
}
```

The reducer, ReducerForStandardComparator.java, does not need any changes, but the ActiveRanges class, which provides the hit method, does. In Listing 9-19, we simplify it to make it aware of the PartitionedTextKeyHelperWithSeparators class, and in that case, to use the real begin and end ranges for a search space request, rather than the per-partition begin and end ranges. If many types of keys are used, this method will quickly become excessively complex. In this case, there is only one type of key, so we can defer that code cleanup to a future that may not come.

Listing 9-19. *Modifications to ActiveRanges.activate to Support the Partition Spanned Search Space Keys*

```
if (helper instanceof PartitionedTextKeyHelperWithSeparators) {
    begin = ((PartitionedTextKeyHelperWithSeparators)helper).getRealRangeBegin();
    end = ((PartitionedTextKeyHelperWithSeparators)helper).getRealRangeEnd();
} else {
    begin = helper.getBeginRange();
    end = helper.getEndRange();
}
```

To provide a secondary sort of the final output, we have the classes DataJoinReduceOutput, DataJoinMergeMapper, and IPv4TextComparator. This set of classes performs a map-side join on all of the reduce output partitions of BruteForceMapReduceDriver, producing a single sorted file as output. The output uses the network begin, end, and name values as secondary sort keys. These also provide an example of how to perform a merge-sort of any reduce task output efficiently using map-side joins.

Listing 9-20 shows the DataJoinReduceOutput method.

Listing 9-20. *DataJoinReduceOutput.java, CustomSetup*

```
ArrayList<String> tables = new ArrayList<String>();
for( String input : inputs ) {
    String []parts = input.split(":");
    if (parts.length==2) {
        Class<? extends InputFormat> candidateInputFormat =
                    conf.getClass(parts[0],null,InputFormat.class);
        if (candidateInputFormat!=null) {
            addFiles(conf,candidateInputFormat, parts[1], tables);
            continue;
        }
    }
    addFiles(conf, KeyValueTextInputFormat.class,input, tables);
}

FileOutputFormat.setOutputPath(conf, new Path(output));
conf.set("mapred.join.expr", "outer(" + StringUtils.join(tables, ",") + ")");
conf.setNumReduceTasks(0);
conf.setMapperClass(DataJoinMergeMapper.class);
conf.setOutputKeyClass(Text.class);
conf.setOutputValueClass(Text.class);
conf.setInputFormat(CompositeInputFormat.class);
//conf.setOutputKeyComparatorClass(IPv4TextComparator.class);
conf.setClass("mapred.join.keycomparator", IPv4TextComparator.class,
      WritableComparator.class);
conf.setJarByClass(DataJoinMergeMapper.class);
```

DataJoinReduceOutput accepts the standard command-line arguments, including the -i path [, path, [path...]] -o output, to set the input datasets and the output path. Unlike a traditional map-side join, where each path item in the input is a table and the matching part-*XXXXX* files of each input path are joined, each individual part-*XXXXX* file is taken as a table, and all of the part-*XXXXX* files are joined together. This causes the map-side join to perform a streaming merge-sort on all of the input data files.

The customSetup() method examines each input in turn. If the input string has a colon (:), it is split and the parts examined:

```
String[] parts = input.split(":");
```

If there are exactly two parts and the first part is a class name that implements InputFormat, that input format is used for loading the directory name in parts[1]. If there is not exactly two parts, the original input is used with KeyValueTextInputFormat. Basically, the input directory can be preceded by a class name and a colon, and the class will be used as the input format for loading files from that input directory.

The addFiles method in shown in Listing 9-21.

Listing 9-21. *DataJoinReducerOutput.addFiles*

```
Path inputPath = new Path(path);
FileSystem fs = inputPath.getFileSystem(conf);
if (!fs.exists(inputPath)) {
    System.err.println(String.format(
                "Input item %s does not exist, ignoring", path));
    return;
}
FileStatus status = fs.getFileStatus(inputPath);
if (!status.isDir()) {
    String composed = CompositeInputFormat.compose(inputFormat, path);
    if (verbose) { System.err.println( "Adding input " + composed); }
    tables.add(composed);
    return;
}
FileStatus[] statai = fs.listStatus(inputPath, new PathFilter() {
    @Override
    public boolean accept(Path path) {
        if (path.getName().matches("^part-[0-9]+$")) {
            return true;
        }
        return false;
    }
}
);
if (statai==null) {
    System.err.println(
         String.format("Input item %s does not contain any parts, ignoring", path));
    return;
}
for( FileStatus status1 : statai) {
    String composed = CompositeInputFormat.compose(inputFormat,
                            status1.getPath().toString());
    if (verbose) { System.err.println( "Adding input " + composed); }
    tables.add(composed);
}
```

This method examines inputPath, constructed from that passed-in path element. If it exists (fs.exists(inputPath)) and is a directory (status.isDir()), the method collects the FileStatus information for each child:

```
FileStatus[] statai = fs.listStatus(inputPath,...
```

The PathFilter restricts the FileStatus entries returned to those that satisfy the accept() method. In this case, the only items accepted have file names that match the regular expression ^part-[0-9]+$, our standard reduce output file format. Rather than try to manage the map-side join table format, the following call builds the table format for the input file:

```
String composed =
  CompositeInputFormat.compose(inputFormat, status1.getPath().toString());
```

All of the individual table entries are aggregated in the ArrayList tables.

The actual join command is built ("outer(" + StringUtils.join(tables, ",") + ")") and stored in the configuration under the key mapred.join.expr. This by itself will merge-sort all of the input data into a single output file. The new piece, the specialty sorting of the input records before the map method, is triggered by the following line:

```
conf.setClass("mapred.join.keycomparator", IPv4TextComparator.class,
              WritableComparator.class);
```

This tells the map-side join framework to use IPv4TextComparator (Listing 9-23) as the key comparator when performing the merges.

The mapper, shown in Listing 9-22, provides a secondary sort by network for the matched requests.

Listing 9-22. *DataJoinMergeMapper.java*

```
reporter.incrCounter("DataJoinReduceOutput", "Input Keys", 1);
/** The number of tables in the join. */
final int size = value.size();
/** Allocate the values array if needed. a null indicates end,
  * so one extra allocated */
if (values==null) {
    values = new Text[size+1];
    outputText = new Text[size];
    /** Make some {@link Text} items, just in case. These probably aren't needed
      * but are made only once. */
    for (int i = 0; i < size; i++ ) {
        outputText[i] = new Text();
    }
}
/** For each table, check to see if it has a value for the key.
 * If it does, store it in values, possibly converting it to a text object by
 * calling {@link Text#set(String)} with the
 * with the string conversion.
```

```
     */
    /** The current index to store into values. */
    int valuesIndex = 0;
    for (int i = 0; i < size; i++) {
        if (value.has(i)) {
            Writable outputValue = value.get(i);
            if (outputValue instanceof Text) {
                values[valuesIndex] = (Text) outputValue;
            } else {
                /** Force a text conversion to simplfy life later. */
                outputText[valuesIndex].set(outputValue.toString());
                values[valuesIndex] = outputText[valuesIndex];
            }
            valuesIndex++;
            reporter.incrCounter("DataJoinReduceOutput", "Output Keys", 1);
        }
    }
    values[valuesIndex] = null;
    if (valuesIndex>1) {
        /** If only one, no reason to bother sorting. */
        Arrays.sort( values, 0, valuesIndex, comparator );
    }
    for ( int i = 0; i < valuesIndex; i++ ) {
        if( LOG.isDebugEnabled()) {
           LOG.debug( String.format( "Output of %d of %d, %s %s",
                        i, size, key, values[i]));
        }
        output.collect( key, values[i] );
    }
```

Each table is checked for a value (`value.has(i)`) and each table value (`Writable outputValue = value.get(i)`) accumulated in the `values` array. Just as a safety check, the values are converted to `Text` objects when needed (`outputText[valuesIndex].set(outputValue.toString())`), and the converted value stored (`values[valuesIndex] = outputText[valuesIndex]`).

If more than one table has a value for this key, the accumulated table values are sorted via `Arrays.sort( values, 0, valuesIndex, comparator )`, using the comparator `TabbedNetRangeComparator` (shown later in Listing 9-24). Once any required sorting is completed, the records are output (`output.collect( key, values[i] )`).

The actual input and output will be detailed in `HADOOP_CLASSPATH=/tmp/commons-lang-2.4.jar hadoop jar /tmp/hadoopprobook.jar com.apress.hadoopbook.examples.ch9.DataJoinReduceOutput -libjars /tmp/hadoopprobook.jar,/tmp/commons-lang-2.4.jar -jt cloud9:8021 -fs hdfs://cloud9:8020  -v -del -i range_join -o merged_range_join`.

The `IPv4TextComparator`, shown in Listing 9-23, provides a binary comparator that handles keys that are IPv4 addresses in the standard dotted-octet format, such as 192.168.0.1. It attempts to operate at the byte level and to minimize object allocation. This class is used in the map-side join to force the correct ordering of the input keys, as the lexical ordering is not what is expected.

Listing 9-23. *IPv4TextComparator.java*

```
public IPv4TextComparator()
{
    super(Text.class);
}
/** Compare the serialized form of two text objects containing IPv4 addresses
 * of the form 0.0.0.0 through 255.255.255.255.
 * @see org.apache.hadoop.io.RawComparator#compare(byte[], int, int, byte[],➥
 int, int)
 */
@Override
public int compare(byte[] b1, int s1, int l1, byte[] b2, int s2, int l2) {
     long a1 = unpack( b1, s1, l1 );
     long a2 = unpack( b2, s2, l2 );
     if (a1<a2) {
         return -1;
     }
     if (a1>a2) {
         return 1;
     }
     return 0;
}

/** Given a byte buffer that contains a standard decimal dotted octet IPv4 address
 * (ie: 0.0.0.0 through 255.255.255.255), as a byte stream, return the long value
 * of the ip address
 *
 * @param buf The byte buffer containing the bytes.
 * @param s The start address in <code>buf</code>.
 * @param l The length of data in <code>buf</code> to use.
 * @return the numeric value of the address 0 -> 2^32, or -1 for parse errors.
 */
public static long unpack( final byte []buf, int s, int l) {
    long result = 0;
    long part = 0;
    l += s;
    for( ; s < l; s++ ) {
        byte b = buf[s];
        switch(b) {
        case '.':
            result <<= 8;
            result += part;
```

```
                part = 0;
                continue;
            case '0': case '1': case '2': case '3': case '4':
            case '5': case '6': case '7': case '8': case '9':
                part *= 10;
                part += Character.getNumericValue((int)b);
                continue;
            default:
                return -1;
            }
        }
        result <<= 8;
        result += part;
        return result;
    }

    /* (non-Javadoc)
     * @see org.apache.hadoop.io.WritableComparator#compare ➥
    (org.apache.hadoop.io.WritableComparable, org.apache.hadoop.io.WritableComparable)
     */
    @Override
    public int compare(WritableComparable a, WritableComparable b) {
        // TODO Auto-generated method stub
        if (a instanceof Text && b instanceof Text) {
            return compare((Text)a, (Text)b);
        }
        return super.compare(a, b);
    }

    /** Compare to text objects that are IPv4 addresses in dotted octet notation.
     * @see org.apache.hadoop.io.RawComparator#compare(Object, Object)
     */
    @Override
    public int compare(final Object a, final Object b) {
        if (a instanceof Text && b instanceof Text) {
            return compare((Text)a, (Text)b);
        }
        return super.compare(a, b);
    }
```

The comparator in Listing 9-24 expects input lines of the form:

```
IP tab IP tab Network Name tab other data
```

It will do a primary sort using the first IP address, secondary on the second IP address, and tertiary on the network name. If at any point there is a parse failure, the element that the parse failed on is considered greater. The parsing is deferred as long as possible in the hopes that it won't be needed. This code tries very hard to work at the byte level and not convert items back into strings.

Listing 9-24. *DataJoinMergeMapper.TabbedNetRangeComparator*

```
public static class TabbedNetRangeComparator implements Comparator<Text> {

    /** The comparator from the {@link Text} class, used for comparing
      * the network names. */
    Text.Comparator comparator = new Text.Comparator();


    /** This expects and requires the value to be IPv4TABIPv4TaBnetworkTABline.
     * the the comparison order is addr1, add2, network
     *
     * @param a Text value 1
     * @param b Text value 2
     * @return -1 1 or 0 less, greater or equal, the first item with
     * a parse failure is considered greater.
     */
    @Override
    public int compare( Text a, Text b ) {
        if( LOG.isDebugEnabled()) {
           LOG.debug( String.format("Comparing %s and %s", a, b));
        }

        /** Do the basic check on <code>a</code>, see if we find the first bit. */
        final byte[] ab = a.getBytes();
        final int al = a.getLength();
        final int at1 = findTab( ab, 0, al );
        if (at1==-1) {
            if( LOG.isDebugEnabled()) {
               LOG.debug(String.format("a %s failed to find first tab", a));
            }
            return 1;
        }

        /** Do the basic check on <code>b</code>, see if we find the first bit. */
        final byte[] bb = b.getBytes();
        final int bl = b.getLength();
        final int bt1 = findTab( bb, 0, bl );
        if (bt1==-1) {
            if( LOG.isDebugEnabled()) {
               LOG.debug(String.format("b %s failed to find first tab", b));
            }
            return -1;
        }
```

```
/** Get the first ip address from <code>a</code>. */
final long aip1 = IPv4TextComparator.unpack( ab, 0, at1 );
if (aip1==-1) {
    if( LOG.isDebugEnabled()) {
       LOG.debug(String.format("a %s failed to unpack %s",
         a, new String( ab, 0, at1)));
    }
    return 1;
}

/** Get the first ip address from <code>b</code>. */
final long bip1 = IPv4TextComparator.unpack( bb, 0, bt1 );
if (bip1==-1) {
    if( LOG.isDebugEnabled()) {
       LOG.debug(String.format("b %s failed to unpack %s", b,
         new String( bb, 0, bt1)));
    }
    return -1;
}
if( LOG.isDebugEnabled()) {
   LOG.debug(String.format("a %x b%x", aip1, bip1));
}

/** Do the ip address comparison on the first IP,
  * if they are different, this routine is done.
 * Since we have longs and the result is int, a simple
 * subtraction may not work as the result may not be an int.
 */
if (aip1<bip1) {
    return -1;
}
if (aip1>bip1) {
    return 1;
}


/** Check the second IP address in <code>a</code> and <code>b</code> */

final int at2 = findTab( ab, at1+1, al);
if (at2==-1) {
    if( LOG.isDebugEnabled()) {
       LOG.debug(String.format("a %s failed to find second tab", a));
    }
    return 1;
}
```

```
final long aip2 = IPv4TextComparator.unpack( ab, at1+1, at2 );
if (aip2==-1) {
    if( LOG.isDebugEnabled()) {
       LOG.debug(String.format("a %s failed to unpack %s", a,
         new String( ab, at1+1, at2)));
    }
    return 1;
}

final int bt2 = findTab( bb, bt1+1, bl);
if (bt2==-1) {
    if( LOG.isDebugEnabled()) {
       LOG.debug(String.format("b %s failed to find second tab", b));
    }
    return -1;
}
final long bip2 = IPv4TextComparator.unpack( bb, bt1+1, bt2 );
if (bip2==-1) {
    if( LOG.isDebugEnabled()) {
       LOG.debug(String.format("b %s failed to unpack %s", b,
         new String( bb, bt1+1, bt2)));
    }
    return -1;
}
if (aip2<bip2) {
    return -1;
}
if (aip2>bip2) {
    return 1;
}

/** At this point both pairs of IP addresses are the same.
  * Pass the network names off to Text, which knows how to compare
  * utf-8 bytes. */
final int at3 = findTab( ab, at2+1, al);
if (at3==-1) {
    if( LOG.isDebugEnabled()) {
       LOG.debug(String.format("a %s failed to find third tab", a));
    }
    return 1;
}

final int bt3 = findTab( bb, bt2+1, al);
if (bt3==-1) {
    if( LOG.isDebugEnabled()) {
       LOG.debug(String.format("b %s failed to find second tab", b));
    }
```

```
            return -1;
        }
        if( LOG.isDebugEnabled()) {
            LOG.debug(String.format("a %s b %s",
              new String( ab, at2+1, at3), new String( bb, bt2+1, bt3)));
        }
        return comparator.compare( ab, at2+1, at3, bb, bt2+1, bt3 );

    }

    @Override
    public boolean equals(Object o) {
        if (o==null) {
            return false;
        }
        if (o==this) {
            return true;
        }
        if (o instanceof TabbedNetRangeComparator) {
            return true;
        }
        return false;
    }

}
```

Listing 9-25 shows the commands used to generate the output. These commands use the machine `cloud9` on port 8021 for JobTracker services and `cloud9` port 8020 for HDFS services. Your local installation will be different.

Listing 9-25. *The Commands Used to Generate the Output*

```
hadoop jar /tmp/hadoopprobook.jar ➥
com.apress.hadoopbook.examples.ch9.BruteForceMapReduceDriver -jt cloud9:8021 ➥
 -fs hdfs://cloud9:8020 -libjars /tmp/hadoopprobook.jar ➥
 -D mapred.reduce.tasks=10 -v --deleteOutput --input searchspace.txt ➥
 access_log.txt -o range_join
HADOOP_CLASSPATH=/tmp/commons-lang-2.4.jar hadoop jar /tmp/hadoopprobook.jar ➥
 com.apress.hadoopbook.examples.ch9.DataJoinReduceOutput -libjars ➥
 /tmp/hadoopprobook.jar,/tmp/commons-lang-2.4.jar -jt cloud9:8021 ➥
 -fs hdfs://cloud9:8020  -v -del -i range_join -o merged_range_join
```

The first command runs the `BruteForceMapReduceDriver`, passing in the JAR file included with the book examples, and specifies that ten reduce tasks are to be run:

```
-D mapred.reduce.tasks=10
```

Most of our later examples accept the arguments `-v -deleteOutput`, enabling verbose logging and causing the job output directory to be deleted if the directory exists. The two input files are a file of network ranges with names, `searchspace.txt`, shown in Listing 9-26, and some Apress.com access log data, `access_log.txt`, shown in Listing 9-27. The first output directory is `range_join`, which will be the input directory of the next command. The second line runs the command `DataJoinReduceOutput` to take the ten partition files and produce a single file that is sorted in IP address order, with secondary sorts on the network begin and end addresses and the network name. The actual output is listed in Table 9-4.

Listing 9-26. *searchspace.txt, Search Space Network Ranges*

```
72810800:72810fff   InTech Online
747d8000:747d8fff   HANANET INFRA
74480000:744bffff   HATHWAY NET
76000000:760fffff   OCN
77ea0000:77eaffff   SINGTELMOBILE
0c000000:0cffffff   ATT
79f00000:79f7ffff   TATACOMM IN
79fec000:79fec1ff   KIDC INFRA SERVERROOM DAUM
79080000:790fffff   CHINANET GD
796150c0:796150cf   BAYAN_REDMAP AP
7aa42000:7aa43fff   ABTS TN DSL 9111 chn
7aa70000:7aa77fff   ABTS KK DSL 9102 blr
7aa98000:7aa9bfff   ABTS AP DSL 9112 hyd
7aea0000:7aeaffff   CHINANET ZJ HZ
7b644000:7b647fff   MAXNET NZ
7c720000:7c73ffff   CHINANET SN
7c512a00:7c512aff   CMTSBDG IM2 HFC ID
7d11ab00:7d11abff   BTNL CHN DSL
80d20000:80d2ffff   PURDUE CCNET
836b0000:836bffff   MICROSOFT
```

Listing 9-27. *First 20 access_log.txt Lines, with the lines truncated for clarity*

```
116.125.47.43 - - [15/Nov/2008:22:07:47 -0800]...
116.125.162.223 - - [15/Nov/2008:22:07:47 -0800]...
193.238.120.192 - - [15/Nov/2008:22:23:13 -0800]...
193.238.192.10 - - [15/Nov/2008:22:23:13 -0800]...
193.238.186.77 - - [15/Nov/2008:22:23:14 -0800]...
193.238.83.101 - - [15/Nov/2008:22:23:14 -0800]...
193.47.137.43 - - [15/Nov/2008:22:25:34 -0800]...
193.47.58.78 - - [15/Nov/2008:22:25:34 -0800]...
193.252.4.14 - - [15/Nov/2008:22:56:05 -0800]...
193.252.144.172 - - [15/Nov/2008:22:56:05 -0800]...
208.80.221.10 - - [15/Nov/2008:23:07:05 -0800]...
208.80.179.99 - - [15/Nov/2008:23:07:05 -0800]...
208.80.168.223 - - [15/Nov/2008:23:14:49 -0800]...
208.80.57.128 - - [15/Nov/2008:23:14:49 -0800]...
66.233.41.9 - - [15/Nov/2008:23:29:13 -0800]...
```

```
66.233.254.37 - - [15/Nov/2008:23:29:13 -0800]...
66.233.78.158 - - [15/Nov/2008:23:29:14 -0800]...
66.233.212.119 - - [15/Nov/2008:23:29:14 -0800]...
66.233.242.121 - - [15/Nov/2008:23:29:17 -0800]...
66.233.220.139 - - [15/Nov/2008:23:29:17 -0800]...
```

Table 9-4. *The First 20 Job Output Lines*

Log IP	Network Start	Network End	Network Name	Log Line
12.6.90.96	12.0.0.0	12.255.255.255	ATT	- - [19/Nov/2008:17:01:18 -0800] "GET / HTTP/1.1" 404 293 "-" "Mozilla/4.0 (compatible; MSIE 6.0; Windows NT 5.1; SV1; InfoPath.1; .NET CLR 1.1.4322; .NET CLR 2.0.50727; .NET CLR 3.0.04506.30)"
12.6.127.77	12.0.0.0	12.255.255.255	ATT	- - [19/Nov/2008:17:03:00 -0800] "GET / HTTP/1.1" 404 293 "http://www.dtsearch.com/CS_Apress_SuperIndex.html" "Mozilla/4.0 (compatible; MSIE 6.0; Windows NT 5.1; SV1; InfoPath.1; .NET CLR 1.1.4322; .NET CLR 2.0.50727; .NET CLR 3.0.04506.30)"
12.6.233.199	12.0.0.0	12.255.255.255	ATT	- - [19/Nov/2008:17:03:00 -0800] "GET / HTTP/1.1" 404 293 "http://www.dtsearch.com/CS_Apress_SuperIndex.html" "Mozilla/4.0 (compatible; MSIE 6.0; Windows NT 5.1; SV1; InfoPath.1; .NET CLR 1.1.4322; .NET CLR 2.0.50727; .NET CLR 3.0.04506.30)"
12.6.239.45	12.0.0.0	12.255.255.255	ATT	- - [19/Nov/2008:17:01:18 -0800] "GET / HTTP/1.1" 404 293 "-" "Mozilla/4.0 (compatible; MSIE 6.0; Windows NT 5.1; SV1; InfoPath.1; .NET CLR 1.1.4322; .NET CLR 2.0.50727; .NET CLR 3.0.04506.30)"
12.30.22.148	12.0.0.0	12.255.255.255	ATT	- - [19/Nov/2008:10:28:55 -0800] "GET /favicon.ico HTTP/1.0" 404 304 "-" "Mozilla/5.0 (Windows; U; Windows NT 5.1; en-US; rv:1.9.0.4) Gecko/2008102920 Firefox/3.0.4"
12.30.31.111	12.0.0.0	12.255.255.255	ATT	- - [19/Nov/2008:10:28:55 -0800] "GET /favicon.ico HTTP/1.0" 404 304 "-" "Mozilla/5.0 (Windows; U; Windows NT 5.1; en-US; rv:1.9.0.4) Gecko/2008102920 Firefox/3.0.4"
12.30.136.50	12.0.0.0	12.255.255.255	ATT	- - [19/Nov/2008:10:28:55 -0800] "GET / HTTP/1.0" 404 293 "-" "Mozilla/5.0 (Windows; U; Windows NT 5.1; en-US; rv:1.9.0.4) Gecko/2008102920 Firefox/3.0.4"
12.30.180.46	12.0.0.0	12.255.255.255	ATT	- - [19/Nov/2008:10:28:55 -0800] "GET / HTTP/1.0" 404 293 "-" "Mozilla/5.0 (Windows; U; Windows NT 5.1; en-US; rv:1.9.0.4) Gecko/2008102920 Firefox/3.0.4"

Log IP	Network Start	Network End	Network Name	Log Line
12.69.76.145	12.0.0.0	12.255.255.255	ATT	- - [18/Nov/2008:13:07:55 -0800] "GET / HTTP/1.1" 404 293 "-" "Mozilla/4.0 (compatible; MSIE 7.0; Windows NT 5.1; .NET CLR 1.1.4322; .NET CLR 2.0.50727; InfoPath.1; .NET CLR 3.0.04506.30; .NET CLR 3.0.04506.648)"
12.69.85.24	12.0.0.0	12.255.255.255	ATT	- - [18/Nov/2008:13:07:30 -0800] "GET / HTTP/1.1" 404 293 "-" "Mozilla/4.0 (compatible; MSIE 7.0; Windows NT 5.1; .NET CLR 1.1.4322; .NET CLR 2.0.50727; InfoPath.1; .NET CLR 3.0.04506.30; .NET CLR 3.0.04506.648)"
12.69.130.223	12.0.0.0	12.255.255.255	ATT	- - [18/Nov/2008:13:07:30 -0800] "GET / HTTP/1.1" 404 293 "-" "Mozilla/4.0 (compatible; MSIE 7.0; Windows NT 5.1; .NET CLR 1.1.4322; .NET CLR 2.0.50727; InfoPath.1; .NET CLR 3.0.04506.30; .NET CLR 3.0.04506.648)"
12.69.229.167	12.0.0.0	12.255.255.255	ATT	- - [18/Nov/2008:13:07:55 -0800] "GET / HTTP/1.1" 404 293 "-" "Mozilla/4.0 (compatible; MSIE 7.0; Windows NT 5.1; .NET CLR 1.1.4322; .NET CLR 2.0.50727; InfoPath.1; .NET CLR 3.0.04506.30; .NET CLR 3.0.04506.648)"
12.167.105.82	12.0.0.0	12.255.255.255	ATT	- - [18/Nov/2008:08:21:28 -0800] "GET / HTTP/1.1" 404 293 "-" "Mozilla/4.0 (compatible; MSIE 7.0; Windows NT 6.0; WOW64; SLCC1; .NET CLR 2.0.50727; .NET CLR 3.0.04506; .NET CLR 3.5.21022)"
12.167.179.22	12.0.0.0	12.255.255.255	ATT	- - [18/Nov/2008:08:21:28 -0800] "GET / HTTP/1.1" 404 293 "-" "Mozilla/4.0 (compatible; MSIE 7.0; Windows NT 6.0; WOW64; SLCC1; .NET CLR 2.0.50727; .NET CLR 3.0.04506; .NET CLR 3.5.21022)"
12.216.40.118	12.0.0.0	12.255.255.255	ATT	- - [17/Nov/2008:11:37:59 -0800] "GET / book/errataSubmit.html?bID=10187 HTTP/1.1" 404 311 "-" "Opera/7.23 (Windows 98; U) [en]"
12.216.48.187	12.0.0.0	12.255.255.255	ATT	- - [17/Nov/2008:11:37:59 -0800] "GET / book/errataSubmit.html?bID=10187 HTTP/1.1" 404 311 "-" "Opera/7.23 (Windows 98; U) [en]"
12.229.67.237	12.0.0.0	12.255.255.255	ATT	- - [18/Nov/2008:14:08:59 -0800] "GET / HTTP/1.1" 404 293 "-" "Mozilla/4.0 (compatible; MSIE 7.0; Windows NT 5.1; .NET CLR 1.0.3705; .NET CLR 1.1.4322; Media Center PC 4.0; .NET CLR 2.0.50727; .NET CLR 3.0.4506.2152; .NET CLR 3.5.30729)"

Continued

Table 9-4. *Continued*

Log IP	Network Start	Network End	Network Name	Log Line
12.229.91.253	12.0.0.0	12.255.255.255	ATT	- - [18/Nov/2008:14:08:59 -0800] "GET / HTTP/1.1" 404 293 "-" "Mozilla/4.0 (compatible; MSIE 7.0; Windows NT 5.1; .NET CLR 1.0.3705; .NET CLR 1.1.4322; Media Center PC 4.0; .NET CLR 2.0.50727; .NET CLR 3.0.4506.2152; .NET CLR 3.5.30729)"
58.68.24.75	58.68.0.0	58.68.127.255	DWL NET	- - [20/Nov/2008:00:47:53 -0800] "GET / HTTP/1.1" 302 315 "-" "Mozilla/4.0 (compatible; MSIE 8.0; Windows NT 5.1; Trident/4.0; .NET CLR 2.0.50727; .NET CLR 3.0.04506.30; .NET CLR 3.0.04506.648; .NET CLR 3.5.21022)"
59.92.23.105	59.88.0.0	59.99.255.255	BSNLNET	- - [19/Nov/2008:12:38:42 -0800] "GET / HTTP/1.1" 302 309 "-" "Mozilla/5.0 (Windows; U; Windows NT 5.1; en-US; rv:1.9.0.4) Gecko/2008102920 Firefox/3.0.4"

Design 3: Future Possibilities

Two possibilities come to mind for this sample MapReduce job:

An indexed map file of search requests in the reduce task: For each search request key, the `configure()` method will open the relevant search space map file—either the full map file for the entire search space or a partitioned file—where the partition contains the networks that keys in this reduce task partition could match. The `MapFile.getClosest()` method would be used to find search space records that could match.

Map-side join of the presorted search requests and a presorted search space: This method requires presorting the search request records and the search space records, and then using the map-side join techniques discussed in Chapter 8 and the classes for working with the IP address described in this chapter.

Both require that the search space records be presorted. Also, in both cases, the search space records can either be partitioned as the search request records are partitioned, or the entire search space be present in each task, in Google Bigtable style (see `http://labs.google.com/papers/bigtable.html`).

There are trade-offs between prepartitioning versus full replicas. The partitioned case reduces the data volume that must be scanned. Even with indexes, the amount of data that needs to be fetched from disk will be smaller in the partitioned case. The downsides are that search space needs to be repartitioned if the number of reduce tasks for the search requests is changed, and there is additional (though small) code complexity to ensure that the correct search space map file is opened in each search request reduce task.

Both techniques lose the data being local for at least the search space records, and neither seem worth the bother at present, as it is not clear that there would be any performance gain.

They also require the search request records to be sorted, and the search space is expected to be relatively small.

Summary

This chapter has walked you through the design and implementation of a nontrivial real-world Hadoop application. In the process, you have seen a number of design decisions made that become invalid as understanding arrives. The design and development process was deliberately oriented to provide initial functionality quickly so that this understanding could arrive sooner, rather than after a large and costly development cycle.

A number of the advanced features, such as chaining and map-side joins, were used in the application, and a partitioner and several comparators were written.

The tight coupling between the custom partitioner and the comparator allowed the application to perform range-based matching very efficiently using MapReduce techniques.

The techniques that you have learned will allow you to efficiently and effectively tackle very complex problems that do not appear to fit the MapReduce framework, but in fact are ideally suited for MapReduce.

Particularly in the rapidly evolving environment of today, you will never have time to build the perfect application—just an application that works for yesterday's goals. Someone else will come along later and modify the application until it meets the new goals. Be kind to that person by leaving comments, testing, and keeping it simple. The person doing those future modifications may be you!

CHAPTER 10

Projects Based On Hadoop and Future Directions

People use Hadoop to solve many types of problems, and a number of teams have built packages on top of Hadoop Core to address an even larger scope of problems. This chapter will walk through some of the many tools being built on top of Hadoop and one tool that can be built into Hadoop. Hadoop Core is an evolving project: over the time of writing this edition of the book, Hadoop 0.19.0 and Hadoop 0.19.1 came out, and Hadoop 0.20.0 became available in May 2009. (You'll see a section on changes later in this chapter.)

Hadoop Core–Related Projects

The main web site for Hadoop Core, `http://hadoop.apache.org/core`, provides a list of related projects and subprojects: HBase, Hive, Pig, Mahout, and Hama. The top-level Hadoop project, `http://hadoop.apache.org/`, also includes ZooKeeper. This section will provide an overview of them and, when feasible, show a quick example of how to set up and use them (as well as what problems users might encounter).

DISCLAIMER

I have little to no experience with most of the projects listed in this chapter, so the information in this chapter is gleaned from reading the project or company web site and/or trying the examples from a current release.

HBase: HDFS-Based Column-Oriented Table

The project description describes HBase as the Hadoop database an open source, column-oriented structured datastore based on the Google BigTable paper, `http://labs.google.com/papers/bigtable.html`. The earlier versions of HBase used the Hadoop MapFile as the underlying storage mechanism and managed updates by maintaining overlay MapFiles. When there were sufficient updates, a merged file was reconstructed, and the overlays were discarded. To speed access and distribute access, each individual MapFile is responsible for only a specific range of data in a table column, and if the MapFile grows past a specified size, it is split into multiple MapFiles. More recent versions of HBase also provide a memcached-based intermediate layer between the user and the MapFiles (`http://www.danga.com/memcached/`).

Prior to the addition of the memcached layer, HBase suffered terrible performance for random reads and writes, primarily because HDFS is not optimized for low latency random access. Ordered reads and writes perform at near-HDFS speed.

HBase has a number of server processes, a single HBaseMaster that manages the HBase cluster and a set of HRegionServers, each of which is responsible for a set of MapFiles containing column regions.

HBase suffers terribly from the inability of applications to flush file data to storage before the file is closed, and a crash of any portion of the HBase servers or a service interrupting crash of HDFS will result in data loss.

In prior chapters there was a discussion of problems caused by applications or server processes attempting to exceed the system-imposed limit on the number of open files; HBase also has this problem. The problem is substantially aggravated because each Hadoop MapFile is actually two files and a directory in HDFS, and each HDFS file also has a hidden checksum file. Setting the per-process open file count very large is a necessity for the HBase servers. A storage file format, HFile, is under development and due for Hbase version 0.20.0, and is expected to solve many of the performance and reliability issues.

HBase relies utterly on a smoothly performing HDFS for its operation; any stalls or DataNode instability will show up as HBase errors. There are HDFS tuning parameters suggested in the troubleshooting section on the HBase wiki: `http://wiki.apache.org/hadoop/Hbase/Troubleshooting`. In particular, if the underlying HDFS cluster is experiencing a slow block report problem, `https://issues.apache.org/jira/browse/HADOOP-4584`, HBase is not recommended.

HBase servers, particularly the version using memcached, are memory intensive and generally require at least a gigabyte of real memory per server; any paging will drastically affect performance. Java Virtual Machine (JVM) garbage collection thread stalls are also causing HBase failures.

HBase generally provides downloadable release bundles that track the Hadoop Core distributions. HBase is not part of the Hadoop Core distribution.

Hive: The Data Warehouse that Facebook Built

Hive provides a rich set of tools in multiple languages to perform SQL-like data analysis on data stored in HDFS. The wonderful people at Facebook have contributed Hive to the Apache project. As of the publication of this book, Hive is undergoing active development. Compiled versions of Hive are part of the contrib subtree of the Hadoop Core distribution.

Cloudera, discussed later in this chapter, provides online training for Hive.

Setting Up and Running Hive

The following four lines are required before attempting to start Hive (your installation might already have the `/tmp` and `/user/hive/warehouse` directories present):

```
hadoop fs -mkdir       /tmp
hadoop fs -mkdir       /user/hive/warehouse
hadoop fs -chmod g+w   /tmp
hadoop fs -chmod g+w   /user/hive/warehouse
```

The only issue I encountered when running Hive was a problem with a missing JAR because of an error I introduced into the `conf/hadoop-env.sh` file (see Listing 10-1).

Listing 10-1. *Hive Configuration Error*

```
jason@cloud9:~/src/hadoop-0.19/contrib/hive$ bin/hive
```

```
java.lang.NoClassDefFoundError: org/apache/hadoop/hive/conf/HiveConf
    at java.lang.Class.forName0(Native Method)
    at java.lang.Class.forName(Class.java:247)
    at org.apache.hadoop.util.RunJar.main(RunJar.java:158)
    at org.apache.hadoop.mapred.JobShell.run(JobShell.java:54)
    at org.apache.hadoop.util.ToolRunner.run(ToolRunner.java:65)
    at org.apache.hadoop.util.ToolRunner.run(ToolRunner.java:79)
    at org.apache.hadoop.mapred.JobShell.main(JobShell.java:68)
Caused by: java.lang.ClassNotFoundException: org.apache.hadoop.hive.conf.HiveConf
    at java.net.URLClassLoader$1.run(URLClassLoader.java:200)
    at java.security.AccessController.doPrivileged(Native Method)
    at java.net.URLClassLoader.findClass(URLClassLoader.java:188)
    at java.lang.ClassLoader.loadClass(ClassLoader.java:306)
    at java.lang.ClassLoader.loadClass(ClassLoader.java:251)
    at java.lang.ClassLoader.loadClassInternal(ClassLoader.java:319)
    ... 7 more
```

I modified the `conf/hadoop-env.sh` file to set the `HADOOP_CLASSPATH` (see Listing 10-2) when I was testing the scheduler services in Chapter 8. The `contrib/hive/bin/hive` script sets `HADOOP_CLASSPATH` with the set of JARs that Hive requires and then invokes the `bin/hadoop` script to start the Hive command-line interpreter.

Listing 10-2. *Incorrect Modification of the HADOOP_CLASSPATH Setting in conf/hadoop-env.sh*

```
# Extra Java CLASSPATH elements.  Optional.
export HADOOP_CLASSPATH=${HADOOP_HOME}/contrib/capacity-scheduler/➥
hadoop-0.19-capacity-scheduler.jar
```

I corrected the error (see Listing 10-3), and Hive started correctly (see Listing 10-4).

Listing 10-3. *Corrected Setting for HADOOP_CLASSPATH in conf/hadoop-env.sh*

```
# Extra Java CLASSPATH elements.  Optional.
export HADOOP_CLASSPATH=${HADOOP_HOME}/contrib/➥
capacity-scheduler/hadoop-0.19-capacity-scheduler.jar:${HADOOP_CLASSPATH}
```

Listing 10-4. *Hive Starts Correctly After Constructing the Required HDFS Path Elements with the Correct Permissions*

```
jason@cloud9:~/src/hadoop-0.19/contrib/hive$ bin/hive
```

```
hive>
```

The examples listed in the wiki page http://wiki.apache.org/hadoop/Hive/GettingStarted did not work particularly well for me (they might be updated by the time you read this chapter).

Pig, the Other Latin: A Scripting Language for Dataset Analysis

Pig provides a high-level language for writing SQL-like operations that apply to datasets. The language is named Pig Latin, and the Pig project provides a compiler that produces MapReduce jobs from a Pig Latin script. Pig is not distributed with Hadoop Core, and is mature enough that the project has releases. At the time of writing, Pig 0.2.0 has been released. Pig also provides grunt, an interactive shell, for running Pig Latin commands directly. Cloudera, listed later in this chapter, provides online training for Pig.

The site http://www.apache.org/dyn/closer.cgi/hadoop/pig provides the main distribution page. At present, it appears that the stock Pig distribution requires the underlying cluster to run Hadoop 0.17.0 or Hadoop 0.18.0.

The setup is as simple as unpacking the distribution and setting the environment variable PIG_CLASSPATH to the directory that contains the hadoop-site.xml file that defines your cluster. The following should work:

```
export PIG_CLASSPATH=${HADOOP_HOME}/conf
```

Mahout: Machine Learning Algorithms

The Mahout project aims to build scalable machine learning algorithms. Its plan is to build libraries for the ten machine learning algorithms listed in http://www.cs.stanford.edu/people/ang//papers/nips06-mapreducemulticore.pdf. As of the time of writing, the first release, 0.1, has been made available for download. The Taste project (a recommendation engine) has become a part of Mahout and is included in the 0.1 release. There is a tutorial available at http://lucene.apache.org/mahout/taste.html.

Mahout requires Maven for operation, and it is not clear from the documentation how to run the examples, including the Taste examples, without Maven.

Mahout also provides a number of distributed clustering algorithms, including k-means, dirichlet, mean-shift, and canopy. There are also two Bayesian classifiers: the naive and the complementary naïve. An implementation of watchmaker is provided for building evolutionary algorithms and support for matrix and vector operations.

Hama: A Parallel Matrix Computation Framework

At the time of writing, Hama is an incubation project. It requires HBase as an underlying storage framework. The project is intended to be used for large-scale numerical analyses and data mining. The project will provide matrix-vector and matrix-matrix multiplication, linear equation solving, tools for working with graphs, data sorting, and methods of finding eigenvalues and eigenvectors. The project is undergoing development and is pre–release 0.1.

ZooKeeper: A High-Performance Collaboration Service

ZooKeeper provides a framework for building high-performance collaborative services. ZooKeeper maintains a shared namespace that looks very similar to a hierarchical file system. Applications rendezvous on entries in the namespace. Each of these namespace entries may have data associated with it. The entry data is accessed atomically, and changes are ordered. In addition, ZooKeeper provides an ephemeral node, an entry that vanishes when the service holding the entry open disconnects. The ephemeral nodes are used to establish service masters and sets of backup servers. Ephemeral nodes are used to support redundant servers with hot failover.

ZooKeeper has been designed to be very reliable and very fast in environments in which data is primarily read.

The examples at `http://hadoop.apache.org/zookeeper/docs/current/recipes.html` provide ZooKeeper recipes for two-phase commit, leader election, barriers, queues, and locks.

Lucene: The Open Source Search Engine

The Lucene project, `http://lucene.apache.org/java/docs/`, provides the standard open source package used for search engines. The Lucene core provides the ability to take in documents in a variety of formats and build inverted indexes out of the terms found in the documents. Lucene also provides a query engine that takes incoming queries, searches the indexes, and returns the documents that match.

Hadoop Core provides a contrib package that manages indexes that are stored in HDFS: `contrib/index/hadoop-<rel>-index.jar`. The main class, `org.apache.hadoop.contrib.index.main.UpdateIndex`, is specified in the JAR. The contrib package supports distributed indexes, shards, and unified indexes.

SOLR: A Rich Set of Interfaces to Lucene

The SOLR project, `http://lucene.apache.org/solr/`, is a stand-alone, enterprise-grade search service built on top of Lucene. SOLR provides XML/HTTP and JSON APIs.

Katta: A Distributed Lucene Index Server

The Katta project, `http://katta.sourceforge.net/`, describes itself as Lucene in the Cloud, a scalable, fault-tolerant, distributed indexing system capable of serving large replicated Lucene indexes at high loads. Katta uses ZooKeeper to coordinate among the individual servers of the Katta cloud. Katta supports storing shards on the local server file system, HDFS, and in Amazon's S3. Katta also provides a distributed scoring service, allowing for the search results from multiple indexes to be merged together.

Thrift and Protocol Buffers

Thrift (`http://incubator.apache.org/thrift/`) and Protocol Buffers (`http://code.google.com/p/protobuf/`) provide a mechanism for using arbitrarily complex data types as keys or values within Hadoop. The core concept is that of defining a type in a text file and having a tool generate per-language APIs for accessing the data structure and for serializing and deserializing the data structure. As of Hadoop 0.17.0, the framework supports using any type that provides serialization services as a key or a value.

Cascading: A Map Reduce Framework for Complex Flows

Cascading, `http://www.cascading.org/`, describes itself as a rich API for handling complex scale-free workflows reliably on a MapReduce cluster. The Cascading package allows the rapid wiring of components together into workflows that support flow control statements. Cascading's metaphor is that the incoming data flows through a series of functions and filters that allow the data to be split into multiple streams and then joined together again as needed. An acyclic-directed graph is built by the framework, out of the functions and filters.

CloudStore: A Distributed File System

CloudStore, `http://kosmosfs.sourceforge.net/` (formerly known as the Kosmos file system), provides an alternative file system for use within a MapReduce cluster. Unlike HDFS, CloudStore is implemented in C++.

Hypertable: A Distributed Column-Oriented Database

The Hypertable project, `http://www.hypertable.org/`, provides a distributed database conceptually similar to HBase and BigTable. The Hypertable site is clear that the project is at a 0.9 release. Currently, the core servers for Hypertable, the Master server and Hyperspace server, are single points of failure. Hypertable does not provide ready-to-run distributions and must be built from source. There are build instructions for CentOS 5.1 and CentOs 5.2, Fedora Core 8 32bit, Gentoo 2007.0, Ubunto 8.10 Intrepid Ibex 32-bit, Max OS X 10.5 Leopard, and Mac OS X 10.4 Tiger.

Hypertable provides HQL, a SQL-like language for running queries.

Greenplum: An Analytic Engine with SQL

Greenplum, `http://www.greenplum.com/`, provides petabyte-scale, scalable database analytics. It provides a download link to allow you to try its software. It also provides an in-database MapReduce that interoperates with SQL.

CloudBase: Data Warehousing

The CloudBase project, `http://cloudbase.sourceforge.net/`, provides a high-performance, data warehousing system built on top of MapReduce, with an ANSI SQL API. The project is developed by `business.com` to speed terabyte scale web log analysis. The current release version is 1.3. CloudBase is released under GLP 2.0. The web site provides detailed instructions for running CloudBase instances on Amazon's elastic compute (EC2) service.

Hadoop in the Cloud

Sometimes you need additional compute resources for only a short time, you want to experiment with particular configurations, or you just don't want to manage your own hardware. Cloud service vendors provide the ability to spin up clusters of almost arbitrary size and capacities for short to long durations. The best-known cloud server provider at the time of writing is Amazon, and there is direct support for running Hadoop in its cloud.

Amazon

Amazon, `http://aws.amazon.com`, provides a large set of cloud computing services:

- Its simple storage S3 service, `http://aws.amazon.com/s3/`, provides large persistent data storage.
- Its EC2 service, `http://aws.amazon.com/ec2/`, provides on-demand computing clusters built of virtual computers with a variety of capacities and operating systems.
- The SimpleDB, `http://aws.amazon.com/simpledb/`, provides a production-grade, distributed, column-oriented database.
- The Elastic Block Store (EBS), `http://aws.amazon.com/ebs/`, provides persistent storage within EC2 and is ideal for longer-running HDFS clusters.
- The Elastic MapReduce service provides on-demand Hadoop clusters, using S3 as the job input and output file system.

The one significant downside to Hadoop in the Amazon cloud is that there is no real data locality something Hadoop works hard to achieve.

■ **Caution** Anything stored on an EC2 machine instance vanishes when the instance is shut down. Do not use EC2 instances for valuable data. Use the EBS or S3 for persistent storage.

Cloudera

Cloudera, `http://www.cloudera.com/`, provides a supported Hadoop distribution. At the time of writing, the base was Hadoop 0.18.3, with important fixes and features back ported from later versions, including unreleased versions. This is an ideal distribution for production use because it provides minimal API changes while providing bug fixes and some new features.

Training

Cloudera also provides a graduated series of training, from basic to advanced. It provides free online basic Hadoop training at `http://www.cloudera.com/hadoop-training-basic`, Hive training at `http://www.cloudera.com/hadoop-training-hive-introduction`, and Pig training at `http://www.cloudera.com/hadoop-training-pig-introduction/`. There is also a session on using Eclipse with Hadoop at `http://www.cloudera.com/blog/2009/04/20/configuring-eclipse-for-hadoop-development-a-screencast/`.

Supported Distribution

Cloudera provides a freely downloadable version of its distribution at `http://www.cloudera.com/hadoop` and a vmware image for training purposes at `http://www.cloudera.com/hadoop-training-virtual-machine`. The virtual machine has an Eclipse installation set up for use with its Hadoop distribution.

■ **Note** I used the Cloudera training virtual machine to work up some of the examples in this book.

Cloudera also provides ready-to-use Amazon EC2 machine images (AMIs) at `http://www.cloudera.com/hadoop-ec2`. The EC2 image has Hive and Pig installed and ready to use.

Paid Support

Cloudera also provides support contracts for installations using its Hadoop distribution.

Scale Unlimited

Scale Unlimited, `http://www.scaleunlimited.com/`, provides Hadoop Core training and consulting. The principals are the Cascading project lead and the Katta project lead. From `http://www.scaleunlimited.com/consulting`:

> *Our consultants' experience does not end with Map Reduce patterns and Hadoop Distributed File System deployment models; but also spans over a wide set of related open source technologies like HBase, ZooKeeper, Cascading, Katta, Pig, Mahout, Casandra, and CouchDB.*

Scale Unlimited also sponsors a live CD image of a Solaris installation with a three-node Hadoop cluster in zones (`http://opensolaris.org/os/project/livehadoop/`).

■ **Note** A live disk is a CD or DVD that boots as a running instance, not requiring any changes to the local machine's hard disk. An image is an `.img` file that most CD/DVD burner applications can burn directly to writable media.

API Changes in Hadoop 0.20.0

Hadoop 0.20.0 introduces a number of new features and changes. At the time of writing, it is becoming clear that it is not ready for production use. This section hopes to whet your appetite for these new features and help you plan for their arrival.

Vaidya: A Rule-Based Performance Diagnostic Tool for MapReduce Jobs

Vaidya processes the log file data of previously run jobs and provides suggestions on how to improve performance.

At the time of writing, Vaidya checks the following:

- How evenly the data is partitioned between the reduce tasks
- Whether map task failure and re-executions are affecting the overall job performance
- Whether reduce task failure and re-executions are affecting the overall job performance
- Whether the `io.sort.space` size is sufficient to prevent the map tasks outputs from being spilled to disk during the map-side sort phase
- Whether substantial data, other than the key/value pairs, is being read from HDFS during the map or reduce tasks

Service Level Authorization (SLA)

The SLA package provides the access control lists for the control APIs of the various Hadoop Core servers, providing some assurance that any client connecting to a server with SLA enabled is an authorized client.

Removal of LZO Compression Codecs and the API Glue

For licensing reasons, the LZO codec interface files were removed. There are plans to bring in another LZO-like codec with a license the Apache Foundations will accept.

New MapReduce Context APIs and Deprecation of the Old Parameter Passing APIs

The core of this change is that a `Mapper` or a `Reducer Context` object is passed to the `Mapper` and `Reducer` classes, in place of the `JobConf`, to `configure()`, and the `Reporter` and `OutputCollector` to `map()` and `reduce()`. The `Mapper` and `Reducer` classes now have a `setup()`, `cleanup()`, and `run()` method in place of the `configure()` and `close()`methods.

Additional Features in the Example Code

As an aid for my development of the example code in this book, I used a number of tools. This section covers the tools I find most useful.

Zero-Configuration, Two-Node Virtual Cluster for Testing

The class `com.apress.hadoopbook.RunVirtualCluster` in `test/src` of the examples will start and run a mini–Hadoop cluster that provides a near-full Hadoop Core installation. This is ideal for use when developing and testing MapReduce jobs that need more than a single reduce task and therefore cannot be run using the `local` JobTracker.

To run it, change to a directory that will be used as the virtual cluster local storage, and run the following:

```
java -jar hadooppro.jar com.apress.hadoopbook.RunVirtualCluster➥
saved_configuration.xml
```

The cluster will be started, information about the web GUI URLs will be printed to stdout, and a configuration file that defines the relevant parameters for this virtual cluster will be written to the file `saved_configuration.xml`. Any Hadoop program that uses the `GenericOptionsParser` may be passed a `-conf saved_configuration.xml` argument, which will cause the program to load the configuration parameters in `saved_configuration.xml`, and to use the virtual cluster for MapReduce and HDFS services.

I find this particularly handy for debugging jobs when I am on the road because the HDFS data persists after the debugger has exited, and I can examine the job status via the web GUIs. The only problem I have is that the per-task log files are not available via the web GUI, and the HDFS files are not available via the web GUI because of issues inside the Hadoop-supplied `MiniMRCluster` code. The following command lists the files in the virtual HDFS:

```
bin/hadoop dfs -conf saved_configuration.xml -ls
```

This came into being when I was trying to work on the unit tests while on the road, using a machine with Windows XP as the host operating system. The virtual clusters would periodically not start, and I became very frustrated. I wrote this and after it started, it stayed running, and I could use it reliably for multiple tests. The ability to examine the data files in HDFS and to interact with the web interfaces was a pleasant discovery.

Eclipse Project for the Example Code

The example code was developed in Eclipse 3.4, and the project and class path files are part of the download, enabling you to load up, experiment with, and run the example code.

Summary

Hadoop is powerful tool for large-scale data processing. Many people and organizations are leveraging the power of Hadoop MapReduce and providing domain-specific package tools. Distributed column-oriented databases are the current mantra of the scalable web services community; and HBase and Hypertable provide them. Data mining, extracting, transforming, and loading without having to write custom MapReduce jobs are provided with Hive and Pig. Machine learning and recognition are provided by Mahout and Hama, and distributed search is provided by the Katta project.

I am partial to the Cloudera Hadoop distribution because it has good support, back ported fixes, training, is free, and is responsive to community needs. Try the various packages discussed in this chapter explore and enjoy.

APPENDIX A

The JobConf Object in Detail

Everything in a job is controlled via the JobConf object; it is the center of the universe for a MapReduce job. The framework will take the JobConf object and render it to XML; then all the tasks will load that XML when they start. This section will cover all the relevant methods (as of Hadoop Core 0.19.0) and provide some basic usage examples.

The JobConf class inherits from the Configuration class. Because the JobConf object is the primary interface between the programmer and the framework, I'll detail all methods available to the user of a JobConf without distinguishing which methods come from the Configuration base class. I suggest that you create and use only JobConf objects. By default, a new JobConf object loads and merges the hadoop-default.xml and hadoop-site.xml files, as shown in Figure A-1.

The default files hadoop-default.xml and hadoop-site.xml, and any additional user-specified XML resources specified by the AddResource() method, are found in the Java Virtual Machine (JVM) classpath and merged into the configuration data in the order added. Configuration values that are loaded as resources are stored separately from the values that are set via setter calls. The values that were loaded via resources are removed by a call to reloadConfiguration(), whereas all the values are removed by a call to clear(). When looking for a value, a value set by a setter call takes precedence over a value loaded from a resource. The lookup process is described in Figure A-1.

Each configuration item is a name and value pair with an optional final parameter. These parameters tell the Hadoop framework code how to contact the cluster, are defaults for various attributes, and allow for passing arbitrary values to the tasks. The conf/hadoop-default.xml file has a list of most of the Hadoop Core framework parameters. Other parameters are found only by reading the source code.

You can set arbitrary names for value pairs in the configuration, and these name-value pairs are made available to MapReduce tasks. Values that are objects are serialized and then deserialized by each MapReduce task when tasks start.

The naming convention for configuration parameters is usually area.subarea.specific name. The parameters that configure the distributed file system start with dfs, and the parameters that configure the MapReduce framework start with mapred.

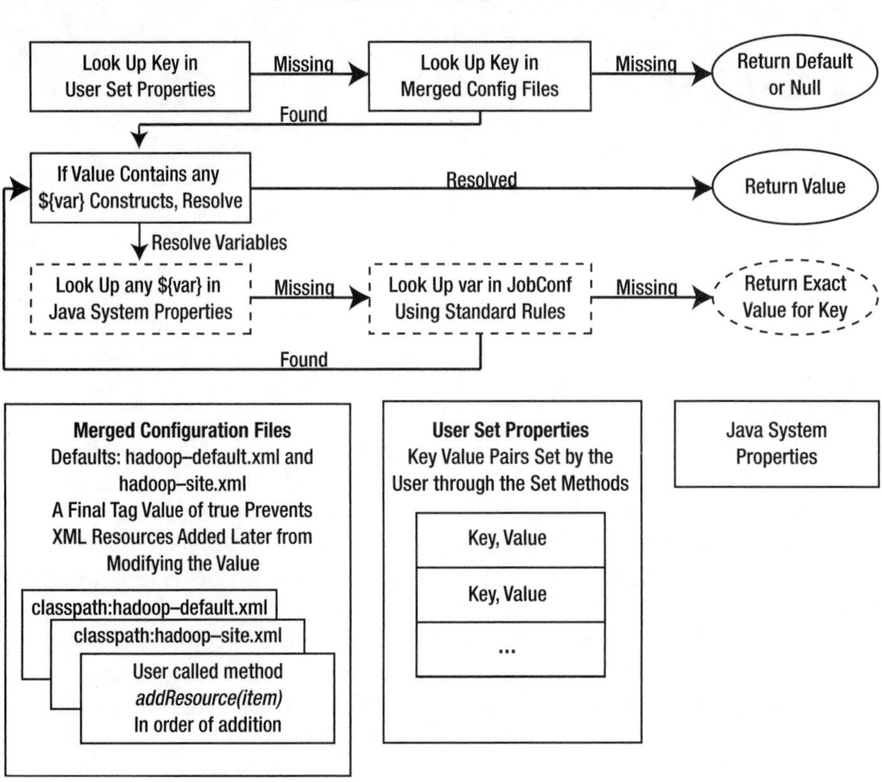

Figure A-1. *How configuration data is loaded into the JobConf object and resolved*

JobConf Object in the Driver and Tasks

The `JobConf` object has two roles. In the job driver, the `JobConf` object is constructed with all the parameters for the job. At job runtime, required data, the `JobConf` object, JAR files, archives, and other resources are stored in the Hadoop Distributed File System (HDFS) in a job-specific directory.

In the task, the `JobConf` object is reconstituted and localized, and it is given a set of directories between the paths defined in `mapred.local.dir`. Any items that must be referenced from the local file system, such as the job JAR file or other items passed via the `DistributedCache`, are unpacked into these local directories and the path references to items in the configuration are adjusted to be the task local path. The classpath for the JVM that the task will run in is also set up for the task to include the location on the local file system that the classpath resources were unpacked into.

JobConf Is a Properties Table

The JobConf instances maintain a table of key/value pairs for all the configuration parameters. The values are all stored as String objects and are serialized if they are objects. At the lowest level, operations get a value for a key or store a value for a key.

Variable Expansion

The JobConf object performs variable expansion on values when raw returned values have special text embedded in them. The syntax is ${key}, which will be replaced by the value of key.

In the configuration files you will often see values in this form: <value>${key}something</value>. If key exists in the System.properties or in the current configuration, a get method will replace ${key} with the value of key.

Note in Listing A-1 that values with ${key} have the key resolved against System.properties. If there is no value found, the value is resolved against the configuration in the JobConf object. This expansion is recursive in that if the expansion contains another ${item} reference, the ${item} is expanded. This process continues until there are no items that are candidates for expansion or there are no items that can be expanded.

Listing A-1. *XML File Used in the Variable Expansion Example: variable-expansion-example.xml*

```
<?xml version="1.0"?>
<?xml-stylesheet type="text/xsl" href="configuration.xsl"?>
<configuration>
   <property>
       <name>no.expansion</name>
       <value>no.expansion Value</value>
   </property>
   <property>
       <name>expansion.from.configuration</name>
       <value>The value of no.expansion is ${no.expansion}</value>
   </property>
   <property>
       <name>java.io.tmpdir</name>
       <value>failed attempt to override a System.properties value
        for variable expansion</value>
   </property>
   <property>
       <name>order.of.expansion</name>
       <value>The value of java.io.tmpdir
        from System.properties: ${java.io.tmpdir}</value>
   </property>
```

```
    <property>
        <name>expansion.from.JDK.properties</name>
        <value>The value of java.io.tmpdir from
           System.properties: ${java.io.tmpdir}</value>
    </property>
    <property>
        <name>nested.variable.expansion</name>
        <value>Will expansion.from.configuration's
  value have substition: [${expansion.from.configuration}]</value>
    </property>
</configuration>
```

The code example in Listing A-2 looks up keys defined in Listing A-1. The first key examined is `no.expansion`; in Listing A-1, the value is defined as `no.expansion Value`, which is the result printed. The value of `no.expansion` is `[no.expansion Value]`.

The next item demonstrating simple substitution is `expansion.from.configuration`, which is given the value of `The value of no.expansion is ${no.expansion}` in Listing A-1. The expanded result is `The value of no.expansion is no.expansion Value`, showing that the `${no.expansion}` was replaced by the value of `no.expansion` in the configuration.

The item for `expansion.from.JDK.properties` demonstrates that the key/value pairs in the `System.properties` are used for variable expansion. The value defined in Listing A-1 is `The value of java.io.tmpdir from System.properties: ${java.io.tmpdir}`, and the result of the expansion is `The value of java.io.tmpdir from System.properties: C:\DOCUME~1\Jason\LOCALS~1\Temp\]`. Note that the actual system property value for `java.io.tmpdir` is used, not the value stored in the configuration for `java.io.tmpdir`, `failed attempt to override a System.properties value for variable expansion.`

The final example demonstrates that the variable expansion results are candidates for further expansion. The key `nested.variable.expansion` has a value of `Will expansion.from.configuration's value have substition: [${expansion.from.configuration}]`, `expansion.from.configuration` has a value of `The value of no.expansion is ${no.expansion}`, and `no.expansion` has the value of `no.expansion Value`. As expected in Listing A-2, the `conf.get("expansion.from.configuration")` returns `The value of no.expansion is no.expansion Value]`.

Listing A-2. *Example of Variable Expansion: the Key is Defined in the JDK System Properties VariableExpansion.java*

```
package com.apress.hadoopbook.examples.jobconf;
import java.io.File;
import java.io.FileOutputStream;
import java.io.IOException;
import java.io.OutputStreamWriter;
import java.io.Writer;

import org.apache.hadoop.mapred.JobConf;
```

```
/** Simple class to demonstrate variable expansion
 * within hadoop configuration values.
 * This relies on the hadoop-core jar, and the
 * hadoop-default.xml file being in the classpath.
 */
public class VariableExpansion {
    public static void main( String [] args ) throws IOException {
        /** Get a local file system object, so that we can construct a local Path
         * That will hold our demonstration configuration.
         */
        /** Construct a JobConf object with our configuration data. */
        JobConf conf = new JobConf( "variable-expansion-example.xml" );
        System.out.println( "The value of no.expansion is [" +
           conf.get("no.expansion") + "]" );
        System.out.println( "The value of expansion.from.configuration is [" +
           conf.get("expansion.from.configuration") + "]");
        System.out.println( "The value of expansion.from.JDK.properties is ["
           + conf.get("expansion.from.JDK.properties") + "]");
        System.out.println( "The value of java.io.tmpdir is [" +
           conf.get("java.io.tmpdir") + "]" );
        System.out.println( "The value of order.of.expansion is [" +
           conf.get("order.of.expansion") + "]" );
        System.out.println( "Nested variable expansion for nested." +
          "variable.expansion is [" +
      conf.get("nested.variable.expansion") +"]");
    }
}
```

```
The value of no.expansion is [no.expansion Value]
The value of expansion.from.configuration is ➥
[The value of no.expansion is no.expansion Value]
The value of expansion.from.JDK.properties is ➥
[The value of java.io.tmpdir from System.properties:  ➥
C:\DOCUME~1\Jason\LOCALS~1\Temp\]
The value of java.io.tmpdir is ➥
[failed attempt to override a System.properties value for variable expansion]
The value of order.of.expansion is ➥
[The value of java.io.tmpdir from System.properties: ➥
C:\DOCUME~1\Jason\LOCALS~1\Temp\]
Nested variable expansion for nested.variable.expansion is ➥
[Will expansion.from.configuration's value have substition: ➥
[The value of no.expansion is no.expansion Value]]
```

Final Values

The Hadoop Core framework gives you a way to mark some keys in a configuration file as final. The stanza `<final>true</final>` prevents later configuration files from overriding the value specified. The `<final>` tag does not prevent the user from overriding the value via the set method. The example in Listing A-3 creates several XML files in the temporary directory: the first file, `finalFirst`, contains the declaration of a configuration key, `final.first`, which has the value `first final value` declared final via `<final>true</final>`. The second file, `finalSecond`, also defines `final.first` with the value `This should not override the value of final.first`. After loading the two resource files via `JobConf conf = new JobConf( finalFirst.toURI().toString() );` and `conf.addResource( finalSecond.toURI().toString());`, the value of the key `final.first` is gotten via `conf.get("final.first")` and found to be `first final value`. The next example calls `conf.set("final.first", "This will override a final value, when applied by conf.set");` to demonstrate that the setter methods will override a value marked `final`.

Listing A-3. *Sample Code Showing DemonstrationOfFinal.java*

```
package com.apress.hadoopbook.examples.jobconf;

import java.io.File;
import java.io.FileOutputStream;
import java.io.IOException;
import java.io.OutputStreamWriter;
import java.io.Writer;

import org.apache.hadoop.mapred.JobConf;
import org.apache.hadoop.conf.Configuration;

/** Demonstrate how the final tag works for configuration files and is ignored
 *  by the {@link Configuration#set(java.lang.String,java.lang.String)} operator
 * This relies on the hadoop-core jar, and the
 * hadoop-default.xml file being in the classpath.
 */
public class DemonstrationOfFinal {
    /** Save xml configuration data to a temporary file,
     * that will be deleted on jvm exit
     *
     * @param configData The data to save to the file
     * @param baseName The base name to use for the file, may be null
     * @return The File object for the file with the data written to it.
     * @throws IOException
     */
```

```
    static File saveTemporaryConfigFile( final String configData,
            String baseName ) throws IOException {
        if (baseName==null) {
            baseName = "temporaryConfig";
        }
        /** Make a temporary file using the JVM file utilities */
        File tmpFile = File.createTempFile(baseName, ".xml");
        tmpFile.deleteOnExit(); /** Ensure the file is deleted
                                 * when this jvm exits. */
        Writer ow = null;
        /** Ensure that the output writer is closed even on errors. */
        try {
            ow = new OutputStreamWriter( new FileOutputStream( tmpFile ), "utf-8");
            ow.write( configData );
            ow.close();
            ow = null;
        } finally {
            if (ow!=null) {
                try {
                    ow.close();
                } catch (IOException e) {
                //  ignore, as we are already handling the real exception
                }
            }
        }
        return tmpFile;

    }
    public static void main( String [] args ) throws IOException {
        /** Get a local file system object, so that we can construct a local Path
         * That will hold our demonstration configuration.
         */

        File finalFirst = saveTemporaryConfigFile(
                "<?xml version=\"1.0\"?>\n" +
                "<?xml-stylesheet type=\"text/xsl\" ➥
href=\"configuration.xsl\"?>\n" +
                "<configuration>\n" +
                "    <property>\n" +
                "        <name>final.first</name>\n" +
                "        <value>first final value.</value>\n" +
                "        <final>true</final>\n" +
                "    </property>\n" +
                "</configuration>\n",
                "finalFirst" );
```

```
        File finalSecond = saveTemporaryConfigFile(
                "<?xml version=\"1.0\"?>\n" +
                "<?xml-stylesheet type=\"text/xsl\" ➥
href=\"configuration.xsl\"?>\n" +
                "<configuration>\n" +
                "   <property>\n" +
                "       <name>final.first</name>\n" +
                "       <value>This should not override the ➥
value of final.first.</value>\n" +
                "   </property>\n" +
                "</configuration>\n",
                "finalSecond" );


        /** Construct a JobConf object with our configuration data. */
        JobConf conf = new JobConf( finalFirst.toURI().toString() );
        /** Add the additional file that will attempt to overwrite
          * the final value of final.first. */
        conf.addResource( finalSecond.toURI().toString());
        System.out.println( "The final tag in the first file will " +
          "prevent the final.first value in the second configuration file "
     +"from inserting into the configuration" );
        System.out.println( "The value of final.first in the " +
         "configuration is [" + conf.get("final.first") + "]" );
        /** Manually set the value of final.first to demonstrate
          * it can be overridden. */
        conf.set("final.first", "This will override a final value,➥
          when applied by conf.set");
        System.out.println( "The value of final.first in the configuration"
         + " is [" + conf.get("final.first") + "]" );

    }
}
```

```
The final tag  in the first file will prevent the final ➥
first value in the second configuration file from inserting into the configuration
The value of final.first in the configuration is [first final value.]
The value of final.first in the configuration is ➥
[This will override a final value when applied by conf.set]
```

Constructors

All code that creates and launches a MapReduce job into a Hadoop cluster creates a `JobConf` object. The framework provides several methods for creating the object.

public JobConf()

This is the default constructor. This constructor should not be used because it doesn't provide the framework with information about the JAR file that this class was loaded from.

public JobConf(Class exampleClass)

This common use case constructor is the constructor you should use. The archive that the `exampleClass` was loaded from will be made available to the MapReduce tasks. The type of `exampleClass` is arbitrary; `exampleClass` is used only to find the classpath resource that the `exampleClass` was loaded from. The containing JAR file will be made available as a classpath item for the job tasks. The JAR is actually passed via the `DistributedCache` as a classpath archive. `exampleClass` is commonly the mapper or reducer class for the job, but it is not required to be so.

■ **Tip** The task JVMs are run on different physical machines and do not have access to the classpath or the classpath items of the JVM that submits the job. The only way to set the classpath of the task JVMs is to either set the classpath in the `conf/hadoop-env.sh` script or pass the items via the `DistributedCache`.

---

public JobConf(Configuration conf)

This constructor is commonly used when your application already has constructed a `JobConf` object and wants a copy to use for an alternate job. The configuration in *conf* is copied into the new `JobConf` object.

It is very handy when unit testing because as the unit test can construct a standard `JobConf` object, and each individual test can use it as a reference and change specific values.

If your driver launches multiple MapReduce jobs, each job should have its own `JobConf` object, and the pattern described previously for unit tests is ideal to support this.

public JobConf(Configuration conf, Class exampleClass)

Construct a new `JobConf` object that inherits all the settings of the passed-in `Configuration` object `conf`, and make the archive that `exampleClass` was loaded from available to the MapReduce tasks.

Classes that launch jobs that may have unit tests or be called as part of a sequence of Hadoop jobs should provide a run method that accepts a `Configuration` object and calls this constructor to make the `JobConf` object for that class's job. This way, the unit test or calling code can preconfigure the configuration, and this class can customize its specific variables and launch the job.

Listing A-4. *Sample Code Fragment of a Class run Method*

```
/** Code Fragment to demonstrate a run method that can be called
 *from a unit test, or from a driver that launches multiple
 * Hadoop jobs.
 *
 * @param defaultConf The default configuration to use
 * @return The running job object for the completed job.
 * @throws IOException
 */
public RunningJob run( Configuration defaultConf ) throws IOException
{
   /** Construct the JobConf object from the passed-in object.
    * Ensure that the archive that contains this class will be
    * provided to the map and reduce tasks.
    */
    JobConf conf = new JobConf( defaultConf, this.getClass() );
    /**
     * Set job specific parameters on the conf object.
     *
     */
    conf.set( "our.parameter", "our.value" );

    RunningJob job = JobClient.runJob(conf);
    return job;
}
```

public JobConf(String config)

Construct a `JobConf` object with configuration data loaded from the file that `config` is a path to.

public JobConf(Path config)

Construct a `JobConf` object and load configuration values from the XML data found in the file `config`. This constructor is used by the TaskTracker to construct the `JobConf` object from the job-specific configuration file that was written out by the Hadoop framework.

public JobConf(boolean loadDefaults)

This method is identical to the no-argument constructor unless the `loadDefaults` value is `false`. If `loadDefaults` is `false`, `hadoop-site.xml` and `hadoop-default.xml` are not loaded.

Methods for Loading Additional Configuration Resources

The methods described in this section load an XML configuration file resource and store it in the `JobConf` parameter set. The order in which these methods are called is important because the contents specified by the most recent call will override values supplied earlier.

If a specified resource cannot be loaded or parsed as valid configuration XML, a `RuntimeException` will be thrown unless quiet mode is enabled via a call to `setQuietMode (true)`.

Each call to one of these methods results in the complete destruction of the configuration data that resulted from the loading and merging of the XML resources. There are no changes made to the configuration parameters that have been created via the set methods. The entire set of XML resources is reparsed and merged on the next method call that reads or sets a configuration parameter.

These resource items follow the same rules as with the `hadoop-default.xml` and `hadoop-site.xml` files, and a parameter in a resource object can tag itself as `final`. In this case, resource objects loaded later may not change the value of the parameter.

Listing A-5. *Sample Final Parameter Declaration*

```
<property>
   <name>my.final.parameter</name>
   <value>unchanging</value>
   <final>true</final>
</property>
```

public void setQuietMode(boolean quietmode)

If `quietmode` is `true`, no log messages will be generated when loading the various resources into the configuration. If a resource cannot be parsed, no exception will be thrown.

If `quietmode` is `false`, a log message will be generated for each resource loaded. If a resource cannot be parsed, a `RuntimeException` will be thrown.

public void addResource(String name)

Load the contents of `name`. The parameter is loaded from the current classpath by the JDK `ClassLoader.getResource` method. `name` can be a simple string or a URL.

The default configuration has two `addResource( String name )` calls: one for `hadoop-default.xml` and the other for `hadoop-site.xml`.

Caution The first `hadoop-default.xml` file and the first `hadoop-site.xml` file in your classpath are loaded. It is not uncommon for these files to accidentally be bundled into a JAR file and end up overriding the cluster-specific configuration data in the `conf` directory. A problem often happens with jobs that are not run through the `bin/hadoop` script and do not have a `hadoop-default.xml` or `hadoop-site.xml` file in their classpath.

---

public void addResource(URL url)

This method explicitly loads the contents of the passed-in URL, `url`, into the configuration.

public void addResource(Path file)

This method explicitly loads the contents of `file` into the configuration.

public void addResource(InputStream in)

Load the XML configuration data from the supplied `InputStream in` into the configuration.

public void reloadConfiguration()

Clear the current configuration, *excluding* any parameters set using the various `set` methods, and reload the configuration from the resources that have been specified. If the user has not specified any resources, the default pair of `hadoop-default.xml` and `hadoop-site.xml` will be used.

This method actually just clears the existing configuration, and the reload will happen on the next get or set.

Basic Getters and Setters

The methods in this section get and set basic types:

- In general, if the framework cannot convert the value stored under a key into the specific type required, a `RuntimeException` will be thrown.
- If the value being retrieved is to be a numeric type, and the value cannot be converted to the numeric type, a `NumberFormatException` will be thrown.
- For boolean types, a value of `true` is required for a `true` return. Any other value is considered `false`.
- For values that are class names, if the class cannot be instantiated, or the instantiated class is not of the correct type, a `RuntimeException` will be thrown.

The framework stores sets of things as comma-separated lists. There is no mechanism currently to escape a comma that must be a part of an individual item in a list.

Under the covers, all data is stored as a java `String` object. All items stored are serialized into a `String` object, and all values retrieved are deserialized from a `String` object. The user is required to convert objects into `String` representations to store arbitrary objects in the configuration and is responsible for re-creating the object from the stored `String` when retrieving the object.

public String get(String name)

This is the basic getter: it returns the `String` version of the value of `name` if `name` has a value or if the method returns `null`. Variable expansion is completed on the returned value. If the value is a serialized object, the results of the variable expansion may be incorrect.

public String getRaw(String name)

Returns the raw `String` value for `name` if `name` exists in the configuration; otherwise returns `null`. No variable expansion is done. This is the method to use to retrieve serialized objects.

public void set(String name, String value)

Stores the `value` under the key `name` in the configuration. Any prior value stored under `name` is discarded, even if the key was marked final.

public String get(String name, String defaultValue)

This method behaves as the `get()` method does: it returns `defaultValue` if `name` does not have a value in the configuration. This method is ideal to use for `get()` operations that must return a value and there is a sensible default value.

public int getInt(String name, int defaultValue)

Many properties stored in the configuration are simple integers, such as the number of reduces, `mapred.reduce.tasks`. If the underlying value for `name` is missing or not convertible to an `int`, the `defaultValue` is returned. If the value starts with a leading `0x` or `0X`, the value will be interpreted as a hexadecimal value.

public void setInt(String name, int value)

Stores the `String` representation of `value` in the configuration under the key `name`. Any prior value associated with `name` will be lost.

public long getLong(String name, long defaultValue)

Many properties stored in the configuration are simple long values, such as the file system block size `dfs.block.size`. If the underlying value for `name` is missing or not convertible to a long, the `defaultValue` is returned. If the value starts with a leading `0x` or `0X`, the value will be interpreted as a hexadecimal value.

public void setLong(String name, long value)

Stores the `String` representation of `value` in the configuration under the key `name`. Any prior value associated with `name` will be lost.

public float getFloat(String name, float defaultValue)

Some properties stored in the configuration are simple floating-point values. You might want to pass a float value to the mapper or reducer, which would use this method to get the float value. If the underlying value for `name` is missing or not convertible to a float, the `defaultValue` is returned.

public boolean getBoolean(String name, boolean defaultValue)

Many properties stored in the configuration are simple boolean values, such as the controlling speculative execution for map tasks, mapred.map.tasks.speculative.execution. If the underlying value for name is missing or not convertible to a boolean value, the defaultValue is returned. The only acceptable boolean values are true or false. The comparison is case sensitive, so a value of True will fail to convert, and the defaultValue will be returned.

public void setBoolean(String name, boolean value)

Convert the boolean value to the String true or the String false and store it in the configuration under the key name. Any prior value associated with name will be lost.

RANGES

The configuration supports storing basic types such as various numbers, boolean values, text, and class names. The configuration also supports a type called a Range. It is two integer values, in which the second integer is larger than the first. An individual range is specified by a String containing a -, a dash character that can also have a leading and trailing integer.

If the leading integer is absent, the first range value takes the value 0. If the trailing integer is absent, the second range value takes the value of Integer.MAX_VALUE.

The simplest range is -, which is the range 0 to Integer.MAX_VALUE. The range -35 parses as 0 to 35. The range 40- parses as 40 to Integer.MAX_VALUE. The range 40-50 parses as 40 to 50. Multiple ranges may be separated by ,: a comma character such as 1-5,7-9,13-50.

public Configuration.IntegerRanges getRange(String name, String defaultValue)

getRange is a relatively unusual method and obtains a named range. It is currently not widely used. As of Hadoop 0.19.0 it is used only to determine which MapReduce tasks to profile. As of Hadoop 0.19.0 there is no corresponding set method, and the base set(String name, String value) is used to set a range The value has to be the valid String representation of a range or later calls to the getRange method for name will result in an exception being thrown.

The defaultValue must be passed in as a valid range. String null may not be passed as the default value, or else a NullPointerException will be thrown.

This method looks up the value of name in the configuration, and if there is no value, the defaultValue will be used. The resulting value will then be parsed as an IntegerRanges object and that result returned. If the parsing fails, an IllegalArgumentException will be thrown.

■ **Note** If there is a value for name in the configuration and it cannot be parsed as an IntegerRanges object, the defaultValue will be ignored, and an IllegalArgumentException will be thrown.

---

public Collection<String> getStringCollection(String name)

The JobConf and Configuration objects (at least through Hadoop 0.19.0) handle parameters that are sets of String objects by storing them internally as comma-separated lists in a single String. There is no provision for escaping the commas.

getStringCollection will get the value associated with name in the configuration and split the String on commas and return the resulting Collection.

Listing A-6. *Sample Use of public Collection<String> getStringCollection(String name)*

```
conf.set( "path.set", "path1,path2,path3,path4");
Collection<String> pathSet = conf.getStringCollection("path.set");
for( String path : pathSet ) {
    System.out.println( path );
}
```

```
path1
path2
path3
path4
```

---

public String[] getStrings(String name)

The JobConf and Configuration objects (at least through Hadoop 0.19.0) handle parameters that are sets of String objects by storing them internally as comma-separated lists in a single String. There is no provision for escaping the commas.

This method gets the value associated with name in the configuration, splits the String on commas, and returns the resulting array (see Listing A-7).

Listing A-7. *Sample Use of public String[] getStrings(String name)*

```
conf.set( "path.set", "path1,path2,path3,path4");
String[] pathSet = conf.getStrings("path.set");
for( String path : pathSet ) {
    System.out.println( path );
}
```

```
path1
path2
path3
path4
```

JAVA 1.5 AND BEYOND VARAGS SYNTAX

As of Java 1.5, variable argument lists are supported for method calls. The declaration of the last parameter may have an ellipsis between the type and the name, `type...name`. The caller can place an arbitrary number of objects of `type` in the method call, and the member method will receive an array of type with the elements from the caller's call. For the method `X(String .... strings)`, a call of the form `X("one","two","three")` would result in the variable `strings` being a three-element array of `String` objects containing `"one"`, `"two"`, `"three"`.

For more details, please visit `http://java.sun.com/j2se/1.5.0/docs/guide/language/varargs.html`.

public String[] getStrings(String name, String... defaultValue)

The `JobConf` and `Configuration` objects (at least through Hadoop 0.19.0) handle parameters that are sets of `String` objects by storing them internally as comma-separated lists in a single `String`. There is no provision for escaping the commas.

This method will get the value associated with `name` in the configuration and split the `String` on commas and return the resulting array (see Listing A-8). If there is no value stored in the configuration for `name`, the array built from the `defaultValue` parameters will be returned.

Listing A-8. *Sample Use of public String[] getStrings(String name, String... defaultValue)*

```
JobConf empty = new JobConf(false); /** Create an empty configuration to
 * ensure the default value is used in our getStrings example.*/
String[] pathSet = conf.getStrings("path.set", "path1", "path2", "path3", path4");
for( String path : pathSet ) {
    System.out.println( path );
}
```

```
path1
path2
path3
path4
```

---

public void setStrings(String name, String... values)

Stores the set of `Strings` provided in `values` under the key `name` in the configuration, deleting any prior value (see Listing A-9). The set of `String` objects defined by `values` is concatenated using the comma (,) character as a separator, and the resulting `String` is stored in the configuration under `name`.

Listing A-9. *Sample Use of public void setStrings(String name, String... values)*

```
conf.setStrings( "path.set", "path1", "path2, "path3", "path4");
String[] pathSet = conf.getStrings("path.set");
for( String path : pathSet ) {
    System.out.println( path );
}
```

```
path1
path2
path3
path4
```

---

public Class<?> getClassByName(String name) throws ClassNotFoundException

It attempts to load a class called `name` by using the `JobConf` customized class loader. If the class is not found, a `ClassNotFoundException` is thrown.

By default, the class loader used to load the class is the class loader for the thread that initialized the `JobConf` object. If that class loader is unavailable, the class loader used to load the `Configuration.class` is used.

Note This method does not look up the value of `name` in the configuration; `name` is the value passed to the class loader.

---

public Class<?>[] getClasses(String name, Class<?>... defaultValue)

If `name` is present in the configuration, parses it as a comma-separated list of class names and construct a class object for each entry in the list. If a class cannot be loaded for any entry in the list, a `RuntimeException` is thrown. If `name` does not have a value in the configuration, this method returns the array of classes passed in as the `defaultValue`.

By default, the class loader used to load the class is the class loader for the thread that initialized the `JobConf` object. If that class loader is unavailable, the class loader used to load the `Configuration.class` is used.

public Class<?> getClass(String name, Class<?> defaultValue)

If `name` is present in the configuration, it attempts to load the value as a class using the configuration's class loader. If a value exists in the configuration and a class cannot be loaded for that value, a `RuntimeException` is thrown. If `name` does not have a value, the class `defaultValue` is returned.

By default, the class loader used to load the class is the class loader for the thread that initialized the `JobConf` object. If that class loader is unavailable, the class loader used to load the `Configuration.class` is used.

public <U> Class<? extends U> getClass(String name, Class<? extends U> defaultValue, Class<U> xface)

If `name` is present in the configuration, attempts to load the value as a class using the configuration's class loader. If a value exists in the configuration, and a class cannot be loaded for that value, a `RuntimeException` is thrown. The loaded class must derive from or implement `xface`, or else a `RuntimeException` will be thrown. If no value is present for `name` in the configuration, the `defaultValue` will be returned.

This `getClass` method returns the result of the call `theClass.asSubclass(xface);` where `theClass` is the class constructed from the class name stored under `name` or the `defaultValue` if there is no value stored under `name`.

By default, the class loader used to load the class is the class loader for the thread that initialized the `JobConf` object. If that class loader is unavailable, the class loader used to load the `Configuration.class` is used.

public void setClass(String name, Class<?> theClass, Class<?> xface)

The class name for `theClass` is stored in the configuration under the key `name`. If `theClass` does not derive from or implement `xface`, a `RuntimeException` is thrown.

Note The name of the class, `theClass.getName()`, is stored, not a serialized version of the class.

Getters for Localized and Load Balanced Paths

The framework provides the capability for multiple local directories to be specified for task temporary files. Multiple locations are allowed to load balance the I/O over multiple devices. The framework will attempt to select one location set either at random or in sequential order. The ordering used will be given in the method description.

Each of the methods described in this section is called with a trailing path component that includes a final file name. The return value will be the full path to the file; all the intermediate directory components will be constructed if needed.

The first found complete path that can be created or exists will be returned by the method. The intermediate directories may be constructed in locations that do not allow the complete construction of the path. If so, those intermediate directories that have been created will not be removed. If no path can be constructed, an `IOException` will be thrown.

These throwing IOException paths are explicitly constructed on the local (host) file system. See Listing A-10.

Listing A-10. *Samples of public Path getLocalPath(String dirsProp, String path) Throwing IOException*

```
conf.setStrings("path.set", "dira/a/a/a/","dirb/b/b/b","dirc/c/c/c");
Path random = conf.getLocalPath( "path.set", "trailing/path/file")
```

The path candidates are as follows:

- `dira/a/a/a/trailing/path`
- `dirb/b/b/b/trailing/path`
- `dirc/c/c/c/trailing/path`

A result in this example might be `dirc/c/c/c/trailing/path/file`.

■ **Note** This method leaves partial paths in place that were constructed during its operation. The pseudo-random method does not guarantee that all possible path candidates will be tried; only that no more than the count of path candidate elements will be tried (as of Hadoop 0.19.0). Also as of Hadoop 0.19.0, the method does not fail if the path candidate is a file, not a directory.

---

public Path getLocalPath(String dirsProp, String pathTrailer) throws IOException

Load balances access to a set of directories that reside on different devices. The goal is to return a resultant path composed of `pathTrailer` as the trailing component and one element out of the set of directories stored under `dirsProp` as the path leader. If `dirsProp` is unset, an `IOException` is thrown. This method uses the Hadoop `LocalFileSystem` object for all path operations. The paths defined by `dirsProp` are searched in a pseudo-random order.

public File getFile(String dirsProp, String pathTrailer) throws IOException

This method is used to load balance access to a set of directories that reside on different devices. The goal is to return a resultant path composed of `pathTrailer` as the trailing component and one element out of the set of directories stored under `dirsProp` as the path leader. If `dirsProp` is unset, an `IOException` is thrown.

This method uses the `java.io.File` methods to create directory paths and test for directory existence. The paths defined by `dirsProp` are searched in a pseudo-random order.

public String[] getLocalDirs() throws IOException

This method looks up the key `mapred.local.dir` in the configuration. The value is expected to be a set of file system paths separated by commas. If present, the value is split on comma characters and the resulting array of `String` objects is returned. If there is no value present, a `null` is returned. This is used by the TaskTracker to find the set of directories to use for per-task local storage. The TaskTracker uses a round robin strategy to allocate a task directory for a new task.

Note The value stored in the configuration under the key `mapred.local.dir` is the set of local file system locations to be used by MapReduce tasks for temporary file storage. This parameter is generally only used directly by the framework.

---

public void deleteLocalFiles() throws IOException

This method deletes all the directory trees stored in the configuration under the key `mapred.local.dir`. The value is parsed as a comma-separated list of paths. This is used by the framework to clean up the local machine temporary areas on TaskTracker start and TaskTracker exit.

public void deleteLocalFiles(String subdir)throws IOException

This method deletes `subdir` from all the directories that are stored in the configuration under the key `mapred.local.dir`. The value is parsed as a comma-separated list of paths. This is used by the framework to clean up the local machine temporary files for a particular task.

public Path getLocalPath(String pathString) throws IOException

This method looks up the key `mapred.local.dir` in the configuration and parses the value as a comma-separated list of local file system paths. For each directory in the resulting list, an attempt is made, in pseudo-random order, to create the path portion of `pathString`, including any leading directory elements. If after this creation attempt that directory exists, the file name portion of `pathString` is appended to the directory path and the resulting path is returned.

public String getJobLocalDir()

This method looks up the key `job.local.dir` in the configuration and returns the value. The value will be the task-specific shared directory for each job on each TaskTracker. This parameter is set only in the `JobConf` object passed to each task.

The returned value will be the path fragment `taskTracker/jobcache/JobId/work`, prefixed by one of the directories specified in the set of directories stored under the key `mapred.local.dir`.

This is used by the framework when setting up the per-job task environment on a TaskTracker node. Tasks can use this method to find the path to the task-specific directory on the local file system, which may be used for temporary file storage.

Note In Hadoop 0.19.0, each task for a job on the same TaskTracker may get a distinct local `dir` if multiple directories are specified in the `mapred.local.dir` value.

Methods for Accessing Classpath Resources

The framework provides a way for tasks to access resources from the task-specific classpath objects.

public URL getResource(String name)

Returns the URL for the resource `name`, found by searching the configuration's class loader. If the resource is not found, `null` is returned.

By default, the class loader used to load the class is the class loader for the thread that initialized the `JobConf` object. If that class loader is unavailable, the class loader used to load the `Configuration.class` is used.

■ **Note** This method does not look up the value of `name` in the configuration; `name` is the value passed to the class loader.

---

public InputStream getConfResourceAsInputStream (String name)

Returns the `java.io.InputStream` resulting from opening the URL for the resource `name`, found by searching the configuration's class loader. If the resource is not found, `null` is returned.

By default, the class loader used to load the class is the class loader for the thread that initialized the `JobConf` object. If that class loader is unavailable, the class loader used to load the `Configuration.class` is used.

■ **Caution** This method does not look up the value of `name` in the configuration; `name` is the value passed to the class loader.

---

public Reader getConfResourceAsReader(String name)

Returns the `java.io.Reader` resulting from opening the URL for the resource `name`, found by searching the configuration's class loader. If the resource is not found, `null` is returned.

By default, the class loader used to load the class is the class loader for the thread that initialized the `JobConf` object. If that class loader is unavailable, the class loader used to load the `Configuration.class` is used.

This method does not look up the value of `name` in the configuration. The name is used directly as a Java resource name. This is approximately equivalent to `new InputStreamReader (System.getClassLoader().getResourceAsInputStream(name));` with error checking and using the `ClassLoader` member variable of the configuration.

Caution `getConfResourceAsReader` does not look up the value of `name` in the configuration; `name` is passed directly to the class loader.

Methods for Controlling the Task Classpath

These methods ensure that the objects referenced are distributed to the task nodes and made available in the classpath of the tasks.

public String getJar()

This method is a shortcut for the call `get("mapred.jar")`. The key `mapred.jar` is the JAR to use for the MapReduce job.

The `mapred.jar` key's value is set by the `setJar(String jar)` method, the `setJarByClass(Class cls)` method, and the `JobConf` constructors that take a `Class` value as a parameter.

If the `mapred.jar` key has been set in the configuration, the value will be returned.

public void setJar(String jar)

Stores the `String jar`, which should be the path to the JAR that contains the map and reduce classes for this job into the configuration under the `mapred.jar` key. Any prior value stored under the key `mapred.jar` will be discarded. This archive will be distributed to the task nodes and placed in the classpath for the map and reduce tasks.

public void setJarByClass(Class cls)

Looks for the first JAR file in the classpath that contains class `cls`. If found, stores the path to it under the `mapred.jar` key in the configuration. If no JAR file is found that contains `cls`, a `RuntimeException` is thrown.

Note This method will look only in JAR files, not in zip files or in directory trees.

Methods for Controlling the Task Execution Environment

These methods control the setup and cleanup of the individual task environment.

public String getUser()

This method returns the value stored in the configuration under the key `user.name`. This parameter is generally initialized to the name of the user that launched the job, but this is not enforced.

Caution Usernames may be overwritten with a different username by any user. This is not a security feature, and Hadoop permissions are not a security feature. Through at least Hadoop 0.19 any user may claim to be any other Hadoop user and act fully as if they are that user, including the removal of files or the scheduling of jobs. There is way to prevent this; you have to trust the users who have access to your cluster because any user can override any Hadoop level permission restrictions placed on that user.

---

public void setUser(String user)

The value of `user` is stored in the configuration under the key `user.name`. This value is used for HDFS permission checking.

public void setKeepFailedTaskFiles(boolean keep)

Stores the value of `keep` in the configuration under the key `keep.failed.task.files`. This value configures the framework to save or not save the intermediate output files of tasks that fail. It is set to `true` to when the task output is needed to debug a failing job.

public boolean getKeepFailedTaskFiles()

Returns the value stored in the configuration under `keep keep.failed.task.files` converted to a `boolean`. If no value is found, or the value is not exactly `true` or `false`, the value `false` is returned.

public void setKeepTaskFilesPattern(String pattern)

Stores a Java regular expression `String pattern` into the configuration under the key `keep.task.files.pattern`. If the task id of a task matches this regular expression, its temporary files will not be removed The file names are written as `*_[mr]_[jobid]_[tasknumber]`. The job id is 0-padded on the left. The pattern `*_m_000027_5` would match the fifth map task of job `00027`. The pattern `*_r_000027_5` would match the fifth reduce task of job `00027`.

This is used to aid in debugging the framework.

public String getKeepTaskFilesPattern()

Returns the value stored in the configuration under the key `keep.task.files.pattern`. The framework calls this in the TaskTracker before cleaning up temporary files after a task completes. If the task id matches the pattern, the temporary files are not removed. This is a framework debugging aid.

public void setWorkingDirectory(Path dir)

This method builds a path by concatenating `dir` with the value of `getWorkingDirectory()` and stores that value in the configuration under the key `mapred.working.dir`. The user calls it and if it has not been called by job submission time, the `JobClient` object will initialize it to the current working directory of the process submitting the job.

■Note If the working directory has not been initialized by the time this method is called, the default working directory for the default file system will be used. For HDFS, it is generally `/user/USERNAME`. The default file system is the file system defined by the configuration key `fs.default.name`.

public Path getWorkingDirectory()

Returns the value stored in the configuration under `mapred.working.dir`. If this value is unset, this method first sets the value for the key to the default working directory for the default file system.

The default file system is defined by the configuration parameter `fs.default.name`; for HDFS, the default working directory is `/user/USERNAME`.

public void setNumTasksToExecutePerJvm(int numTasks)

Prior to Hadoop 0.19.0, a new JVM was created for each task run by the TaskTracker. As of Hadoop 0.19.0, the TaskTracker has the capability to reuse the task JVM for additional tasks. The configuration key `mapred.job.reuse.jvm.num.tasks`'s value is the number of times that a JVM may be reused. This method stores `numTasks` in the configuration under the key `mapred.job.reuse.jvm.num.tasks`.

■Note Calling `setNumTasksToExecutePerJvm` with a value that is `<= 0` will result in erroneous behavior.

public int getNumTasksToExecutePerJvm()

Looks up the value of `mapred.job.reuse.jvm.num.tasks` in the configuration and converts the value to an integer. If the value does not exist or if the value cannot be converted to an integer, it returns `1`.

Prior to Hadoop 0.19.0, a new JVM was created for each task run by the TaskTracker. As of Hadoop 0.19.0, the TaskTracker has the capability to reuse the task JVM for additional tasks. The value stored in the configuration under `mapred.job.reuse.jvm.num.tasks` is the number of times to use a JVM for a task.

Methods for Controlling the Input and Output of the Job

The methods described in this section are used to configure how the jobs' input and output will be handled. This includes how the input is parsed and presented to the framework, the compression of intermediate and final output, and how the output is written.

public InputFormat getInputFormat()

This method looks up the value of the key `mapred.input.format.class` in the configuration and instantiates a class of that name. If the value is missing, a `TextInputFormat.class` will be returned. If the class name cannot be instantiated, or if the instantiated class is not an instance of `InputFormat`, a `RuntimeException` will be thrown.

The returned class will be used by the framework to read the input data set for the job. The key/value pairs that the class extracts from the input will be passed to the map method of the mapper class in the map tasks. There will be one instance created per map task, and that instance will receive the input split for that map task as input.

public void setInputFormat(Class<? extends InputFormat> theClass)

This method stores the class name of `theClass` in the configuration under the key `mapred.input.format.class`. An instance of this class will be instantiated in each map task to convert the input split data into a set of key/value pairs for the map method of the mapper class. If `theClass` does not implement the interface `InputFormat`, a `RuntimeException` will be thrown.

public OutputFormat getOutputFormat()

This method looks up the value of the key `mapred.output.format.class` in the configuration and instantiates a class of that name. If the value is missing, a `TextOutputFormat.class` will be returned. If the class name cannot be instantiated or if the instantiated class is not an instance of `OutputFormat`, a `RuntimeException` will be thrown.

The returned class will be used by the framework to write each of the key/value pairs output by the `reduce()` method of the reducer class, and if not explicitly configured, each of the key/value pairs output by the map method of the mapper class.

There will be one instance of this class created for each reduce task. By default, one instance of this class is created for each map task.

public void setOutputFormat(Class<? extends OutputFormat> theClass)

This method stores the class name of `theClass` in the configuration under the key `mapred.output.format.class`. This class transforms the key/value pairs passed to the output in the `reduce()` method of the reducer into the output format. The default value is `TextOutputFormat`. If `theClass` does not implement the `OutputFormat` interface, a `RuntimeException` will be thrown.

public OutputCommitter getOutputCommitter()

The framework provides a unique output directory for each task and stores this directory in the per-task configuration under the key `mapred.output.dir`. As of Hadoop 0.18, this key is set via the `FileOutputFormat.setOutputPath` static method.

As of 0.19.0, the `OutputCommitter` object is used to process the files in the per-task temporary area on successful task completion, and is responsible for deciding which output files are moved to the actual output area. Prior to this, any files present were moved to the user-specified output path.

This method retrieves the value stored in the configuration under the key `mapred.output.committer.class`. If the retrieved value is `null`, the `FileOutputCommitter` class will be returned. If the retrieved value is not `null`, the method will attempt to instantiate a class using the value as the class name. If the class name cannot be instantiated or if the instantiated class is not derived from the class `OutputCommitter`, a `RuntimeException` will be thrown.

public void setOutputCommitter(Class <? extends OutputCommitter> theClass)

This method will store the class name of `theClass` in the configuration under the key `mapred.output.committer.class`. If `theClass` does not implement the `OutputCommitter` interface, a `RuntimeException` will be thrown. See `getOutputCommitter` for a description of what the `OutputCommitter` is used for.

public void setCompressMapOutput(boolean compress)

This method stores the `String` equivalent of the value of `compress` in the configuration under the key `mapred.compress.map.output`. If the stored value is `true`, the map output data that will be consumed by the reduce phase will be compressed, using either the default compression codec or the codec specified by the method `setMapOutputCompressorClass`.

MAP TASK OUTPUT COMPRESSION

During a Hadoop job that has a reduce phase, the map phase produces intermediate output that will be further processed by the framework. This output will eventually become the input to the reduce phase. This output may be compressed to reduce transitory disk space requirements and network transfer requirements. The call `setCompressMapOutput(true)` will enable this compression. To enable map output compression when the job will not have a reduce phase, the call `FileOutputFormat.setCompressOutput(conf, true)` must be made.

Having the map output compressed can save substantial time because the amount of data that must traverse the network between the map and the reduce phase may be substantially reduced.

Having the job output compressed may also save substantial time because the amount of data to be stored in HDFS may be substantially reduced, greatly reducing the amount of network traffic for the replicas.

public boolean getCompressMapOutput()

This method returns the value stored in the configuration under the key `mapred.compress.map.output`. If the value is unset or is not one of `true` or `false`, the value `false` will be returned. If this value is `true`, map task output that will be reduced will be compressed using the compression defined for `SequenceFiles`.

public void setMapOutputCompressorClass(Class <? extends CompressionCodec> codecClass)

This method stores the class name of `codecClass` in the configuration under the key `mapred.map.output.compression.codec`. An instance of this class will be used to compress the map task output that is to be passed to the reduce tasks if the configuration key `mapred.compress.map.output` has the value of `true`. This key may be set by the `JobConf` method `setCompressMapOutput(boolean)`. If `codecClass` does not implement the `CompressionCodec` interface, a `RuntimeException` will be thrown.

public Class<? extends CompressionCodec> getMapOutputCompressorClass(Class<? extends CompressionCodec> defaultValue)

This method looks at the value of the key `mapred.output.compression.codec` in the configuration. If the value is not found, `defaultValue` is returned. If the value cannot be instantiated as a class that is derived from `CompressionCodec`, a `RuntimeException` will be thrown.

OUTPUT KEY AND VALUE CLASSES

The Hadoop framework is responsible for loading the job input and converting that input into key/value pairs that are passed to the map method of the mapper, passing the key/value pairs output by the map method of the mapper to the `reduce()` method of the reducer, and taking the key/value pairs output by the `reduce()` method of the reducer and writing them to the job output.

The class that loads and transforms the input into key/value pairs is derived from `InputFormat` and requires that the type of the key and the type of the value be specified. The class that handles loading the input is responsible for producing keys and values of the correct type. A commonly used class is the `KeyValueTextInput` class, which parses the input as text files, with each record on a single line and the key and value separated by the first tab character. The key type is `org.apache.hadoop.io.Text`, and the value type is `org.apache.hadoop.io.Text`. If the job does not explicitly configure the map output class as `org.apache.hadoop.io.Text` or the job output class as `org.apache.hadoop.io.Text`, the reduce will fail with a key type mismatch error.

An `InputFormat` object has an associated `RecordReader` object. The `RecordReader` must provide `createKey` and `createValue` objects. The types of these objects will be used to define the mapper class input key and value types.

The class to receive and transform the output key/value pairs is derived from `OutputFormat`. The default value for the key class is `LongWritable`, and the default value for the value class is `org.apache.hadoop.io.Text`. The job may specify different classes via the `setOutputKeyClass` and `setOutputValueClass` methods, respectively. By default, the expected map output types are the same as the expected reduce input and output types.

The job may specify that the map output key type and or the map output value type is different from the job output key and value type. The setMapOutputKeyClass method allows the job to specify the map output key class and the reduce input key class as being different from the job output key class. The setMapOutputValueClass method allows the job to specify the map output value class and reduce input value class as being different that the job output key class.

The class specified under the key map.sort.class in the configuration will be used to sort the key objects if a reduce has been requested by the job. The default value for this key is org.apache.hadoop.util.QuickSort, an implementation of org.apache.hadoop.util.IndexedSorter.

The key and value classes can be any type as long as the framework is provided with serializer classes and deserializer classes that implement org.apache.hadoop.io.serializer.Serializer and org.apache.hadoop.io.serializer.Deserializer, and the class names are added to the list stored in the configuration under the key io.serialization.

public void setMapOutputKeyClass(Class<?> theClass)

This method stores the class name of theClass in the configuration under the key mapred.mapoutput.key.class. The type of this class will be used as the type of the map method output keys and the Reducer.reduce() method input keys. This class must be serializable and deserializable by a class defined in the list of serializers specified in the value of the configuration key io.serializations. theClass must be sortable by the class returned by getOutputKeyComparator().

public Class<?> getMapOutputKeyClass()

This method looks up the value of the key mapred.mapoutput.key.class in the configuration. If the value is unset, null is returned. If the value cannot be instantiated as a class, a RuntimeException is thrown. This class will also be the Reducer.reduce() method input key class. The default class for this is the job output key class, getOutputKeyClass(), and the default for it is LongWritable.

public Class<?> getMapOutputValueClass()

This method looks up the value of the key mapred.mapoutput.value.class in the configuration. If the value is unset, the value of getOutputValueClass() is returned. If the value cannot be instantiated as a class, a RuntimeException is returned. This class will also be the reduce() method input value class. The default value for the output value class is org.apache.hadoop.io.Text.

public void setMapOutputValueClass(Class<?> theClass)

This method stores the class name of theClass in the configuration under the key mapred.mapoutput.value.class, which will be the class for the Mapper.map() value output and the Reducer.reduce ()values. The default is the type used for the reduce output value, org.apache.hadoop.io.Text. This class must be serializable by a class defined in the list of serializers specified in the value of the configuration key io.serializations.

public Class<?> getOutputKeyClass()

This method looks up the key mapred.output.key.class in the configuration. If the value is unset, the class object org.apache.hadoop.io.LongWritable will be returned. If the value is set and a class of that name cannot be instantiated, a RuntimeException will be thrown. This class is the key class that the Reducer.reduce() method will output.

public void setOutputKeyClass(Class<?> theClass)

This method stores the name of theClass in the configuration under the key mapred.output.key. This will be the type of key output by the Reducer.reduce() method. Unless overridden by setMapOutputKeyClass, theClass will also be the Mapper.map() output key. theClass must be sortable by the class returned by getOutputKeyComparator() if it will also be used as the Mapper.map() output key class. theClass class must be serializable by a class defined in the list of serializers specified in the value of the configuration key io.serializations. The default value is org.apache.hadoop.io.LongWritable.

public Class<?> getOutputValueClass()

This method looks up the value of the key mapred.output.value.class in the configuration. If the value is unset, the class org.apache.hadoop.io.Text is returned. The value is instantiated as a class, and the class is returned. If the value cannot be instantiated as a class, a RuntimeException will be thrown.

public void setOutputValueClass(Class<?> theClass)

This method stores the name of theClass in the configuration under the key mapred.output.value.class. This value will be used as the type of the Reducer.reduce() output value; if not overridden by setMapOutputValueClass(), it will be the type of the Mapper.map() output value. If this class is used as a map output value, it must be serializable by a class defined in the list of serializers specified in the value of the configuration key io.serializations.

Methods for Controlling Output Partitioning and Sorting for the Reduce

The partitioner determines which key/value pair is sent to which reduce task. The comparator, the class returned by getOutputKeyComparator(), determines the ordering of the key/value pairs, and the class returned by getOutputValueGroupingComparator() determines which adjacently sorted keys are considered equal for producing a value group to pass to the Reducer.reduce() method. Classes used as comparators must implement the RawComparator interface.

DEFINING OPTIMIZED COMPARATORS

A class used as a key object in Hadoop may define an optimized comparator class. The comparator has to implement the `org.apache.hadoop.io.WritableComparable` interface. The comparator must be registered with the framework by calling `org.apache.hadoop.io.WritableComparator.define(Key.class, ComparatorInstance)`. The common key class `org.apache.hadoop.io.Text` defines a custom comparator that does a byte-wise comparison of the actual serialized text. This avoids having to deserialize the `Text` object and then run `String` comparisons on the data in the reconstituted objects.

public RawComparator getOutputKeyComparator()

This method looks up the value of the key `mapred.output.key.comparator.class` in the configuration. If the value is unset, the class `org.apache.hadoop.io.WritableComparable WritableComparator` will be returned. If the value cannot be instantiated as a class that is an instance of `org.apache.hadoop.io.RawComparator`, a `RuntimeException` will be thrown.

public void setOutputKeyComparatorClass(Class <? extends RawComparator> theClass)

This method stores the class name of `theClass` in the configuration under the key `mapred.output.key.comparator.class`. `theClass` will be used to order the keys being presented to the `Reducer.reduce()` method. The default class is the comparator for the `Mapper.map()` key output class. If *`theClass`* does not implement the `RawComparator` interface, a `RuntimeException` will be thrown.

public void setKeyFieldComparatorOptions(String keySpec)

This method stores the `String` *keySpec* in the configuration under the key `mapred.text.key.comparator.options`. This method also changes the `OutputKeyComparatorClass` (key `mapred.output.key.comparator.class`) to the class `org.apache.hadoop.mapred.lib.KeyFieldBasedComparator`.

The key fields are separated by the character that is the value of the configuration key `map.output.key.field.separator`. If there is no value set for the key `map.output.key.field.separator`, the separator character will be the ASCII tab character.

The key will be split on `map.output.key.field.separarator` characters into pieces. These pieces are numbered from 1.

The *keySpec* `String` is composed of one or more space-separated groups. Each group defines the following items:

- The piece number to start the comparison region.
- The character number in the piece to start the comparison. The first character is `1`; the last character is `0`. This is optional and defaults to position `1`.
- How to sort, either numerically via the `n` option or in reverse order via the `r` option. This is optional and defaults to the standard `String` comparison ordering.

- The piece number to end the comparison region. This is optional and defaults to the starting piece number.
- The character number in the piece to end the comparison. This is optional and defaults to the last character in the `String`: `0`.

The specification `-k1.5nr` specifies numeric reverse-order sorting using the characters from position 5 through the end of the first piece of the key.

The specification `-k2.2,3.4r` specifies reverse `String` comparison using the characters from character 2 in key 2 through to character 4 in piece 3.

Given the line `01234 6789`, key piece 1 would be `01234`, and key piece 2 would be `6789`. The key spec `-k1.2,2.3` would provide a comparison segment of `1234 678`. There is a test class for these key fields in the examples that can be run by giving it three arguments: the key, the key spec for the combiner, and the key spec for the partitioner. The field separator is hard coded as a space: `bin/hadoop jar '\Documents and Settings\Jason\My Documents\Hadoop Source\hadoop-0.19.0\hadoopprobook.jar' com.apress.hadoopbook.examples.jobconf.KeyFieldDemonstrator "01234 6789 abcd" "-K1,2" "-k3,3"`. The summarized output is `Partitioner[key: (abc)] Comparator[key: (01234 6789 abcd), key: (01234 6789 abcd)]`.

Note Changing the output key comparator class via `setOutputKeyComparatorClass` disables field-based key comparisons. The output key comparator class must be `org.apache.hadoop.mapred.lib.KeyFieldBasedPartitioner` or a functional equivalent.

---

public String getKeyFieldComparatorOption()

This method looks up the value of the key `mapred.text.key.comparator.options` and returns the value. Please see `setKeyFieldComparatorOption` for a discussion of the appropriate values.

PARTITIONING

When a job is configured to have a reduce phase, the output will be split into partitions (one partition per reduce task). The framework has a default partitioning strategy of using the hash code of the key, modulus the number of partitions, `key.hashCode() % conf. getNumReduceTasks()`. If your job has three reduces specified, the default partition for a key will be `key.hashCode() % 3`. The user is free to specify a custom partitioning class. The framework provides three partitioning classes:

- `org.apache.hadoop.mapred.lib.HashPartitioner`: default partition based on the key's hash code
- `org.apache.hadoop.mapred.lib.KeyFieldBasedPartitioner`: partition based on a segment of the key
- `org.apache.hadoop.mapred.lib.TotalOrderPartitioner`: partition by absolute range of the keys

A custom partitioning class must implement the interface `org.apache.hadoop.mapred.Partitioner`.

public Class<? extends Partitioner> getPartitionerClass()

This method looks up the value of the key `mapred.partitioner.class` in the configuration. If the value is unset, the class `org.apache.hadoop.mapred.lib.HashPartitioner` is returned. If the value is set, it is instantiated as a class that must be an instance of `org.apache.hadoop.mapred.Partitioner.class`. If the value cannot be instantiated or is not an instance of the `Paritioner` class, a `RuntimeException` will be thrown. `HashPartitioner` simply uses the hash value of the key, modulus the number of reduce tasks, to determine which reduce will receive any given key/value pair.

public void setPartitionerClass(Class<? extends Partitioner> theClass)

This method stores the class name of `theClass` in the configuration under the key `mapred.partitioner.class`. An instance of this class will be created for each map task and used to determine which reduce will receive which key/value pair that the `Mapper.map()` method outputs. If `theClass` does not implement the `org.apache.hadoop.mapred.Partitioner` interface, `RuntimeException` will be thrown.

public void setKeyFieldPartitionerOptions(String keySpec)

This method stores the `String keySpec` in the configuration under the key `mapred.text.key.partitioner.options`. The output partitioning class will also be set to `org.apache.hadoop.mapred.lib.KeyFieldBasedPartitioner` via a call to the `setPartitionerClass()` method.

The portion of the key selected will be hashed, and that hash modulus the number of reduces will be the partition number.

The `keySpec String` is composed of one or more space-separated groups. Each group defines the following items:

- The piece number to start the comparison region.
- The character number in the piece to start the comparison: `1` is the first character; `0` is the last character. This is optional and defaults to position `1`.
- How to sort, either numerically via the `n` option and or in reverse order via the `r` option. This is optional and defaults to the standard `String` comparison ordering.
- The piece number to end the comparison region. This is optional and defaults to the starting piece number.
- The character number in the piece to end the comparison. This is optional and defaults to the last character in the `String`: `0`.

■ **Note** The key parser (at least through Hadoop 0.19.0) has an issue: it doesn't understand that the last piece of the key might not have a separator character after it. If your job generates `ArrayIndexOutOfBounds` exceptions, explicitly end the key piece selection for the second key piece: `-k2.2` explicitly ends the piece at the last character; `-k2` includes the second key piece and the separator after piece `2`.

The specification -k1.5nr specifies numeric reverse order sorting using the characters from position 5 through the end of the first piece of the key.

The specification -k2.2,3.4r specifies reverse String comparison using the characters from character 2 in key 2 to character 4 in piece 3.

Given the line 01234 6789, key piece 1 would be 01234, and key piece 2 would be 6789. The key spec -k1.2,2,3 would provide a comparison segment of 234 678.

public String getKeyFieldPartitionerOption()

This looks up the key mapred.text.key.partitioner.options in the configuration and returns the value. For this value to have an effect, the output partitioner class must be org.apache.hadoop.mapred.lib.KeyFieldBasedPartitioner. See setKeyFieldPartionerOptions for a description of the returned value.

OUTPUT VALUE GROUPING

It is often the case that there is a requirement for grouping output data. Hadoop Core provides a way to group output values that acts very much like a secondary sort on the key data. For this to work in the manner that the user expects, the output partitioner, the output comparator, and the output grouping comparator have to cooperate.

The outputKeyComparator must order the keys using the primary and secondary sort. Because keys that must group together may not be equal in this method, the outputPartitioner has to be able to place keys that must group together into the same partition. The outputValueGroupingComparator must return equality only for those keys that are equal in the primary sort. This will result in a call to the Reduce.reducer method for each group of keys.

public RawComparator getOutputValueGroupingComparator()

This method looks up the value of the key mapred.output.value.groupfn.class in the configuration and attempts to instantiate a class that is an instance of org.apache.hadoop.io.RawComparator. If the value is unset, the comparator class for the Map key class is returned. If the value cannot be instantiated or the resulting class does not implement org.apache.hadoop.io.RawComparator, a RuntimeException is thrown.

public void setOutputValueGroupingComparator(Class <? extends RawComparator> theClass)

This method stores the class name of theClass in the configuration under the key mapred.output.value.groupfn.class. If theClass does not implement the org.apache.hadoop.io.RawComparator interface, a RuntimeException will be thrown.

The use of this method enables a grouping operator on keys and a secondary sort. The user must set both a partitioner and a comparator that cooperate for this to be used. It is common for the default output comparator to be used to force complete sorting of the keys output by the `Mapper.map()` method. The output comparator must compare keys so all keys that are to be grouped together are adjacent in the sort. The partitioner must ensure that all keys that are to be grouped together are sent to the same partition.

The `Reducer.reduce()` method will receive the first key in the group, and the values will be the values from all adjacent keys that the output value grouping comparator considers equal. If keys are of the form `item rank` and the values are of the form `data`, the partitioner must use only `item` to partition. The standard output comparator will sort lexically on `item rank`. The output value grouping operator will use only `item` for comparing keys. The `Reducer.reduce()` method will receive all keys that share `item`, and the values will be lexically sorted by `rank`.

The keys are composed of `item rank`, where the `item` is one of Key1 or Key2, and the `rank` is one of 00, 01, 02. The partitioner would use the `item` for partitioning. The output comparator would fully sort the keys by `item rank`. The output value grouping comparator would use only `item` for comparing keys. (See Table A-1 and Table A-2.)

Table A-1. *Sample Input*

Key	Value	Partitioner Value	Output Comparator Sort
Key1 00	00	Key1	Key1 00
Key1 01	01	Key1	Key1 01
Key1 02	02	Key1	Key1 02
Key2 00	00	Key2	Key2 00
Key2 01	01	Key2	Key2 01
Key2 02	02	Key2	Key2 02

Table A-2. *Reducer.reduce Calls*

Key	Values		
Key1 00	00	01	02
Key2 00	00	01	02

Methods that Control Map and Reduce Tasks

These methods actually specify the class that provides the `Mapper.map()` and `Reducer.reduce()` methods. They specify if the map methods may be run from multiple threads or in a single thread.

They specify if the framework will attempt to run multiple instances of a task to see if one will run faster, and when to consider a task completely failed and a job completely failed.

SINGLE THREADED OR MULTI-THREADED MAPPERS

The framework creates an instance of the mapper class in each map task. By default, a single-threaded map runner is used, and the key/value pairs are passed to the `Mapper.map()` method serially. The user may inform the framework that multiple threads are to run the `Mapper.map()` method. There will be multiple simultaneous calls to the `map()` method of the single instance of the Mapper class, running in the JVM that hosts the map task. The input of key/value pairs are treated as a queue, being serviced by a thread pool, which invokes the `Mapper.map()` method on each pair pulled from the queue.

The user specifies this behavior by setting the map runner class to `org.apache.hadoop.mapred.lib.MultithreadedMapRunner` and by storing the number of threads to run in the configuration under the key `mapred.map.multithreadedrunner.threads`.

public Class<? extends Mapper> getMapperClass()

This method looks up the value of the key `mapred.mapper.class` in the configuration and attempts to instantiate the value as a class of type `org.apache.hadoop.mapred.Mapper`. If the value is unset, the class `org.apache.hadoop.mapred.lib.IdentityMapper` is returned. If the value cannot be instantiated as a class of the correct type, a `RuntimeException` is thrown.

The returned class will provide the map method that all the input data will be passed through.

public void setMapperClass(Class<? extends Mapper> theClass)

This method stores the name of `theClass` class in the configuration under the key `mapred.mapper.class`. An instance of this class will be created in each map task, and each input key/value pair will be passed to `theClass` map method. If `theClass` does not implement the `org.apache.hadoop.mapred Mapper` interface, a `RuntimeException` will be thrown.

public Class<? extends MapRunnable> getMapRunnerClass()

This method looks up the key `mapred.map.runner.class` in the configuration and instantiates the value as a class of type `org.apache.hadoop.mapred.MapRunnable`. If the value is unset, the class `org.apache.hadoop.mapred.lib.MapRunnable` is returned.

public void setMapRunnerClass(Class<? extends MapRunnable> theClass)

This method stores the name of `theClass` in the configuration under the key `mapred.map.runner.class`. This is commonly used when the `Mapper.map()` method is to be threaded, and `theClass` in this case is `org.apache.hadoop.mapred.lib.MultithreadedMapRunner.class`. When this is done, there is usually a `setInt("mapred.map.multithreadedrunner.threads", threadCount)` call.

The multithreaded map runner is very handy when the map method is not blocked waiting on local CPU or IO, such as when the map method is used to fetch URLs.

public Class<? extends Reducer> getReducerClass()

This method looks up the key `mapred.reducer.class` in the configuration and instantiates the value as a class of type `org.apache.hadoop.mapred.Reducer`. If the value is unset, the class `org.apache.hadoop.mapred.lib.IdentityReducer` is returned. If the value cannot be instantiated as a class of the correct type, a `RuntimeException` will be thrown.

public void setReducerClass(Class<? extends Reducer> theClass)

This method stores the name of `theClass` in the configuration under the key `mapred.reducer.class`. If `theClass` does not implement the `Reducer` interface, a `RuntimeException` will be thrown.

One instance of this class will be created in each reduce task. Each unique key will be passed to one instance of the `Reducer.reduce()` method of `theClass`, with all the values that share that key.

COMBINERS: A WAY TO REDUCE INTERMEDIATE DATA

A combiner class is a minireducer that is run in the context of the map task to pregroup key/value pairs that share a key.

Combiners can greatly minimize the amount of output that has to pass between the map and reduce tasks and speed up the job.

The class used for combining must implement the `Reducer` interface, and the class's `reduce()` method will be called to combine map output values that share a key.

If the job's reducer class is being used as a combiner, `reduce()` must not have side effects because there is no constraint on the number of times the `reduce()` method will be called in as a map output combiner. In particular, if the same class is used for combing and reducing, unless care is taken to change the counter names, the counts displayed at job end will be the sum of the combiner and reducer counts. Please see `com.apress.hadoopbook.examples.ch5.CounterExamplesWithCombiner`, and look at the `NaiveReducer` counter values and compare them against the `reducer` and `combiner` counter values.

public Class<? extends Reducer> getCombinerClass()

This method looks up the key `mapred.combiner.class` in the configuration and instantiates the value as a class implementing the `Reducer` interface. If the value is unset, `null` is returned. If the value cannot be instantiated as a class implementing the `Reducer` interface, a `RuntimeException` is thrown.

public void setCombinerClass(Class<? extends Reducer> theClass)

This method stores the name of `theClass` in the configuration under the key `mapred.combiner.class`. If `theClass` does not implement the `Reducer` interface, a `RuntimeException` is thrown.

SPECULATIVE EXECUTION: FRIEND AND FOE

Hadoop has its roots in clusters of heterogeneous machines. In this environment, the amount of wall clock time for any given machine to execute a map or reduce task could vary widely because of differing machine capabilities. In addition, there is no guarantee that any given `InputSplit` will take the same amount of wall clock time to execute.

Speculative execution informs the cluster that any unused task slots may be used to run duplicate instances of an already running task. The first of these duplicates to complete has its results used, and the other task has its output discarded.

If your tasks do not have side effects that Hadoop cannot undo, do not consume resources with some real costs or load your machines so that other tasks run slower. Speculative execution is your friend.

■ **Note** Hadoop only knows how to discard task output that is in the form of job counters or output that is placed in the per-task output directory. Ensure that speculative execution is disabled if your tasks have output that Hadoop cannot discard or side effects that Hadoop cannot undo.

---

public boolean getSpeculativeExecution()

This method returns `true` if either `getMapSpeculativeExecution()` or `getReduceSpeculativeExecution()` is `true`. The default Hadoop configuration has speculative execution enabled for map tasks and for reduce tasks.

public void setSpeculativeExecution (boolean speculativeExecution)

This method calls `setMapSpeculativeExecution(speculativeExecution)` and `setReduceSpeculativeExecution(speculativeExecution)`. If `speculativeExecution` is `true`, speculative execution will be enabled for both map and reduce tasks. If `speculativeExecution` is `false`, speculative execution will be disabled for both map and reduce tasks.

public boolean getMapSpeculativeExecution()

This method looks up the value of the key `mapred.map.tasks.speculative.execution` in the configuration and converts that value to a `boolean` value, which is then returned. If the value is unset, `true` is returned. If the value is not the `String` `true`, `false` is returned.

public void setMapSpeculativeExecution (boolean speculativeExecution)

This method stores the `String` value of the `boolean` `speculativeExecution` in the configuration under the key `map.tasks.speculative.execution`.

public boolean getReduceSpeculativeExecution()

This method looks up the value of the key mapred.reduce.tasks.speculative.execution in the configuration and converts that value to a boolean value, which is then returned. If the value is unset, true is returned. If the value is not the String true, false is returned.

public void setReduceSpeculativeExecution (boolean speculativeExecution)

This method stores the String value of the boolean speculativeExecution in the configuration under the key mapred.reduce.tasks.speculative.execution.

public int getNumMapTasks()

This method looks up the value of the key mapred.map.tasks in the configuration and returns the value converted to an int. If the value is unset, 1 is returned. If the value cannot be converted to an int, a NumberFormatException is thrown. This value is the suggested number of map tasks to run. The actual number of map tasks will be determined by the number of InputSplits that the framework constructs from the input data. In general, there is at least one InputSplit for each input file. The input format might be able to make multiple InputSplits from a single file. The FileInputFormat set of input formats will split uncompressed files on HDFS block boundaries, which by default are 64MB. Many installations increase this size to 128MB or higher.

public void setNumMapTasks(int n)

This stores the String representation of n in the configuration under the key mapred.map.tasks. The input format will attempt to ensure that this is the maximum number of map tasks, but may not be able to do so if there are more individual files that this in the input directory. In general, tuning this and the split size setInt("mapred.min.split.size ", NUMBER), so map tasks take more than a minute to run is considered optimal.

public int getNumReduceTasks()

This method looks up the value of the key mapred.reduce.tasks in the configuration and returns the value converted to an int. If the value is unset, 1 is returned. If the value cannot be converted to an int, a NumberFormatException is thrown.

Unlike the number of map tasks, this is exactly the number of reduce tasks that will be run.

public void setNumReduceTasks(int n)

This method stores the String representation of n in the configuration under the key mapred.reduce.tasks. Exactly this number of reduce tasks will be run by the framework. If this number is 0, no reduce tasks will be run, and no output partitioning or sorting will be done. There will be one output file per map task, written to the output directory configured for the job.

public int getMaxMapAttempts()

This method looks up the value of the key `mapred.map.max.attempts` in the configuration and returns the value converted to an `int`. If the value is unset, the value `4` is returned. If the value cannot be converted to an `int`, a `NumberFormatException` is thrown.

The framework will attempt to reschedule map tasks that fail up to `getMaxMapAttempts()` times before the job is considered failed.

public void setMaxMapAttempts(int n)

This method stores the `String` representation of `n` in the configuration under the key `mapred.map.max.attempts`. This is rarely changed by the user other than to set it to `0` to disable the retrying of failed jobs.

The framework will attempt to reschedule map tasks that fail up to `getMaxMapAttempts()` times before the job is considered failed.

public int getMaxReduceAttempts()

This method looks up the value of key `mapred.reduce.max.attempts` in the configuration and returns the value converted to an `int`. If the value is unset, the value `4` is returned. If the value cannot be converted to an `int`, a `NumberFormatException` is thrown.

The framework will attempt to reschedule reduce tasks that fail up to this value times before the job is considered failed.

public void setMaxReduceAttempts(int n)

This method stores the `String` representation of `n` in the configuration under the key `mapred.reduce.max.attempts`. This is rarely changed by the user other than to set it to `0` to disable the retrying of failed jobs.

The framework will attempt to reschedule reduce tasks that fail up to this value times before the job is considered failed.

public void setMaxTaskFailuresPerTracker(int noFailures)

This method stores the `String` representation of `noFailures` in the configuration under the key `mapred.max.tracker.failures`. This value is the number of tasks for this job that may fail on a specific TaskTracker before that TaskTracker is considered failed for this job.

public int getMaxTaskFailuresPerTracker()

This method looks up the value of the key `mapred.max.tracker.failures` in the configuration and returns the value converted to an `int`. If the value is unset, the value `4` is returned. If the value cannot be converted to an `int`, a `NumberFormatException` will be thrown. This value is the number of tasks for this job that may fail on a specific TaskTracker before that TaskTracker is considered, failed, for this job.

public int getMaxMapTaskFailuresPercent()

This method looks up the value of the key `mapred.max.map.failures.percent` in the configuration. If the value is unset, `0` is returned. If the value cannot be converted to an `int`, a `NumberFormatException` is thrown.

If this value is not zero, a job may succeed if less than this value as a percentage of the map tasks cannot be successfully completed. So if the job has 100 map tasks, and this returns 1, only 99 of the map tasks have to complete successfully for the job to be considered a success.

Map tasks that do not succeed are retried up to `getMaxMapAttempts()` times before being considered failed.

public void setMaxMapTaskFailuresPercent(int percent)

This method stores the `String` representation of `percent` in the configuration under the key `mapred.max.map.failures.percent`. This is the percentage of map tasks that can fail without the job being marked as a failure. The default value for this parameter is `0`.

A map task that does not succeed is retried `getMaxMapAttempts()` times, which defaults to `4`, before being that task is considered failed.

public int getMaxReduceTaskFailuresPercent()

This method looks up the value of the key `mapred.max.reduce.failures.percent` in the configuration. If the value is unset, `0` is returned. If the value cannot be converted to an `int`, a `NumberFormatException` is thrown.

If this value is not zero, a job may succeed if less than this value as a percentage of the reduce tasks cannot be completed successfully. So if the job has 10 reduce tasks, and this returns `10`, only 9 of the reduce tasks have to complete successfully for the job to be considered a success.

Map tasks that do not succeed are retried up to `getMaxReduceAttempts()` times before being considered failed.

public void setMaxReduceTaskFailuresPercent(int percent)

This method stores the `String` representation of `percentage` in the configuration under the key `mapred.max.reduce.failures.percent`. This is the percentage of reduce tasks that can fail without the job being marked as a failure. The default value for this parameter is `0`.

A reduce task that does not succeed is retried `getMaxReduceAttempts()` times, which defaults to `4` before being that task is considered failed.

Methods Providing Control Over Job Execution and Naming

These methods provide a way to specify a job name and a session identifier as well as to specify a priority for a job. The naming is also helpful for distinguishing jobs in the reporting frameworks.

They also provide a way to enable profiling of specific tasks and of running a debugging script on failed tasks.

public String getJobName()

This method looks up the value of the key `mapred.job.name` in the configuration and returns the result. If the value is unset, an empty `String` is returned.

This is the name that the job will be identified by to the user.

public void setJobName(String name)

This method stores `name` in the configuration under the key `mapred.job.name`. `name` will be used to identify the job in user-reporting mechanisms.

HADOOP ON DEMAND

Hadoop On Demand (HOD) is a package that provides virtual map/red clusters on top of a larger HDFS installation. It is used extensively inside of Yahoo. The use of HOD requires an understanding of torque: `http://www.clusterresources.com/pages/products/torque-resource-manager.php`. The author and the team the author was working with found it too complex for the benefits provided and discontinued using it.

HOD is described on the Hadoop site: `http://hadoop.apache.org/core/docs/r0.19.1/hod_user_guide.html`. HOD has probably improved significantly because the author used it last with Hadoop 0.16.1. The author recommends avoiding HOD unless there is a local torque expert to handle the torque installation and day-to-day operation.

public String getSessionId()

This method looks up the value of the key `session.id` in the configuration and returns it. If the value is unset, an empty `String` is returned. This is primarily used by HOD to distinguish different virtual clusters. The session name may also help distinguish this job in the metrics reporting framework.

public void setSessionId(String sessionId)

This method stores `sessionId` in the configuration under the key `session.id`. This value will be used as a token in the name used to identify any metrics that are reported by this job. This method is primarily intended for use by HOD.

public JobPriority getJobPriority()

This method looks up the value of the key `mapred.job.priority` in the configuration. If the value is unset, `JobPriority.NORMAL` is returned. If the value cannot be parsed as a `JobPriority`, an `IllegalArgumentException` is thrown. (Hadoop versions prior to 0.19 had only this simple mechanism for handling multiple running jobs on a cluster.)

A job with a higher priority has first right of refusal for any map or reduce task slot available on the cluster. If jobs have equal priority, the first requester gets the open task slots. There is no preemption of executing tasks.

Caution Queuing multiple jobs into a cluster with this mechanism can result in a cluster deadlock in which no job can complete.

Hadoop 0.19 also provides a queuing mechanism that provides rich control over how task slots are allocated between multiple competing jobs. (Refer to Chapter 8.)

public void setJobPriority(JobPriority prio)

Store the `String` representation of `prio` in the configuration under the key `mapred.job.priority`.

Jobs with a higher priority have first choice of available task slots when executing in an environment in which multiple jobs are queued into a cluster.

public boolean getProfileEnabled()

This method looks up the value of the key `mapred.task.profile` in the configuration. If the value is unset or not the `String`, `true`, `false` is returned; otherwise, `true` is returned.

If this is `true`, the framework may profile specific tasks by using the results of `getProfileTaskRange()` to select individual tasks to profile. Profiling is performed on both map tasks and reduce tasks if enabled. If only profiling on maps is required, the user must specify a range of reduce values that is not available to the `setProfileTaskRange()` method, with `false` as the first argument. If the number of reduces is `10`, the reduces will be `0` through `9`, and calling `setProfileTaskRange(false, "10")` would effectively disable profiling for reduces. It is harder to absolutely know the number of map tasks, but the same technique applies.

public void setProfileEnabled(boolean newValue)

This method stores the `String` value of `newValue` in the configuration under the key `mapred.task.profile`. If `newValue` is `true`, profiling information will be collected for tasks that match the `getProfileTaskRange()` method.

public String getProfileParams()

This method looks up the value of the key `mapred.task.profile.params` in the configuration, returning that value. If the value is unset, the following `String` is returned:

```
-agentlib:hprof=cpu=samples,heap=sites,force=n,thread=y,verbose=n,file=%s
```

This `String` is passed to the JVM to control how the profiling is performed for the task to be profiled. At runtime, for a profiled task a single `%s` will be substituted in the value with the name of the task-specific `profile.out` file.

public void setProfileParams(String value)

This method stores `value` in the configuration under the key `mapred.task.profile.params`. This value, with a single `%s` substituted with the name of the task-specific profile output file, is passed to the JVM of a task to be profiled.

public Configuration.IntegerRanges getProfileTaskRange (boolean isMap)

This method looks up the value of the key `mapred.task.profile.maps` if `isMap` is `true`, or the value of the key `mapred.task.profile.reduces` if `isMap` is `false`. If the value is unset, the range `0-2` is constructed. If the value cannot be parsed as a set of ranges, an `IllegalArgumentException` is thrown. Ranges are specified as a set of comma-separated values, in which each value is a single positive integer or two positive integers separated by a dash. Some valid ranges include the following:

- `0-2`: tasks 0, 1, and 2
- `2`: task 2 only
- `0-2,5-7`: tasks 0, 1, 2, 5, 6, and 7
- `-7,0,6-11`: tasks 0, 5, 6, 7, 8, 9, 10, and 11 (ordering is not needed, and overlap is allowed)
- `0-3, 9-11,13`: tasks 0, 1, 2, 3, 9, 10, 11, and 13

No checking is performed to ensure that the individual ranges in a comma-separated set do not overlap and ordering is not required. A linear search through the list in the order supplied is performed for each task when profiling is enabled.

public void setProfileTaskRange(boolean isMap, String newValue)

This method stores `newValue` under the key `mapred.task.profile.maps` if `isMap` is `true`, or the key `mapred.task.profile.reduces` if `isMap` is `false`. The value must be a comma-separated list of ranges composed of positive integers. During task setup, the TaskRunner will get this value via `getProfileTaskRange()`, if the value stored in the key is not a valid range, an exception will be thrown and the task will be aborted. (See `Configuration.IntegerRanges getRange()`, earlier in this chapter, for a discussion of range formats.)

Some valid ranges include the following:

- `0-2`: tasks 0, 1, and 2
- `2`: task 2 only
- `0-2,5-7`: tasks 0, 1, 2, 5, 6, and 7
- `5-7,0,6-11`: tasks 0, 5, 6, 7, 8, 9, 10, and 11 (ordering is not needed, and overlap is allowed)
- `0-3, 9-11,13`: tasks 0, 1, 2, 3, 9, 10, 11, and 13

No checking is performed to ensure that the individual ranges in a comma-separated set do not overlap and ordering is not required. A linear search through the list in the order supplied is performed for each task, when profiling is enabled.

public String getMapDebugScript()

This method returns the value of the key `mapred.map.task.debug.script` from the configuration. If the value is unset, `null` is returned. This script will be run for a map task that the framework is going to mark as failed or about to kill.

The value is the script and script arguments to be used to debug failed tasks. The value will be split into tokens using the space character as a separator. Five additional arguments are added:

- The path to the task standard output file
- The path to the task standard error file
- The path to the task syslog output file
- The path to the file containing the XML representation of the `JobConf` object for the task
- The program name if this is a pipes job or empty `String`

All the tokens are passed to the shell to be executed as a command. The input of the command will be connected to `/dev/null`, and the standard and error output collected in a single stream.

The script is run with the current working directory as the task local directory. If the script is not resident on all the TaskTracker nodes and normally executable, it must be distributed via the `DistributedCache` and symlinked.

The following code fragment arranges for the executable program that is on the local file system at `LocalFileSystemPathToDebugScript` to be distributed to all tasks and made available for execution as `./MyDebugScript`. In Listing A-11, the URI fragment `#MyDebugScript` informs the framework to create a symbolic link named `MyDebugScript` between the task local copy of `LocalFileSystemPathToDebugScript` and the current working directory of the task.

Listing A-11. *Adding a Debug Script to the DistributedCache*

```
Job.setMapDebugScript("./MyDebugScript map argument2 argument3" );
DistributedCache.createSymlink(job);
DistributedCache.addFile("HDFSFileSystemPathToDebugScript#MyDebugScript");
```

The script will be invoked in the task local directory via the following shell command:

```
./MyDebugScript map argument1 argument2 argument3 taskStdoutFile taskStderrFile➥
taskSyslogFile taskJobConfXmlFile pipesProgramName➥
< /dev/null 2>&1 > ./debugout
```

The user can specify how many lines to keep from the output by setting an `int` value on the key `mapred.debug.out.lines`. The default value `-1` keeps all the output lines. The value specified is the number of lines from the tail of the file to keep. If the value is `10`, the last 10 lines of the output file are saved.

This information is made available via the JobTracker web interface in the task detail output.

■ **Caution** Having shell metacharacters in the value of `mapred.map.task.debug.script` may lead to unpredictable results.

---

public void setMapDebugScript(String mDbgScript)

This method stores `mDbgScript` in the configuration under the key `mapred.map.task.debug.script`. (See `getMapDebugScript()` for details on the format and use of `mDbgScript`.)

public String getReduceDebugScript()

This method return the value stored under the key `mapred.reduce.task.debug.script`. If the value is unset, `null` is returned. The usage is the same as the usage of `getMapDebugScript()`, except it reduces tasks.

public void setReduceDebugScript(String rDbgScript)

This method stores `rDbgScript` in the configuration under the key `mapred.reduce.task.debug.script`. (See `getMapDebugScript()` for details on the format and use of `rDbgScript`.) This script will be used for failed or about to be killed reduce tasks.

JOB END NOTIFICATION

If a URL is stored in the configuration under the key job.end.notification.url or via setJobEndNotification(), an HTTP GET will be made on this URL when the job finishes.

The text $jobId and $jobStatus, if present in the URL, is replaced with the job id and the job status, respectively. The job status will be either SUCCEEDED or FAILED.

The parameter job.end.retry.attempts controls the number of retry attempts that will be made if the HTTP GET does not return the numeric status code of 200. The default is 0 retries.

The parameter job.end.retry.interval controls the delay between retry attempts, with a default value of 30,000 msec.

If either parameter is set and the value cannot be converted to an int, a NumberFormatException will be thrown in the context of the JobTracker, which may cause the JobTracker to abort or otherwise behave unpredictably.

public String getJobEndNotificationURI()

This method looks up the value of the key job.end.notification.url in the configuration and returns that value. If the value is unset, null is returned. The value will be used as a URL in an HTTP GET.

public void setJobEndNotificationURI(String uri)

This method stores uri in the configuration under the key job.end.notification.url.

public String getQueueName()

This method looks up the key mapred.job.queue.name in the configuration and returns the value. If the value is unset, default is returned.

Queues, which are new to Hadoop 0.19.0, provide a mechanism to allow multiple jobs to share cluster resources in a specified manner (refer to Chapter 8).

public void setQueueName(String queueName)

This method stores queueName in the configuration under the key mapred.job.queue.name. If queueName is not a valid queue name, the JobTracker behavior is unpredictable.

MEMORY LIMITS FOR TASKS AND THEIR CHILDREN

Hadoop provides a mechanism to control the limit of virtual memory that an individual task and the task's children use.

The user can specify the maximum amount of memory, in kilobytes, in the configuration under the key `mapred.task.maxmemory`; the method `setMaxVirtualMemoryForTask(vmem)` can also be used. The overall default can be specified by storing the value in kilobytes under the key `mapred.task.default.maxmemory`.

When the virtual memory consumption of a task and its children exceed this value, the task is killed by the framework, and marked as failed. This is predicated on the system reporting virtual memory usage for processes in kilobytes.

The default value for `mapred.task.maxmemory` is `-1`. The value of `-1` tells the framework to use the framework limit, which is stored under the key `mapred.task.default.maxmemory`. The default value for this key is `536,870,912` kilobytes (roughly one-half terabyte).

long getMaxVirtualMemoryForTask() {

This method looks up the value of the key `mapred.task.maxmemory` and returns it as a `long`. If the value is unset, `-1` is returned. If the value cannot be converted to a long, a `NumberFormatException` will be thrown.

void setMaxVirtualMemoryForTask(long vmem) {

This method stores the `String` version of `vmem` in the configuration under the key `mapred.task.maxmemory`. If a task and its children's virtual memory usage exceed this value, the task will be killed by the framework.

Convenience Methods

These methods provide convenience functions for accessing the configuration data.

public int size()

This method returns the number of keys in the configuration.

public void clear()

This method completely clears all keys and values from the configuration.

public Iterator<Map.Entry<String,String>> iterator()

This method returns an integrator for to the key/value pairs stored in the configuration.

public void writeXml(OutputStream out) throws IOException

This method serializes the key/value pairs in the configuration to XML in the standard configuration file format and writes the data to `out`.

The destination for this output data can be used as input to the `addResource()` method. This method is used by the framework to serialize the job configuration and store it in HDFS so that the individual tasks load the job configuration at task start.

public ClassLoader getClassLoader()

This method returns the class loader that is used to search for resources that are added via the `addResource()` methods and to instantiate classes when a class is being returned.

public void setClassLoader(ClassLoader classLoader)

This method sets the class loader to be used for locating resources and instantiating classes to `classLoader`.

This is primarily used by the framework when preparing map and reduce tasks to include the `DistributedCache` classpath items in the classpath.

public String toString()

This returns a `String` composed of the names of all the resources that were loaded into this configuration. This method does not return the key/value pairs that are stored in the configuration.

Methods Used to Pass Configurations Through SequenceFiles

The configuration class implements the `Writable` interface, which allows the framework to serialize and deserialize the configuration. These two methods are required for the `Writable` interface. It is not clear that these methods are used by the framework at the current time.

public void readFields(DataInput in) throws IOException

This method will deserialize the configuration key/value pairs. The key value pairs will be read from the `DataInput` stream `in`.

public void write(DataOutput out) throws IOException

This method serializes the configuration into a form that is suitable for use in `SequenceFiles`. The serialized data is written to the `DataOutput` stream `out`.

Index

D

E

I

K

L

M

Q

R

S

T

U